	DATE DUE		

Instruments and the Imagination

Instruments and the Imagination

THOMAS L. HANKINS AND
ROBERT J. SILVERMAN

PRINCETON UNIVERSITY PRESS
PRINCETON, NEW JERSEY

Published by Princeton University Press, 41 William Street, Princeton, New Jersey 08540
In the United Kingdom: Princeton University Press, Chichester, West Sussex

---

Library of Congress Cataloging-in-Publication Data

Hankins, Thomas L.
Instruments and the imagination / Thomas L. Hankins and Robert J. Silverman.
p. cm.
Includes bibliographical references and index.
ISBN 0-691-02997-0
1. Scientific apparatus and instruments—Europe—History.
2. Science—Europe—History. 3. Science—Historiography.
4. Science—Methodology. 5. Creative ability in science.
I. Silverman, Robert J. II. Title.
Q185.H25 1995 502'.8'0940903—dc20 95-4301 CIP

This book has been composed in Sabon

Princeton University Press books are printed on acid-free paper, and meet the guidelines for permanence and durability of the Committee on Production Guidelines for Book Longevity of the Council on Library Resources

Printed in the United States of America by Princeton Academic Press

1 2 3 4 5 6 7 8 9 10

FOR BETSY AND POLLY

CONTENTS

ILLUSTRATIONS

ACKNOWLEDGMENTS

THIS BOOK owes much to numerous friends who have read chapters, helped with translations, provided illustrations, and assisted in more ways than we can possibly describe. We apologize if we have failed to mention any of our benefactors in these acknowledgments. Our gratitude is not diminished by any such oversight. We thank Jennifer Alexander, Susan Alon, Keith R. Benson, Jody Bourgeois, Gunnar Broberg, John Campbell, Geoffrey Cantor, V. K. Chew, James Clowes, Donna Coke, Ruth Schwartz Cowan, Thatcher Deane, Peter Dear, Peter Degen, John and Amanda Ellis, Jacqueline Ettinger, James Evans, Seth Fein, B. Raymond Fink, Bernard S. Finn, Maarten Franssen, Mott T. Greene, Caroline Hannaway, Timothy Heinrichs, Lisa Hendrickson, John Henry, Bruce W. Hevly, Ray Holstein, Bruce Hunt, William Kangas, Ann Koblitz, Brandon Konoval, Pamela Laird, Tom Leahey, Ann LeBar, Mark Levinson, Timothy McMannon, Ernan McMullin, Raimonda Modiano, R. Laurence Moore, James Naiden, Keith Nier, Keith Pickus, Robert C. Post, William B. Provine, Stephen Pumfrey, Joan Richards, Nancy Rockafellar, Joseph Roza, Iris Sandler, Lewis O. Saum, Laurie Sears, Stephanie and Larry Shea, Jody Segal, Stan Smith, Stephen Straker, Peter Sugar, Woodruff T. Sullivan III, Arnold Tamarin, Emily Thompson, John E. Toews, Alice Walters, Deborah J. Warner, L. Pearce Williams, Joella Yoder, Youngsoo Yook, Susanne Young, and Judith Zilczer.

For help with translations we wish to thank Diane Johnson, Pia Friedrich, Robert Elrich, and Paul Pascal.

We are grateful to Kevin Downing, Lauren Lepow, Emily Wilkinson, and the staff at Princeton University Press for their efforts throughout the production of this book.

The following libraries have been most helpful: Bibliothèque inguimbertine, Carpentras, the Burndy Library, the Harvard University Archives and Libraries, the Library of Congress, New York Public Library, the Science Museum, London, the Oliver Wendell Holmes Stereoscopic Research Library, the Princeton University Libraries, the Smithsonian Institution, the University of Washington Libraries, the Washington University Medical School Library, and the Yale Medical History Library.

We wish to express our special thanks to the National Endowment for the Humanities for their support of our research through an Interpretive Research Grant in the Humanities, Science and Technology Program. We would like to thank the Smithsonian Institution for the award of a fellowship that greatly assisted our research. We also owe a debt of gratitude to the Department of History, to the Howard and Frances Keller Fund, to the members of the History Research Group, and to the Program in the History of Science, Technology, and Medicine at the University of Washington for their constant support of our endeavor.

Two chapters from this book have been published previously. Chapter 4 first appeared in *Osiris* 9 (1994): 141–156, © 1994 by The History of Science Society. All rights reserved. Chapter 7 first appeared in *Technology and Culture* 34 (1993): 729–756, © 1993 by the Society for the History of Technology. All rights reserved. We thank the History of Science Society and the Society for the History of Technology for permission to reproduce these chapters.

We have made every effort to obtain permissions to reproduce material that may be under copyright. In a few cases we have been unable to discover who holds copyright or have had no response to our inquiries. If there are any sources that we have not properly acknowledged, we regret our oversight and extend our sincere thanks.

Finally we want to thank our families and colleagues for their support and encouragement during the many hours that went into the making of this book.

Instruments and the Imagination

CHAPTER ONE

Instruments and Images: Subjects for the Historiography of Science

IN THE second aphorism of the *Novum Organum* Francis Bacon argued that "neither the naked hand nor the understanding left to itself can effect much. It is by instruments and helps that the work is done, which are as much wanted for the understanding as for the hand. And as the instruments of the hand either give motion or guide it, so the instruments of the mind supply either suggestions for the understanding or cautions"[1] In this aphorism Bacon identified two wants of natural philosophy—a new method for investigating nature, and new instruments for carrying out that investigation. He succeeded in elaborating a method, but his suggestions about instruments were vague, going little beyond the insistence that real knowledge of nature lay in the hands of the craftsman and not the philosopher.

From our perspective Bacon could not help but be vague, because in 1620, when he wrote, the instruments that made the experimental philosophy possible were just beginning to arrive on the scene. Of course instruments to measure those things that Aristotle called quantities—that is, distance, angle, time, weight—are as old as recorded history. These include rulers, balances, clocks of different kinds, and instruments for surveying, navigation, and astronomy. In the early modern period they were called "mathematical" and were manufactured and employed by "mathematical practitioners."[2] But in the seventeenth century a different kind of instrument made its appearance. The most important of these instruments was the telescope that Galileo used successfully in astronomy for the first time in 1609. Other new instruments in the seventeenth century were the microscope and the air pump—instruments that were to transform natural science. Instead of just measuring length, weight, or time, these instruments distorted nature in some way, either by magnifying it as in the case of the telescope and microscope, or by producing an unnatural condition as in the vacuum created in an air pump. Experiments performed with these instruments were called "elaborate" and were performed in an "elaboratory" or "laboratory." They were called elaborate because they went beyond mere observation and "tortured" nature in order to reveal her secrets.[3] They were also called "philosophical" (as opposed to mathematical) and they were employed by philosophers, whose interests were more intellectual than practical.

Such devices as the telescope and the microscope had existed before the seventeenth century, but not as philosophical instruments. They were instead part of what was called "natural magic." The purpose of the instruments of natural magic was to produce wondrous effects. Natural magic differed from black magic in that the effects were natural rather than supernatural even though they may have *appeared* to be miraculous. As Giambattista Della Porta explained in 1558:

> There are two sorts of magic; the one is infamous, and unhappie, because it hath to do with foul spirits, and consists of inchantments and wicked curiosity; and this is called sorcery; an art which all learned and good men detest; neither is it able to yeeld any truth of Reason or Nature, but stands meerly upon fancies and imaginations, such as vanish presently away and leave nothing behinde them. . . . The other Magick is natural; which all excellent wise men do admit and embrace, and worship with great applause; neither is there anything more highly esteemed, or better thought of by men of learning. . . . I think that [natural] Magick is nothing else but the survey of the whole course of Nature.[4]

The natural magician reveled in his ability to trick the senses of his audience and to conceal the causes of the effects he produced, and he did it with instruments. Della Porta's *Natural Magick* (1558) was loaded with trick mirrors, secret speaking tubes, and automata of all kinds along with recipes for removing spots from clothes, curing diseases, removing pimples, making seeds grow, and other such "secrets."[5] But among his tricks were the germs of the telescope, microscope, barometer, and air pump. It is not coincidental that the earliest known sketch of a telescope is by Della Porta, that Galileo probably got the idea for his thermometer from Cornelis Drebbel's famous perpetual motion machine at the court of James I, that Robert Boyle learned of the air pump from reading the *Mechanica hydraulico-pneumatica* (1657) of the natural magician Gaspar Schott, and that even Newton got his prisms at a fair where they were sold as instruments of natural magic.[6] Most of the "philosophical" instruments, which were the foundation of the experimental philosophy as it developed during the Scientific Revolution, had existed in an earlier version in natural magic.

Experimental philosophers like Robert Boyle and Robert Hooke put their instruments to new and different uses. But one cannot conclude that an enlightened experimental philosophy simply replaced a baleful natural magic in the seventeenth century. In the first place natural magic did not disappear. Athanasius Kircher (about whom we will have much to say in the following pages) happily compiled his enormous Latin tomes completely oblivious to the radical new methods of his contemporaries Descartes, Boyle, and Newton. Nor was Kircher by any means the last of the practitioners of natural magic. Instruments in the natural magic tradition continued to be invented well into the nineteenth century, even though they ceased to be called magical. Natural magic never really disappeared. It was merely subsumed under new categories such as entertainment, technology, and natural science.

One reason for the persistence of natural magic was its practicality. We tend not to think of magic as a practical art, certainly not in a utilitarian sense, but many of the goals of natural magic—creating realistic images where there is no substance, communicating instantly around the globe, imitating and preserving the human voice, revealing hidden sources of power, traveling under the sea, and flying through the air—are technologies we now take for granted. We no longer consider them magic, but in the seventeenth century they were, and their modern "inventors" such as Charles Wheatstone, David Brewster, and Alexander Graham Bell had more than a toehold in natural magic. One can ask, for instance, why Brewster wrote his *Letters on Natural Magic Addressed to Sir Walter Scott* in 1832.[7] In the letters Brewster extols the triumph of modern science over dark superstition, but he is nonetheless captivated by the instruments of the magician, as his own invention of the kaleidoscope demonstrates. Or we can ask whether Wheatstone's telegraph and stereoscope were that far distant from his "enchanted lyre" and "concertina," or Bell's telephone so very different from the speaking machines of von Kempelen and Faber (chapters 7 and 8).

A second reason for the persistence of natural magic comes from its emphasis on instruments. If we approach the Scientific Revolution through a study of experimental method, we recognize an important divergence between the aims of natural magic and those of experimental philosophy—the goal of natural magic was to emulate the wonders of nature and glorify their "wondrousness"; the goal of the experimental philosophy was to establish "matters of fact." If, on the other hand, we study instruments, we see a continuity. Historians and philosophers of science have traditionally debated the relative roles of observation, experiment, and theory in science with the assumption that instruments are made and used in obvious ways in response to the demands of observation and experiment. More recently they have begun to recognize that instruments are much more problematic. Instruments have a life of their own. They do not merely follow theory; often they determine theory, because instruments determine what is possible, and what is possible determines to a large extent what can be thought.[8] In this book we consider a number of instruments that came from the natural magic tradition but also became subjects of debate by experimental philosophers. Because they are part of both traditions, they raise questions about what counts as a scientific instrument, what is the proper method for studying nature, and ultimately, what is natural science.

Rather than trace out a sharp boundary between natural science and other human activity, we show how these instruments moved easily from natural philosophy to art and to popular culture; our investigation will follow the same paths. In so doing, we will use these instruments to consider such problems as the purpose of natural magic, the nature of demonstration, analogies between the senses, distortion versus duplication of the senses, language and signs, images of sight and sound, and alternative views of nature.

Instruments and the Senses

With the exception of the sunflower clock, all the instruments discussed in this book were used to replicate or investigate in some way the phenomena of sight and sound. To those who believed, as Robert Hooke did, that instruments were extensions of the senses, this would not have been surprising.[9] Sight and sound were privileged senses in art, literature, and science. Therefore an instrument was often something with which to see or hear. What was seen or heard, of course, required interpretation, especially if the instrument intentionally magnified or in some other way distorted the image or sound. These were problems of which seventeenth-century investigators were well aware. Francis Bacon, for instance, assumed the existence of four mental faculties in order to explain how the mind learned about nature. These were sense, memory, imagination, and reason. According to Bacon, information coming from the senses had to be organized into images before the reason could operate on it, and this was the function of imagination (see fig. 1.1).[10] The imagination was not limited to the senses, because it could also call up images that had been stored in the memory and could combine parts of images to create new ones. Thus the imagination was the creative faculty, because it had the ability to create entirely new images from old ones, even fantastical images as in dreams. It could also draw comparisons between them. In the *Advancement of Learning* Bacon wrote, "For the mind of man is far from the nature of a clear and equal glass wherein the beams of things should reflect according to their true incidence; nay, it is rather like an enchanted glass, full of superstition and imposture, if it be not delivered and reduced."[11] In this analogy Bacon compared the imagination to a favorite instrument of natural magic and found both of them unreliable. Both instruments and the imagination were essential for creativity in natural science, but both could distort as well as create.

As Bacon planned his project for the reformation of all knowledge, he recognized that the images produced by instruments needed to be "delivered and reduced," just as did the images formed by the imagination. Natural magic produced wonders by tricking the senses, but it was, according to Bacon, "full of error and vanity," and thus he sought a new use for instruments that would correctly "deliver and reduce" the images that they created.[12]

In certain cases a distorted image could be an advantage. The telescope and microscope distorted the image by magnifying it in order to make hidden things visible. But Bacon doubted the usefulness of these instruments, because their application seemed to him to be severely limited. He admitted that Galileo had, indeed, made "noble discoveries" with the telescope, but added that Galileo's "experiment stops with these few discoveries, and many other things equally worthy of investigation are not discovered by the same means."[13] The air pump created a space that might or might not be a

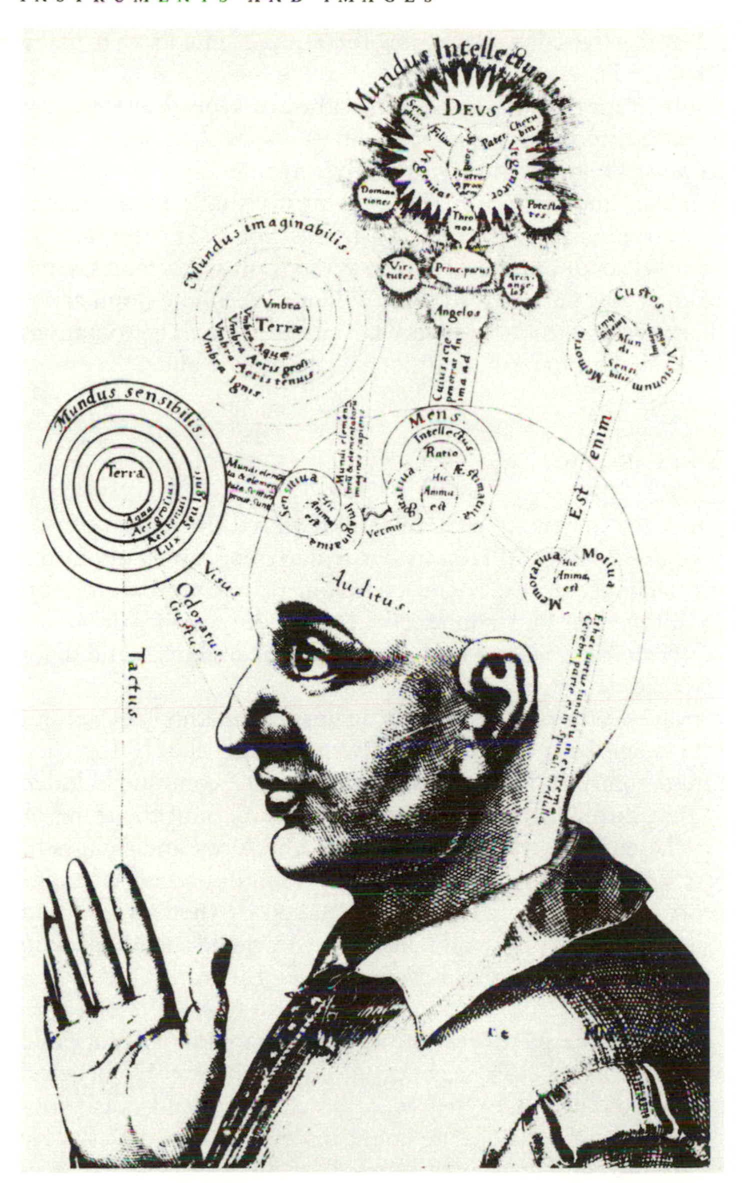

Fig. 1.1. In this drawing by Robert Fludd the faculty of the imagination is seen mediating between the senses and the reason. From Fludd, *Tomus Secundus De Supernaturali*. Courtesy of the University of Washington Libraries.

vacuum, but certainly it was a distortion of ordinary atmospheric space. This question of whether instruments should duplicate human perception exactly or distort it to the philosopher's advantage was not limited to the Scientific Revolution. It was argued as vigorously in the nineteenth century with respect to the stereoscope as it had been in the seventeenth with respect to the telescope, and it remains a question for all instruments that make visual images (chapter 7).

Likewise the substitution of one sense for another, or synesthesia, was an important issue raised by the instruments that we discuss. In some cases, such as graphical recording in acoustics, the only way to analyze sound "scientifically" rather than musically was to represent it visually by an instrument that made a graphic trace (chapter 6). It was not clear how far this substitution of one sense for another could be taken. Louis-Bertrand Castel, the inventor of the ocular harpsichord, and William Jones, the popularizer of the Aeolian harp, wanted to carry it very far indeed, as did the romantics who employed the Aeolian harp in their poetry (chapters 4 and 5).

Instruments and Language

The reduction from the imagination that Bacon required the reason to perform needed to be done carefully. He was particularly concerned about the reduction from images to words. Bacon complained that words often replaced substance altogether. Philosophers, who dealt only with words, lost contact with the objects of sense and built philosophical systems that did not correspond to experience.[14]

Bacon's objection was, in part, a reaction against the methods of natural magic. In natural magic, words and things were bound closely together. Words were more than arbitrary symbols for things; they contained hidden signification, so that through the word one could learn about the thing. A good example is the emblem, which employed both words and images to create an allegory with a hidden meaning. It was a kind of secret language that signified more than the images and words taken by themselves could mean. It was also secret in the sense that only those learned individuals who had been initiated into the meaning of the images and words could understand it.

The natural magician shared this tradition. He operated by allegory and analogy, because that was the only way that he could operate. He believed that real causes were unknowable; they were occult and could not be observed directly, but through analogy he could discern them indirectly. He believed that both words and instruments pointed to the concealed essences of things. Thus Athanasius Kircher's sunflower clock (chapter 2) was more than a timepiece. It was emblematic of (and therefore revealed and made manifest) the occult cosmic magnetic force that was the cause of all change. Because the method of analogy willingly conflated words with things, natural magicians were entranced by both instruments and language.

The "new philosophy" of Bacon, Descartes, Boyle, and Locke turned against this way of looking at the world. If natural philosophers were to use *instruments* that they found in natural magic, they would have to get rid of the hidden sympathies and antipathies associated with them.[15] Just as they called for a reformation of language, they also called for a reformation in the use of instruments—that is, a new experimental method.[16] This created a dilemma for philosophers like John Locke. Locke was especially vigorous in his condemnation of figurative speech. "All the art of rhetoric," he wrote, "all the artificial and figurative application of words eloquence hath invented, are for nothing else but to insinuate wrong ideas, move the passions, and thereby mislead the judgment; and so indeed are perfect cheats."[17] Like Bacon before him, Locke vigorously attacked the confusion between words and things. He believed that the major source of error in our study of the natural world could be found in this misuse of language.

But how should one consider philosophical instruments? Are they more like words, or more like things, or are they halfway in between? They are certainly things, but things whose purpose it is to help us analyze and reason about other things. They are things that we construct to represent and interpret nature. In these capacities they act more like words. If figures expressed in words—that is, analogies like metaphor, simile, comparison, and so forth—are perfect cheats, are the figures presented by instruments cheats also? Would it not be better to depend on our direct unaided senses, rather than allow distorting instruments to come between us and the objects that we observe? Locke finally concluded that the microscope, while not exactly a cheat, would be of little use in studying nature.[18] His criticism of figurative speech was directed especially against the figure of analogy, and to the extent that instruments were used as analogies to nature they were suspect. Not all were convinced, however. Throughout the eighteenth and nineteenth centuries, those seeking alternative approaches to the study of nature attempted to reinstate the magician's identity between words and things, and to reestablish the analogical and symbolical character of instruments (chapters 4 and 5).

The association between instruments and language took a new twist at the beginning of the nineteenth century with the introduction of recording instruments—instruments that wrote down their results in their own "languages." Two new sciences, acoustics and experimental physiology, were made possible or greatly aided by instruments that could detect and record phenomena beyond the reach of the human senses. The new languages of recording instruments (or the new signs that substituted for language) were experimental graphs. The experimental graph first appeared in the second half of the eighteenth century but did not become common until around 1820. Early graphs were often called languages. Édouard-Léon Scott de Martinville designed his phonautograph to assist sound to "write itself in the air." Étienne-Jules Marey called his graphs a new "universal language" to be employed in physiology. The appearance of graphical recording instruments and the rise of the graphical method were for them a new visual language. The study of signs was formalized in the second half of the nineteenth century as semiotics, but it had an important harbinger in J. H. Lambert, who, a century earlier, had created his own semiotics and was the first to make consistent use of experimental graphs (chapter 6).

Whether one wants to call these systems of signs new languages or alternatives to language depends on whether one believes language holds a privileged position among signs—whether words are, on the one hand, mere conventions or, on the other hand, pointers to secret connections with things. The signs recorded by instruments lead us back to the question of the natural magician. What do instruments tell us? Do their inscriptions, like the magician's words, reveal nature's secrets, or do they, like Locke's words, merely state conventions that we have designed into the machinery (chapter 6)? This question was implicit in many of the debates over the validity of philosophical instruments.

Historians of science, who have for the most part been trained either with a strong mathematical orientation, or, like the majority of historians, with a decided literary bent, do their research in libraries filled with words rather than in laboratories filled with instruments. Thus they have not, until recently, really confronted Bacon's problem of the confusion between words and things.[19] Scott and Marey wanted to create instruments that would reduce phenomena to language automatically. To some extent they achieved their goal, but the problem of language still haunts us and we historians continue to live with the tyranny of words that Bacon warned against.

Instruments as Mediators

Just as the imagination mediates between sense and reason (according to Bacon), so instruments mediate between the objective external world and the subjective mind. Some investigators, such as Louis-Bertrand Castel, believed that instruments and their makers quite literally inhabited a world of artifice that mediated between the world of nature and the world of spirit (chapter 4). One could approach this world from either direction. In the case of the telescope, for instance, one could consider it as an extension of the sense of sight (that is, as a way to "see better"), or one could regard it as a revealer of what is "out there" (as a producer of better objects to be viewed). Which way one regards an instrument will depend to some extent on the instrument. We say that we "see through" a telescope or microscope, but we "look at" the output of a mass spectrograph. The general tendency among scientists has been to take the latter position: to regard a scientific instrument as a physical object that produces phenomena which are "detected" by other instruments or by the senses directly. The instrument manipulates nature but not our senses.

Many of the instruments that we discuss seem strange, because they belong to the subjective side of this divide; that is, they were built to explore and imitate human functions—namely, sight, hearing, and speech. This imitative function of instruments is possible because the human organs of sense can themselves be regarded as instruments.[20] Galileo thought of the human eye in this way. He treated it as an optical instrument, and an imperfect one at that.[21] Imitating the senses was a common goal of natural magic; the books of Kircher and Schott are replete with speaking heads, ear trumpets, magic mirrors, and magic lanterns. Natural magicians used their knowledge of these instruments in large part to entertain and to mystify, but through the eighteenth and nineteenth centuries, the quest for ways to imitate human hearing, vision, and speech produced both practical instruments of communication and the "serious" sciences of acoustics and the physiology of vision. While the new instruments lost much of their "magic," they still retained their ability to entertain, as attested by the stereoscope, the cinema, and audio and video recording and display (chapters 7 and 8).

In the history of science, instruments have played manifold mediatory roles. In addition to improving the existing senses, instruments have been called upon to measure things, produce images, model phenomena, and alter the state of nature. In each case the instrument allows the observer to approach nature in a way that would be impossible with only the unaided senses. Although some of these functions have been ascribed solely to the early modern period, they were, in fact, all still active through the nineteenth century. In order to investigate the relationship between nature and its students, we examine instruments as they mediate between the object and the observer.

Instruments as Demonstrators

As mediators between objects and observers, instruments often performed the function of display. They "showed" something, or "demonstrated" something to the observer or observers. This "showing" could be on several different levels. On the first level, the instrument itself could be an object of display. During the seventeenth century, the *cabinet* of a wealthy collector would contain instruments alongside natural history specimens, rare manuscripts, paintings, and antiquities; alternatively, the instruments would be part of a more specialized *cabinet de physique* (chapter 3). Potential patrons received such instruments from would-be clients. The instruments themselves were objects that conferred status and acknowledged rank.

On a second level instruments displayed phenomena. They created effects that did not occur naturally. This was the major purpose of the instruments of natural magic. The instruments "showed" phenomena the causes of which were hidden. In the experimental philosophy instruments established "matters of fact." These were events or deeds performed by or with instruments and testified to by men. As in natural magic, instruments in experimental philosophy displayed unusual events the truths of which were validated by witnesses.

On a third level, instruments confirmed or "demonstrated" theory. In this case the "showing" was neither of the instrument itself nor of the phenomenon that it produced, but of the cause or explanation behind the phenomenon. Thus Newton's prisms "demonstrated" his theory of colors.

The word "demonstration" still carries all three of these meanings, and therefore its use in any particular case has to be inferred from what is demonstrated, whether it be a new species of plant (display of an object), the effects of extremely cold temperatures (display of a phenomenon), or the law of falling bodies (confirmation of theory). The instruments that we discuss in this book, because of their roots in natural magic, typically fall into the first two categories of demonstration—that is, they tend to be vehicles for display of objects and phenomena rather than agents for confirming hypotheses or theory. This gives them more of a carnival character than the instruments that we commonly think of as "scientific."

Scientists may choose to dismiss them as toys, but as historians we cannot, because they raise such issues as the distinction between teaching and research (a distinction that did not exist until the late eighteenth century), the origin of the demonstration lecture (a late-seventeenth-century innovation that resulted in the new discipline of experimental physics), and, more generally, the question of what counts as an experimental demonstration. With the introduction of precision measuring instruments in experimental physics and chemistry in the second half of the eighteenth century the emphasis shifted dramatically from demonstration as display to demonstration as confirmation of theory. Ironically the demonstration lecture enjoyed a revival at the same time and, in its most popular form, eschewed all mathematics and quantitative measure (chapter 3). The instruments that we discuss here can help expose these distinctions precisely because they are not so obviously "scientific."

The reader of this book may feel as though he or she has entered the attic of a very old house and has found in dusty trunks the vestments of an earlier era. The ridiculous hats and the profusion of petticoats seem completely unsuitable as human attire. It is difficult to conceive how such awkward garments could ever have been in fashion.

The instruments we discuss are similarly outdated. They bear little resemblance to modern tools of science or, consequently, to those instruments from the annals of science that seem to have been the most crucial. Modern scholars are apt to view speaking heads and Aeolian harps as utterly marginal to scientific progress. They strike contemporary scientists and historians as too toylike and frivolous to merit consideration as "serious" science. These devices have been relegated to the cellars and attics of historiography. Yet the study of these apparently tangential objects can disclose connections that would otherwise be invisible—such as the links between the development of modern science and the enduring tradition of natural magic, or between the comparative roles of instruments and language. These themes may join other subjects in the history of science, like the role of the occult, which was once rejected as unscientific but is now a favorite subject of study.

Perhaps it is the peculiar status of our instruments—their capacity to bridge the gap between the rigidly scientific and the amusingly nonscientific—that is the source of their alleged marginality. However, "marginality" per se should not be grounds for dismissing an item from the field of historical inquiry. Margins, of course, indicate the penumbrae of boundaries—in this case the boundaries of scientific legitimacy. But margins are also surfaces of contact and connection between and among different themes and entities. Documents concerning the Aeolian harp or the stereoscope, for example, confuse the historian because they present analyses and discussions that are perfectly lucid from the point of view of modern science (the analysis of fluid flow around a cylinder and the geometry of binocular vision) *alongside* issues that are entirely opaque to a modern reader (the analogy between sight and sound, and the instrumental imitation of human speech and vision). This dualism and strangeness, however, should provide a spur to historical investigation, rather than a cause for discarding the subject. "Scientific instruments," including those that seem merely quasi-scientific, are the material indexes of the study of nature. They embody approaches to nature—oftentimes approaches that are unfamiliar to us. The instruments discussed below possess a full measure of this unfamiliarity. Their contextual roles may not have been tailored to the expectations of modern-day tastes. But, upon examination, these devices remind us of the diverse origins of the vast and complex enterprise of modern science and of some thematic threads in its fabric that historians have forgotten or ignored.

CHAPTER TWO

Athanasius Kircher's Sunflower Clock

WHEN Athanasius Kircher arrived in Avignon in 1632 he at last found some relief from the turmoils of the Thirty Years War that had propelled him across Europe. From Paderborn, where in 1618 he had been admitted as a novice to the Jesuit order, he had been driven by the Protestant forces to Köln, Koblenz, Heiligenstadt, Mainz, Würzburg, and finally Avignon. His peregrinations had not, however, prevented him from pursuing his studies, and by the time he arrived at Avignon he was already known for his profound erudition. Claude Fabri de Peiresc and his friend Pierre Gassendi were pleased to have Kircher in their vicinity, especially because his mastery of exotic languages promised to help them decipher the hieroglyphic and Coptic manuscripts that Peiresc had acquired. Kircher also promised to be a useful companion in the astronomical observations that Peiresc and Gassendi were carrying out (see fig. 2.1).[1]

In addition to astronomy and hieroglyphics, Kircher's interests included magnets and clocks. His first book, *Ars Magnesia* (1631), was a study of magnetism, and while at Avignon he wrote another on sundials entitled *Primitiae gnomonicae catoptricae* (1635; dedicated May 10, 1633). Kircher not only wrote about clocks, he also built them. He erected a square sundial at the Jesuit College in Koblenz in 1623, and the next year he built another at Heiligenstadt. In 1631 he made a "pantometrum" for Ferdinand III, archduke of Austria, to measure "length, breadth, heights, depths, areas, of both earthly and heavenly bodies," and at Avignon he built the most complex of all, an elaborate indoor sundial that employed mirrors to bring the sun's rays into the building.

On March 2, 1633, Peiresc alerted Gassendi to an impending visit by Kircher at Aix. Gassendi's presence was needed not only for his instruments, which he, Kircher, and Peiresc would use to observe an upcoming eclipse, but also for his influence in persuading Kircher to forsake his teaching duties at Avignon in favor of astronomy and hieroglyphics at nearby Aix.[2] The effort succeeded, at least in part, because Peiresc wrote to his friends Jacques and Pierre DuPuy on May 21, 1633, that Kircher had been with him for four or five days and that he and Gassendi were negotiating to keep him for an entire year. In addition to Kircher's knowledge of hieroglyphics, Peiresc was taken by the many "beautiful secrets of nature" that Kircher had at his command, especially a clock that was driven by a sunflower seed, which followed the sun just as the blossom does from sunrise to sunset. Moreover, this clock followed the sun even indoors and when the sun was covered by clouds. Kircher had demonstrated the clock "en bonne compagnie en pleine

Fig. 2.1. Claude Fabri de Peiresc, scientific amateur and correspondent. Courtesy of the Burndy Library, Dibner Institute, Cambridge, Mass.

Athanasius Kircher, natural magician and "monster" of erudition. From Kircher, *Athanasii Kircheri e Soc. Jesu China monumentis*, frontispiece. Courtesy of University of Washington Libraries, Special Collections.

table" before the elector of Mainz, who could testify to its success. For Peiresc this was a "great miracle of nature, which merits being seen."[3]

With Peiresc's help and that of Marin Mersenne, news of the sunflower clock circulated rapidly.[4] Godefroid Wendelin at Brussels and René Descartes in Deventer both responded to Mersenne's questions about the clock. Descartes was skeptical but did not consider it impossible:

> If the experiment that you have described to me about a clock without sun is certain, it is indeed curious and I thank you for having written to me about it; but I still doubt the effect, and at the same time I do not judge it impossible. If you have seen it I would appreciate more information about it.[5]

Wendelin added a new twist to the story. He reported that he had seen a similar clock made by another Jesuit, the famous or notorious Father Linus (Francis Hall) at Liège, who later challenged the experiments of both Boyle and Newton. Linus's clock consisted of a sphere filled with a liquid in which floated a smaller sphere. This smaller sphere, with hours inscribed about its equator, rotated under some cosmic influence, and a small stationary fish floating beside the rotating sphere pointed out the hours.[6]

On September 3, 1633, Kircher was again at Aix, and this time Peiresc had an opportunity to see the sunflower clock put to a test. Kircher carried out his demonstration at the Jesuit College at Aix before witnesses; Peiresc wrote up a description of the proceedings.[7] Kircher first demonstrated a magnetic clock of his own devising that consisted of a piece of lodestone, "somewhat larger than a nut," wedged in a groove on a circular disk of cork, which in turn floated in a tub full of water. Around the edge of the tub was a paper scale divided into twenty-four hours, and on the cork was another paper scale also divided into twenty-four meridian lines. As the magnet rotated, it carried the cork around with it and the divisions on the scales gave the time not only at Aix, but also at other major cities around the world. After demonstrating his magnetic clock, Kircher tried out his sunflower clock, but in this case Peiresc could observe no effect. He wrote to Mersenne, "As for the sunflower clock . . . we haven't been able to obtain one and I am of your opinion and don't believe in it any more than you do."[8] Peiresc was disappointed by the failure of the sunflower clock to live up to expectations, but he remained in good relations with "le bon Athanase."

At the end of September Kircher was ordered to go to Trieste, much to Peiresc's annoyance and chagrin. After a dangerous and adventurous journey he reached Rome, where he found that Peiresc had arranged for him to stay there at the Collegium Romanum and for Father Christoph Scheiner to make the trip to Trieste in his stead. Kircher's hasty departure from Avignon bothered Peiresc, not only because he lost a valuable colleague, but also because he had not had time to set up a safe conduit for correspondence.[9] The problem of communication is illustrated by the fact that a letter from Peiresc to Galileo written January 26, 1634, did not reach its destination until after March 18 of that year. It was accompanied by a letter from Gassendi and was transmitted via G. G. Bouchard at Rome and Raffaello Magiotti, both of whom added letters of their own in which they announced to Galileo the arrival of Kircher at Rome. Magiotti wrote:

> There is now at Rome a Jesuit, long in the Orient, who, besides knowing twelve languages and being a good mathematician etc., has with him many lovely things, among them a root which turns as the sun turns, and serves as a most perfect clock. This is affixed by him in a piece of cork, which holds it freely on the water, and on this cork there is a needle of iron that shows the hours, with a scale for knowing what hour it is in other parts of the world.[10]

Apparently Kircher's demonstration was more successful in Rome than in Aix, or else Magiotti was more easily persuaded than Peiresc.[11]

The Magnetic Clock of Father Linus

In December 1634 Peiresc received a visit from the papal nuncio of Köln, Pierluigi Caraffa, accompanied by his confessor, Father Sylvester Pietrasancta. Pietrasancta brought with him his recently published *De symbolis*

Fig. 2.2. The magnetic clock of Father Francis Linus as depicted in Sylvester Pietrasancta's *De symbolis heroicis* (1634), from the workshop of Peter Paul Rubens. From Pietrasancta, *De symbolis heroicis libri IX*, p. 146. Courtesy of the University of Washington Libraries, Special Collections, neg. no. 15408.

heroicis, which contained a description of Linus's clock, the one that Wendelin had mentioned to Mersenne more than a year earlier (see fig. 2.2). Pietrasancta wrote in *De symbolis*:

> Recently at Liège, P. Franciscus Linus, a mathematics instructor in the English College of our Society, devised most successfully this orb, which is placed inside a glass phial, which orb stays in the centre of the surrounding water (just as the Earth stays in the centre of surrounding air) by a secret balancing of its mass. But the orb by an arcane force and as if by a certain love strives after the conversion of the sky from east to west and is driven around altogether in the space of 24 hours. A little fish is placed inside as indicator, and like an expert swimmer, its weight poised, watches the fleeting hours and designates them with its snout, its eyes gazing intently on them. When the phial is moved, if impetus is given to the water,

soon by its own will it regains the path of its orb; and the calculation of time will be wholly unaltered after tranquillity is restored. . . . It will hasten [to the sun] ever so quickly since love knows no delay; and although it may leap back and forth several times, finally it will obtain that position to which, as a comrade and fellow traveller of the Sun, it will return without fail.[12]

Peiresc saw the book, but apparently had not immediately noticed or had not been impressed by what Pietrasancta and Caraffa had told him.

However, further confirmation of the magnetic clock was on its way. The painter Peter Paul Rubens wrote to Peiresc from Antwerp to supply what Peiresc had missed:

I enclose here a folio from the Reverend Father Sylvester de Pietra Sancta's *De Symbolis Heroicis*, on the mysterious clock (or glass globe) in a decanter filled with water. You will see it reproduced in the engraving and described in the text. . . . You need not doubt the authenticity of the thing (the mystery consists in a certain attraction and magnetic power); I have talked with men of ingenuity who have seen and operated it with ease, and have the greatest admiration for it.[13]

Here was more direct evidence that the clock actually worked, and Peiresc now wished that he had queried Pietrasancta and Caraffa more specifically when he had had them at Aix. Rubens had drawn the frontispiece for Pietrasancta's book, and his workshop was almost certainly responsible for the other illustrations, including the drawing of Linus's magnetic clock.[14]

Peiresc wrote Galileo about the clock and said that Pietrasancta and Monsignor Caraffa had both assured Rubens that they had observed the clock themselves; Caraffa had even had it at his house for two days, and it was reliable.[15] A month later Wendelin sent a letter to Gassendi (which was also intended for Peiresc) to say that he had insinuated himself into friendship with Linus and had asked him what was in the sphere that made it go around. Linus had merely shrugged his shoulders, indicating annoyance at Wendelin's uncivil curiosity. Mersenne—apparently having overcome his earlier doubts about the efficacy of the sunflower clock—suspected that it was driven by a sunflower seed. When asked how long the clock would keep going, Linus told Wendelin only three or four hours, which led Wendelin to suspect that the wax ball floating in the liquid became waterlogged after that time.[16] These reports encouraged Peiresc, who continued to write enthusiastically about the clock.

But it was Galileo himself who dampened Peiresc's enthusiasm. He wrote on May 12, 1635:

The water-clock [of Linus] will truly be a thing of extreme marvel if it is true that the globe suspended in the middle of the water goes naturally turning by an occult magnetic force. Many years ago I made a similar invention, but with the aid of a deceptive artifice, and the machine was this. The little globe with 12 meridians for the 24 hours was of copper, hollow within, with a little piece of magnet placed at the bottom, and almost in balance with the density of water; so that placing in the

> vessel some salt water, and then on that some sweet water, the globe stayed between the two waters, that is, in the middle of the vessel, which vessel had a wooden base in which there was concealed a clock made expressly in such a way as to rotate a piece of magnet that was fitted upon it, making one revolution in 24 hours, which motion the other magnet placed in the little globe obeyed, making it turn and show the hour. Thus far went my speculation; but if this one of Father Linus without any artifice makes his globe obey the motion of the heavens, truly it will be a celestial and divine thing, and we shall have a perpetual motion. Your excellency, by those means which you recite, will easily be able to come to a knowledge of the whole matter; I, meanwhile, have wished to indicate my thought in order to have a witness beyond all exception that I have not usurped the invention from Father Linus—if indeed his machine does not have any more to it than mine.[17]

Meanwhile Peiresc waited for a firsthand description of the clock from someone who had actually seen it, either Rubens, who offered to go to Liège to witness its operation, or his friend Dormalius.[18] It was Henri Dormalius who finally came to Aix and on June 18, 1635, delivered an eyewitness account of the clock. As Peiresc informed DuPuy, "We have had the pleasure of hearing from his [Dormalius's] mouth a description of the machine of P. Liny, but the particular details that he has told us about its horizontal movement make me suspect an artifice which would not be natural, and would therefore greatly diminish the admiration of the instrument."[19]

We do not have the report of Dormalius, but probably the fact, reported by Pietrasancta, that the sphere would occasionally move erratically led Peiresc to suspect a hidden mechanical contrivance. Galileo's doubts and Dormalius's report reduced Linus's clock from an important discovery to a curiosity.[20] The suspicion of a "concealed artifice" was greatly strengthened in 1641 when Kircher described an identical clock in his *Magnes* (although claiming it as his own invention) and revealed that it contained a mechanism concealed in its base, as Galileo had suspected (see fig. 2.3). The *Magnes* also contained a description of Kircher's sunflower clock, but for this clock Kircher did not admit any artifice.

Descartes, however, was tired of such tricks. After looking at the *Magnes* he wrote to Constantijn Huygens:

> The Jesuit has lots of tricks [*farfanteries*]; he is more charlatan than savant. He speaks among other things of a substance that he says he purchased from an Arab merchant, which turns day and night toward the sun. If this were true, it would be curious; but he doesn't explain what this material is. Father Mersenne wrote me about it approximately eight years ago, and said it was the seed of a heliotrope; which I don't believe unless this seed has more power in Arabia than it does here, because I tried it when I had some free time and it didn't work.[21]

Descartes was not prepared to accept any clock driven by an occult celestial sympathy.

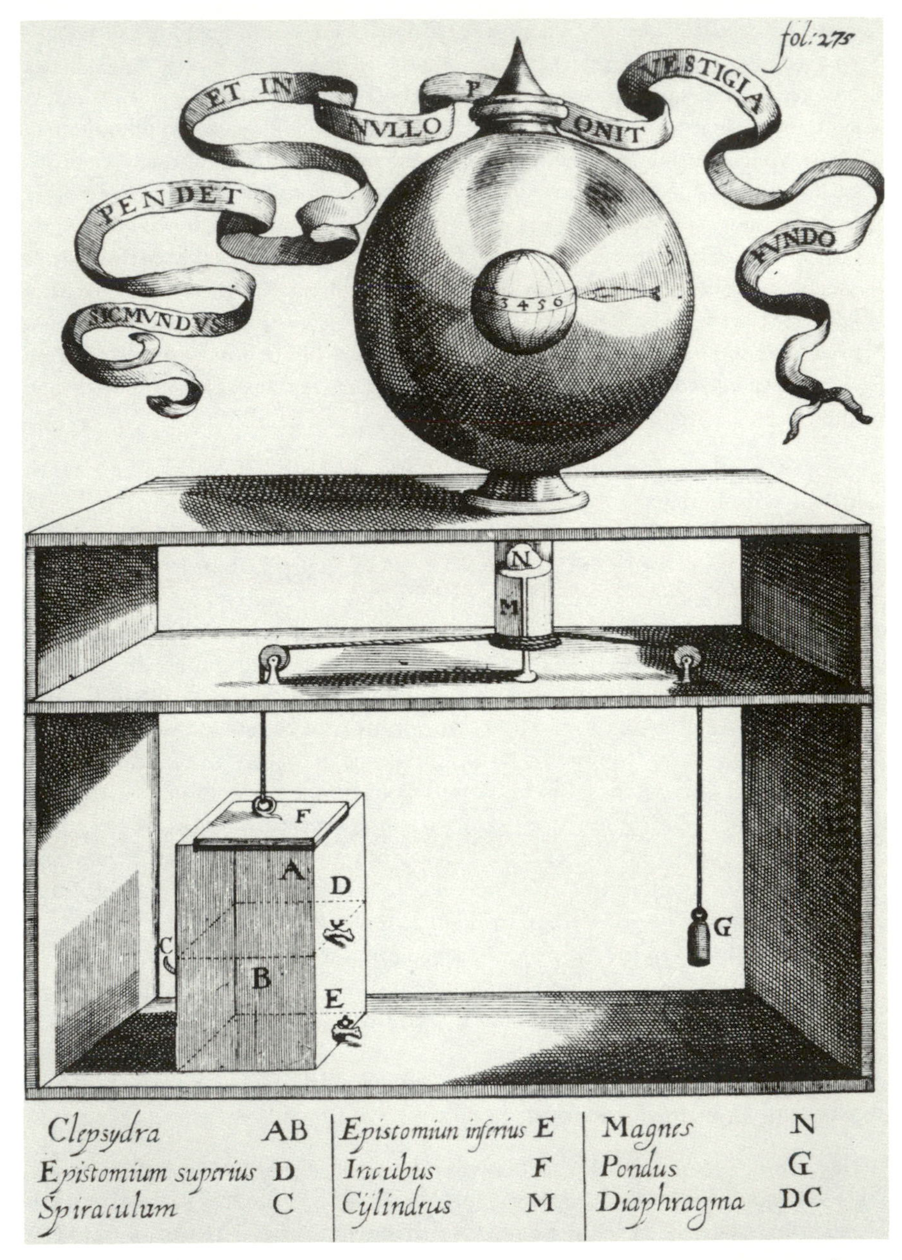

Fig. 2.3. Kircher reveals the secret of the magnetic clock in his *Magnes, sive de arte magnetica* (Rome, 1641). From Kircher, *Magnes, sive de arte magnetica*, p. 311. Courtesy of the University of Washington Libraries.

The Clocks as Instruments

So much for the story of the clocks. How are we to make sense of such strange instruments? Were these clocks hoaxes? Did Linus and Kircher intend to deceive their audiences? And if so, why were they not exposed? Why were they treated with such tolerance? Also, how could Kircher demonstrate his sunflower clock in Mainz, Avignon, Aix, and Rome, and still convince himself and others that it worked? The late-nineteenth-century historian Georges Monchamps, in his analysis of the story, concluded that although "it is fashionable among a number of authors to deny the truth of certain extraordinary claims of Father Kircher . . . it seems incontestable that one cannot reasonably doubt [his claims]."[22] Monchamps accepted Kircher's account! But unfortunately magnets and sunflower seeds do not rotate by themselves, at least not for anyone but Kircher.

We can also ask how Linus, Galileo, and Kircher all came up with the same magnetic clock independently. Or were they independent? It is apparent that these instruments and their demonstrations in the seventeenth century were not being judged as we would judge them today. We need to ask, "What was the purpose of these clocks, and what was the purpose of similar instruments in the seventeenth century?"

The Magnetic Philosophy

One can make a convincing argument that during the first half of the seventeenth century there existed a natural philosophy that was neither scholastic nor "mechanical," but magnetic.[23] It took its lead from William Gilbert's *De Magnete* of 1600 and continued Gilbert's effort to explain natural phenomena in terms of the single fundamental force of magnetism. Gilbert had made magnetism real. It remained the paradigm of an occult cause, that is, a cause that was known only by its effects, but in Gilbert's hands it was also a *natural* cause, not a Hermetic or spiritual one. Also Gilbert's experiments were a model for a new empirical approach to natural philosophy. Kepler and Galileo both took great interest in magnetism, and when Kepler argued that he worked with "real physical forces," not just imagined orbs and epicycles, he meant magnetic forces, which he believed controlled the motions of the planets.[24]

Magnetism was also a popular part of natural magic, which sought to produce prodigies or "wonders" through hidden causes, and there was no cause more hidden than magnetism. One of Della Porta's favorite magnetic tricks had been to move lodestones beneath a table that had been covered with fragments of another crushed lodestone. The fragments would stand up and march around with no visible cause. Della Porta used his magnetic table to imitate armies maneuvering on the field and locked in combat.[25]

The magnetic philosophy also included the study of medicinal herbs and other plants that were believed to operate by magnetic influence, the prime example being heliotropism.[26] The ability of a flower to follow the sun was ascribed to the same cosmic magnetic influence as that which moved the planets and caused the rotation of the earth. Other characteristics of plants suggested polarity like that of a magnet. The root and shoot from a seed always grow in the proper directions no matter how the seed is oriented. A cutting must be grafted to a branch by the correct end; otherwise it will not grow.[27] A willow wand used in dowsing is attracted like a magnet to water. Transplanted shrubs grow better if their north-south orientation is not changed. So Kircher's sunflower clock was not as odd an idea as it might initially seem. If the sunflower was moved by the sun's magnetic influence, there was no reason why it should not move as well at night as during the day. The sunflower added drama to his demonstration because it was a striking plant and had the added mystery of having come relatively recently from the New World.[28] In keeping with the magnetic philosophy, Kircher believed that magnetism was the universal occult cause behind all motion, so it is not surprising to find him trying to construct clocks driven by that mysterious force.

Galileo's Trial

Galileo's trial caused problems for Kircher and his fellow Jesuits. While Kircher seemed inclined to accept the Copernican system in 1632, he most definitely rejected it after it was condemned by the Holy Office.[29] In his *Magnes* of 1641 Kircher led the way in criticizing Gilbert, Kepler, Stevin, and any other philosopher who might employ magnetism to account for a moving earth. In fact, Kircher and his fellow Jesuits argued that magnetism was responsible not for moving the earth, but for keeping it stationary, while the celestial sphere turned about it.[30]

Peiresc, on the other hand, was a convinced Copernican and a great admirer of Galileo. Linus's clock was for him a heaven-sent demonstration of the earth's motion, for if the little sphere inside the liquid-filled glass globe rotated by magnetic influence, then by analogy the sphere of the earth at the center of the celestial sphere would likewise rotate by its magnetism as Gilbert had claimed. The irony of such a demonstration's having been discovered by a Jesuit did not escape Peiresc, as he mentioned to Galileo, and he bent every effort to persuade Cardinal Barberini, the pope's nephew, to use it as a reason for reconsidering the sentence against Galileo.[31] Likewise the sunflower clock was further evidence that there existed a cosmic magnetic influence which rotated objects with a diurnal motion. Kircher's experiments at Aix with a magnet and a sunflower seed were two attempts to detect the same force. Not wishing to contradict his friend and patron, Kircher would not openly oppose Peiresc, but after Peiresc's death in 1637 he inserted in his *Magnes* of 1641 a cutaway drawing of Linus's clock revealing a hidden mechanism in the base and thereby undermining the argument for a moving earth.[32]

Trials and Testimony: Peiresc's Report of the Trial at Aix

We can get some insights into the clocks by looking closely at contemporary descriptions of their use. Peiresc's account of Kircher's demonstration at Aix is a traditional case of "witnessing" wherein the validity of an experiment or instrument is established by a trial before witnesses (usually noblemen), whose word can be trusted and whose patronage can be expected (see fig. 2.4). The manuscript is titled "1633, 3 Sept. at Aix THE CLOCK of P. ATHANASE Kircher made with the Seed or flower of the sunflower or with the lodestone," and the text is as follows:[33]

> The Reverend Father Athanase Kircser, Jesuit, has shown us this third of September 1633 at the Jesuit College the trial that he has made with a clock which shows the hours in a darkened closed chamber by the magnetic virtue alone. Having put in a clay pot full of water . . . a round piece of cork larger than the palm of the hand and as thick as one finger in the middle of which he had made a groove to hold a piece of lodestone larger than a nut, which he had carefully adjusted to its pole and therefore also to the proportion of its true declination from the meridian line. He then attached this piece of cork at the center on the underside to a thread the other end of which was attached to the center of the bottom of the pot in such a way that the cork could turn horizontally to and fro departing scarcely at all from the center of the water surface. . . . He next covered the edge of the clay pot with a circle of paper divided into twenty-four equal parts, and he covered the piece of cork with another circular piece of paper, also divided into twenty-four equal parts on which he marked the names of cities situated on all the meridians in relation to the meridian of the city of Aix, to which he had adjusted a paper pointer [*dent*] which reached to the outside circle and served to show the time. He afterward adjusted a little movable cardboard Alidade {for finding} the required proportion of departure from the meridian line for his geography paper and situated his pot on the true meridian line, which he marked on the table where it was situated.
>
> He then moved the cork and allowed it to return to its natural position. The little marker that was on the meridian for the City of Aix stopped itself exactly at one-third of an hour after two in the evening or afternoon, and at the same time the same paper showed by a definite relation what the time was at Rome, Constantinople, Jerusalem, Babylon, the Indies, China, America, Peru and the Canarys, and also other places.
>
> But it was without doubt necessary for him to turn his vase or the circle mounted on it in order for it to mark other hours. Because the little paper marker that was supposed to give the time [for Aix] ought to remain immobile since its

1633. 3 Sept. a Aix.
L'HOROLOGE DU P. ATHANASE Kircher
tant avec la Graine & fleur du solanum
qu'avec la pierre d'Aymant.

215

Le R. P. ATHANASE KIRCSER, Jesuite nous a faict voir le 3 sept
1633. au College des Jesuites l'essay qu'il avoit faict d'une
Horologe qui monstre les heures a l'ombre dans une
chambre fermee, par la seule vertu Magnetique.
Ayant mis dans un pot de terre remply d'eau de la largeur ou
ouverture d'un pan ou environ un morceau de liege de forme ronde
plus large que la paulme de la main, espois d'un doigt, dans le milieu duquel
il avoit faict un creux pour y placer un morceau d'Aymant
plus large qu'une noix. lequel il avoit bien adjusté à
son pole, et par consequent à la proportion de sa vraye
declinaison de la ligne meridienne. Et avoit attaché
ce morceau de liege par son centre inferieur, à un fillet
attaché, au centre du cul du vase ensorte qu'il se pouvoit
tourner orisontalement ça et là sans s'esloigner de guieres
du centre de la superficie de l'eau, ou de l'ouverture
de la Gueulle du vase. parceque le filet n'estoit pas plus long que
l'espoisseur de l'eau, tellement qu'il faudroit que le Morceau de liege
s'enfonceast dans l'eau pour pouvoir s'esloigner du centre plus
d'un costé que d'aultre. Il avoit par aprez couvert
le bord du pot de terre, d'un cercle de papier divisé en
XXIIII. parties esgales, et toute la superficie de son liege
d'un aultre morceau de papier couppé en rond, et pareillement
divisé en XXIIII. parties esgales, sur lesquelles il avoit marqué
des noms de Villes situees en tous les 24. meridiens
qui respondoient par les longitudes au meridien de la ville
d'Aix auquel il avoit adjousté une dent de papier sortant
hors de la rondeur d'iceluy, pour servir à monstrer l'heure.
Il avoit par aprez adjousté une petite alidade de cartoncin
qui estoit mobile, pour gouverner la proportion requise de l'esloignement
de la ligne meridienne, avec son papier de Geographie.
et avoit situé son vase en sa vraye ligne meridienne
qu'il avoit marquee sur la table ou il estoit situé.

Fig. 2.4. Peiresc's testimony of the trials of Kircher's magnetic and sunflower clocks at Aix on September 3, 1633. Bibliothèque inguimbertine, Archives et Musée Municipal, Carpentras, Peiresc MS 1864, fol. 215. Courtesy of the Archives Communales, Carpentras.

> position is relative to the pole or to the local declination of the magnet. And for [the magnet] to follow the course of the sun, it would be necessary that the circle of hours be moved, which movement could not be regulated unless he had previously marked the principal points on the table or on the edge of the vase, and in fact he admitted to me that in order to prepare his instrument it was necessary for him to be with it for several hours in advance in order to adjust his apparatus [*son faict*] to { } hour, so that he had a way of moving the circle and stopping it at all the points of the hours that he wished to examine with his magnet. Which is not only tiresome, but also useless, since this proportion is what one seeks with the instrument and not what one accommodates to the instrument {at each trial.}

No wonder Peiresc was disappointed: what Kircher had demonstrated was a compass! The floating cork carried a piece of lodestone, which, not surprisingly, lined up with the north and south magnetic poles. The circle of hours around the outside of the pot corresponded to the hours marked on any twenty-four-hour clock, and the small cardboard pointer on the cork (fixed on the meridian line representing Aix) corresponded to the hour hand—except that in this case the hour hand did not move by itself; it stayed fixed on magnetic north. Before the trial began, Kircher had to rotate the pot or the circle of hours, so that during the trial, magnetic north would coincide with the hour that the trial was to take place. During the trial Kircher would move the cork, and as the spectators watched, the cork with its pointer would slowly swing back to precisely the correct time. The hope was that the spectators would be gullible enough not to realize that it was actually indicating not an hour but a direction. Of course, a lodestone is not a magnetic needle, and the line of the poles on a lodestone "slightly larger than a nut" would not be obvious to an observer. Therefore it would not be immediately apparent that the lodestone was seeking magnetic north.

The fancy alidade and meridian lines telling the time at all the major cities of the world were designed to impress the spectator. The relationship between local time at Aix and time anywhere else in the world is a simple matter of the difference in longitude—one hour for every fifteen degrees, just as we understand from our time zones. Kircher was following the tradition of elaborate clocks: the more complex, the more impressive—a tradition that he had followed in building ever more complicated sundials.[34] It was also a way of increasing the wonderment his instrument caused, and it had the added advantage of drawing attention away from the fact that the lodestone always pointed in the same direction.

But Peiresc was not taken in, if, indeed, that had been Kircher's intention. Peiresc had been trained in the law and served as senator in the parliament at Aix. He knew how to examine and take testimony. Note that he describes the instrument in great detail, the date and place of the trial are carefully noted, and he did not hesitate to put Kircher on his word. He forced Kircher to admit that his clock would tell the correct hour only if it were adjusted ahead of time so that the pointer would come out in the right place, which, of course, defeated the purpose of the clock. The report of the trial continues:

> He [Kircher] said that he had demonstrated his instrument at Mainz before the elector, then living, and at Avignon before three fathers of his company where he had put on his cork in a little hollow canal a quantity of sunflower seed [*solanum montanum*] of the kind that is { } while the seed used in the experiment at Avignon was from the Alps. And that this material followed the sun by a sympathy similar to that followed by the flower both east and west as well as meridional. . . .
>
> But what made me doubt the certitude of his experiment and of his words was the fact that he would not swear that the sunflower seed alone was sufficient for the demonstration; thus, without actually saying it, he left me with the understanding that he required some other unknown ingredient that he did not wish to declare, and which I guess to be his magnet. Now I do not find this to be a miracle of any kind or even a useful convenience because it is necessary to assure oneself of the position of the instrument in advance in order to make it seem that it finds the correct time, in order to trick the spectators.

In this version of the experiment Kircher used a sunflower seed in place of the magnet and a different kind of pivot for the cork float. It is not obvious that Kircher actually demonstrated this clock before Peiresc, but Peiresc asked the right question. He wanted to know if there was anything more to the apparatus than Kircher had described, and from Kircher's refusal to swear that the sunflower seed alone made the effect, he concluded that Kircher's instrument contained a concealed magnet. His suspicion is strengthened by Magiotti's report to Galileo of Kircher's successful trial at Rome. At that later demonstration Kircher employed "a root which turns as the sun turns, and serves as a most perfect clock. This is affixed by him in a piece of cork, which holds it freely on the water, and on this cork there is a needle of iron that shows the hours, with a scale for knowing what hour it is in other parts of the world."[35] An iron needle has replaced the cardboard pointer, and one cannot help but suspect that it was magnetized in order to "help" the sunflower root turn the cork float. Peiresc's hope, as expressed to DuPuy the previous May, that he would see a "great miracle of nature" had been sadly diminished.[36] But Kircher had one more instrument to try:

> He then took a piece of cork in the form of a spoon in the hollow of which he put some white seeds of the sunflower of Mr. Robin in order to try its effect, and he wanted to fix it to the right of the meridian of the city of Aix on his cork so that the hollow of the spoon would face directly into the sun according to the inclination of the ecliptic as do the flowers of the sunflower, but this was useless and without effect.
>
> He said further that the effect of this sunflower seed clock was more sensitive when exposed to the rays of the sun than when covered and shadowed from the rays.

> And that he had made this first experiment [with the sunflower seed] in Germany in winter because there were no flowers and that the flowers {worked} better.

If the trial took place at approximately 2:20 in the afternoon, which was the time on Kircher's magnet clock, then the sun would have been to the right of the southern meridian line and the sunflower seeds should have had their greatest effect if they were inclined to the horizontal by the degree of the sun's altitude and could thereby face directly toward the sun. The sunlight could not strike the seeds, because the experiments were done indoors, but the supposed magnetic influence of the sun would, of course, penetrate the walls and the sunflower seed should turn to face it. But in this case, as Peiresc reported, there was no effect. And in conclusion Peiresc reports:

> He [Kircher] is persuaded that the flower henceforth will be capable of telling time if one were able to keep it planted on the cork floating on water, because then it would not require as great a force to turn itself to follow the sun as when its stem is fixed immobile to the root and it has to twist its neck violently from sunrise to midday to sunset. And if it can do this with some ease [when fixed to the root and planted in the ground], it would appear that in aiding or facilitating the sympathy or natural inclination of this plant, it would be able to produce its effect in a more regular way and therefore be more capable of showing the hours.

Kircher reasons that if the magnetic influence is sufficient to turn the blossom of a rooted plant toward the sun, then it *must* be sufficient to turn the plant poised on a frictionless pivot both during the day and at night.

Kircher's Account

In the *Magnes* (1641) Kircher has an elaborate engraving of a sunflower in full blossom, floating on a cork and pointing out the hours by a stylus attached to the center of the blossom (see fig. 2.5). Because the scale of hours to which the stylus points is held up by mysterious hands protruding from clouds (symbolizing, perhaps, the magnetic powers of the sun and the moon), one is forced to conclude that the drawing is idealized. On the other hand, there is little doubt that Kircher built some sort of clock with a complete sunflower plant. As he describes it in the *Magnes*, the flower is supported by a cork disk, just as Kircher had described it to Peiresc seven years earlier. He instructs anyone trying the experiment to wrap the roots of the sunflower with bands of wool, which serve as a wick to bring moisture to the plant. And he concludes, as Peiresc says in his report, that "since the stem is not a resisting force that could divert the flower from the sun's attraction, the sun easily causes the flower to turn with it by its force of attraction."[37] The logic is compelling if one believes that the sunflower follows the sun by a force of attraction. If this is so, then mounting the flower

Fig. 2.5. The sunflower clock in full flower. From Kircher, *Magnes, sive de arte magnetica*, p. 644. Courtesy of the University of Washington Libraries.

on a frictionless pivot would make it swing to face the sun much more easily than if it were fixed in the ground—so much more easily, in fact, that it ought to work even where the sun's influence is weakened, in the shadow or at night.

Kircher tells us that the clock using an entire flower did not work well because it was impossible to enclose it in a glass case and every breeze deflected it from its true position. Also "when the sunlight was weak, and itself was as if withered and worn out, it ran a little slow, seeking rest. Added to this is the fact that a clock of this sort can barely last one month, even though cared for with the greatest effort; thus nothing is perfect in every aspect."[38] In the *Magnes* Kircher tells us that in his effort to improve his clock, he was led by some divine spirit to encounter an Arab merchant at Marseille in 1633. After talking about matters dealing with Arabia and the Red Sea, Kircher checked his signet-ring watch to see if it was time to return home. The Arab "appeared to derive an extraordinary pleasure from the use of a clock so suitable and convenient. Therefore, when I asked certain pertinent questions concerning the clocks widely used in Arabia, he replied that astronomers were in the habit of using various instruments to tell time; among the others was a very famous doctor who, with the help of a kind of material that constantly turns toward the sun, would find what time it was both night and day. The merchant said that not only did he know about material of this sort, but had even brought some with him among his aromatic wares and was prepared to exchange some of that material of his for my signet-ring watch. No sooner said than done. I gave him my watch in exchange for the stuff. First at Aix, then at Avignon repeating my experiment with it, I ascertained that it was more genuine than I supposed."[39] The Arab is here presented as the origin of the idea that sunflower seed or root could tell time as well as the flower. At Aix, however, he had made no mention of the Arab and told Peiresc that he had looked for a heliotropic effect in sunflower seeds because in Germany, where he first tried the experiment, it was winter and there was no plant available. This would mean that the first experiment with the seed preceded his meeting with the mysterious Arab. One suspects the Arab was conjured up to add mystery to the report.

Gassendi's Account

One would have to conclude that the demonstration at Aix had not been a great success. Peiresc, who hoped for much, saw little that suggested a cosmic magnetic force capable of moving clocks, to say nothing of the earth. It is strange, then, that Gassendi, in his *Life of Peiresc*, did not mention the failure of Kircher's demonstrations. He wrote from firsthand knowledge, both of Peiresc and of Kircher, and should have known Peiresc's opinions.

The focus of Gassendi's account, however, was on Linus's clock and the visit of Caraffa and Pietrasancta to Aix in December 1634, more than a year after Kircher's demonstration at Aix. Gassendi reports that the information from Pietrasancta

> was confirmed to him, both by the Letters of *Rubens*, and the *Relation of Dormalius*, who returning into Italy towards the end of Spring, and being detained certain daies at Aix, described the thing according as himself had seen it. Wherefore *Peireskius* praised that wonderful invention; and began to cast divers waies with himself, what power of Nature could effect such a thing. . . . But he chiefly called to mind, that which *Kircherus* had told him two years before, how he had stuck certain seeds of the Flower of the Sun into a piece of cork, which following the course of the Sun, as the flowers use to do, did turn about the floating Cork, and by a certain hand annexed, point out the hours, which were marked upon the Vessels.[40]

Whatever doubts Peiresc had had about Kircher's demonstration in 1633, they were dispelled, at least for a while, by the accounts of Linus's instrument in late 1634 and 1635.[41] Kircher's instruments had not convinced Peiresc, but others might succeed where Kircher had failed. Testimony was strong that Linus's clock had succeeded. Caraffa and Rubens were men of substance whose words carried conviction. Therefore Peiresc again took heart and believed that he had evidence for a moving earth.

It is remarkable how important the authority of witnesses was to validating or invalidating these instruments. Peiresc sought testimony from his friends Rubens and Dormalius in addition to the testimony from Caraffa and Pietrasancta, and he himself wrote out a detailed report of the trial at Aix. None of them appears to have handled or operated the clocks. Their testimony was based on observing the instruments, witnessing the trial, and putting questions to the makers.[42] Peiresc did not accuse Kircher of fraud, although he suspected a trick behind Kircher's clocks. Nor did he demand a complete explanation. Kircher, in turn, was careful to respond truthfully, if evasively, to Peiresc's questions. Although the trial seemed to destroy Kircher's claims, Peiresc was not of one mind about the clocks. According to Gassendi's narrative, Peiresc still appreciated the "marvel" even though he suspected that it was produced by a trick.

The trial had the character of a stage performance and a legal proceeding. These are two of the areas where we still practice and experience the art of rhetoric, while we supposedly reject it in the natural sciences. Therefore Kircher's magic tricks seem strange to us and totally "unscientific." Kircher, on the other hand, was using his instruments to make a rhetorical point—that the world operates by occult forces that, though hidden from us, can be exhibited analogically by such instruments as the sunflower clock.

How Many Clocks?

It is curious that Linus, Kircher, and Galileo all claimed to have invented the same instrument. There was probably more than one magnetic clock. Linus told Wendelin that the globe in his clock was made of wax, but Linus's fellow Jesuits said it was made of bronze or copper. Galileo's instrument had a copper globe floating between fluids of different density. Kircher's copper globe was suspended by an invisible thread.[43] All of this suggests that others besides Linus made spherical magnetic clocks. Dormalius reports that other Jesuits were attempting to duplicate Linus's clock, and we know that Peiresc acquired copies of both Linus's magnetic clock and Kircher's sunflower clock, because they were listed in the inventory of Peiresc's *cabinet* drawn up after his death. It is reasonable to assume that by 1635 magnetic clocks existed in several different versions.

There was also a tactical reason for Galileo and Kircher to claim the magnetic clock as their own. Peiresc, a man of great prominence and a friend of the pope's nephew, had come to Galileo's aid not without some danger to himself. He had urged Cardinal Barberini to order Caraffa, Pietrasancta, and even Linus himself to Rome in order to verify the clock, because he thought it would decide the question of the earth's motion in Galileo's favor. Galileo did not want to offend Peiresc or Linus, a Jesuit, but he also did not want his chance for freedom to depend on a doubtful instrument. Therefore he told Peiresc that he had long ago made essentially the same instrument, but with the aid of a deceptive artifice. After describing his clock he concluded, "Your Excellency, by those means which you recite, will easily be able to come to a knowledge of the whole matter."[44] Without accusing anyone of fraud or gullibility Galileo had alerted Peiresc to the probability that Linus's clock contained an artifice.

Kircher employed essentially the same technique. In his *Magnes* of 1641 he placed his drawing of Linus's clock on page 311 along with a description of the clepsydra in the base that drove it, but did not mention Linus or Pietrasancta. On page 739, he wrote that he had abandoned sunflower clocks because of their corruptibility and had turned instead to magnetic clocks, which were impervious to rot. Then he added, "I discover among the *Symbolis Heroicis* of Sylvestris Pietrasancta a similar sympathetic clock put together in a glass sphere by a certain English priest of our order, Francis Linnius; but since I have not been permitted to know the manner and method of its construction, I have judged that it is not fitting for me to interpose my judgment."[45] Like Galileo, he described how *his* clock worked, thereby exposing Linus's artifice without accusing him of fraud. He probably obtained details of the inner workings of Linus's clock from Pietrasancta, who, after his visit with Peiresc, returned to Rome where he would have been near Kircher.[46] Kircher was in an awkward position. He had to argue that the earth was immobile, which required him to expose Linus's artifice, and yet he did not want to contradict members of his own order. Describing the clock as his own invention was one way to accomplish this difficult task (although it should be understood that Kircher frequently claimed as his own instruments that he had only seen or heard described).

Kircher's Purpose

One suspects that Kircher would have preferred to keep the artifice of Linus secret, and that he exposed it only because he needed to argue against the natural motion of the earth. Since such hidden artifices drove most of his instruments, he would not willingly have confessed to them. By not confessing to them was Kircher consciously and willingly practicing fraud? The answer to that question is not simple. He certainly knew how his instruments worked—in fact, he was annoyed at any suggestion to the contrary—but an experiment for him was a demonstration more in the sense of an illustration than a test. Kircher introduced his description of the mechanism driving the magnetic clock in the following words:

> All those artificers who have fitted their activities to an unchangeable model of nature know that they have found a true key to the innumerable secrets hiding themselves in Nature's bosom. Therefore, since every motion has been dependent upon some first unmoved mover—as Aristotle in Book 4 of the *Physics* [states]—it is just to presuppose [the existence of] some first mover of every movement of bodies that Nature's ape, the artificer, with remarkable industry brings about, by which, when moved, the remaining bodies are moved in an orderly way. We effect things of this kind by the pouring of sand, or of water, or of another liquid, in the following way.[47]

And then follows his description of the hidden mechanism.

According to Kircher, all motion in nature is caused by a mover and that mover by a previous mover until we are carried back to the first mover, which is undoubtedly hidden from us. As art imitates nature, the artificial instrument must also have an artificial first mover that replicates the occult or hidden first cause in nature. Kircher's instruments are analogies to nature. They mimic nature rather than test it or probe it. After describing the sunflower and magnetic clocks he wrote: "In the nature of things much lies hidden that transcends all human intellectual capacity. Who would ever be induced to believe those wonders of the magnet which we demonstrated in the above books if we did not recognize through sensible and palpable demonstrations that these things are most true?"[48] An instrument for Kircher illustrates by imitation a wondrous effect in nature whose cause is occult. This is the purpose of natural magic. Both the wondrous effect and the artificial instrument illustrating it are "natural" in the sense that they employ no supernatural or spiritual powers, although they may well appear miraculous to the untrained observer. In fact, the ability of the instrument to mimic nature demonstrates that the phenomena of nature are not miraculous.

To reveal the hidden mechanism in an instrument would destroy the analogy between it and the natural effect that it imitated. The clocks illustrated a supposed natural phenomenon, in this case the effects of celestial magnetism. Kircher and Linus built them with the conviction that celestial magnetism really existed, even though they knew that they were using tricks to produce the motions of their clocks. When the Holy Office pronounced the Copernican system false, they realized that they had been wrong in assuming that magnetism could cause the rotation of the earth. Their clocks, therefore, had been based on a false analogy. The trial at Aix was no *experimentum crucis* that could be understood in one way and no other. Kircher repudiated the moving earth and continued to demonstrate his sunflower clock at the Collegium Romanum without any apparent conflict.[49]

The fact that Kircher did not perform experiments as we understand them does not mean that he lacked instruments. The Jesuits, and Kircher in particular, were the envy of other natural philosophers in the seventeenth century including the English experimenters like Robert Boyle and Robert Hooke. The Jesuits had the best cabinets of instruments and the greatest patronage for their work.[50] This was particularly true in the middle of the century before the scientific societies at London and Paris became active.

Kircher's books and instruments should not be regarded as "research" in the modern sense. His work was encyclopedic in the medieval tradition. He sought universal causes for natural phenomena, not by exploring certain effects in detail, but by encompassing all known examples. Magnetism is a good case. Kircher believed that magnetism was the ultimate cause of all motion, and therefore he attempted to demonstrate his theory by describing literally hundreds of instruments and unusual natural phenomena.[51] His instruments served rhetorical and didactic purposes. They were instruments of demonstration, and also instruments of patronage and education. They added to his reputation as a "monster of erudition," a man who knew everything. Kircher was not a philosophical rebel. He rested his arguments on the authority of Aristotle, but his use of instruments and his love for the exotic and prodigious made him an Aristotelian of a new kind.

Instruments and Occult Qualities

Peiresc no doubt understood Kircher's reasoning and shared his love of prodigies. Descartes, on the other hand, declared him a charlatan. Descartes understood the purpose and method of natural magic, but denied that it had any value. The contrast between Kircher's and Descartes's approaches to magnetism illustrates very clearly the different meanings of "occult" in the seventeenth century.[52] For Kircher the cause of magnetism was beyond all human capacity to understand. He therefore made it an experiential axiom in his philosophy. Magnetism was an occult cause that could not be explained and was therefore fundamental. For Descartes magnetism was an occult cause to be explained away. In his program all such supposed causes were produced mechanically by the interaction of particles that could not be sensed directly. Kircher, following in the Aristotelian tradition, could not accept such an explanation, because for Aristotle, anything beyond the range of human sense was unintelligible. One could not "imagine" such things as atoms because there was no way that the faculty of the imagination could form an image of something that was not in the senses or in the memory. If the imagination could not form an image, then there would be nothing for the reason to work upon. For Kircher the occult cause of magnetism was both insensible and unintelligible. There was no way that it could be approached by the intellect, and therefore it could only be taken as fundamental.

Descartes, on the other hand, insisted that even though magnetism was occult, it was possible for the imagination to create for it a mechanism of subsensible particles that was intelligible. He imagined how sensible objects might work the same effect on a large scale, and then mentally reduced the mechanism to a subsensible scale. Like Kircher he relied on analogy to create his particulate model of magnetism, but he limited his analogies to the particular kind called comparison.[53] By restricting his analogies to mechanical comparisons between his particles and objects of sense he hoped to avoid the errors caused by the loose analogies Kircher employed.

Kircher, for his part, specifically rejected the corpuscularian approach of Descartes.[54] Because they were out of the range of the senses, Descartes's particles could be only vain imaginings based on no evidence of the senses. Instead Kircher followed his own "mechanical philosophy." His mechanical cause was the hidden mechanism of the artificer that in each instrument aped the insensible first mover of nature. Kircher's mechanical analogies were real instruments, not drawings of imaginary particles streaming out into space and screwing themselves into magnets and pieces of iron (see fig. 2.6). Descartes, the mechanical philosopher, made machines only in his mind. Kircher, the natural magician, made real machines by the hundreds—enough to fill an entire museum.[55]

There is a special significance in the fact that the instruments we have been discussing were clocks. Otto Mayr has demonstrated in great detail the extraordinary pervasiveness of the clock metaphor in the seventeenth century. Clocks, during Kircher's lifetime, were more models of the cosmos than they were timepieces. The mechanical clock, like Kircher's magnetic and sunflower clocks, provided an analogy to the secret workings of nature. The poet Georg Philipp Harsdörffer wrote, "Just as we see the hand of the clock and read the hours from its turning without having insight into the ingenious workings of its complex gears, so we can observe the blessings and punishments of God without knowing and understanding their secret causes."[56] Descartes, as is well known, made extensive use of the clock anal-

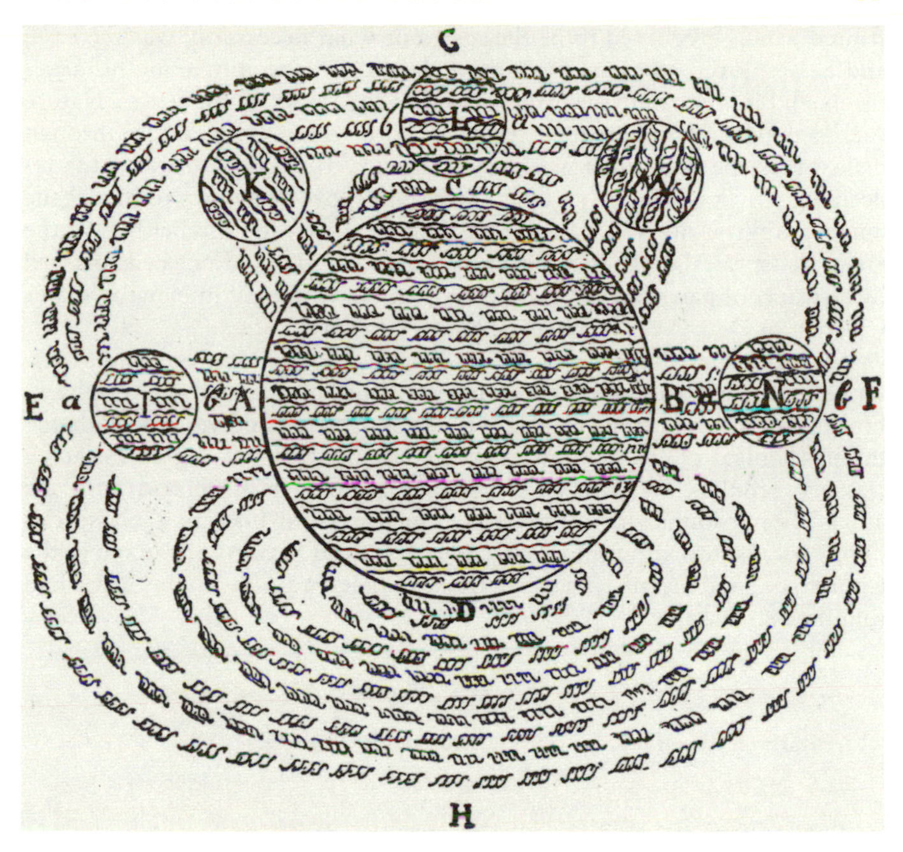

Fig. 2.6. Descartes's mechanical explanation of magnetic attraction. Because the magnet possesses polarity, the particles streaming out from the magnet must be asymmetrical. Descartes imagines them to be threaded with either left- or right-hand threads. From Descartes, *Oeuvres*, 8:288. Courtesy of the University of Washington Libraries.

ogy. He sought to make the "secret causes" of nature intelligible by attributing them to the mechanical interaction of tiny particles of matter. But the clock analogy did not necessarily imply a particulate universe. Kircher's imagery was similar to that of Johannes Kepler, who stated in 1605: "It is my goal to show that the celestial machine is not some kind of divine being, but rather like a clock. . . . In this machine nearly all the various movements are caused by a single, very simple magnetic force, just as in a clock all movements are caused by a simple weight."[57] Kepler, like Kircher, emphasized not the mechanical linkage in the clock but the hidden first cause of its motion.

Of course Kircher's clocks mimicked only what he *thought* was a true effect. On the basis of the authority of the ancients, the testimony of others, and his own general experience, he believed that the sunflower followed the sun because of some occult magnetic influence. His sunflower clock demonstrated what he believed to be the case, not what necessarily was the case, and he employed a certain amount of sleight of hand in making his arguments. But it is not only the magician who plays tricks on the senses. Nature can do the same thing, and Kircher's purpose was to demonstrate through his artifices the wondrous workings of nature. Beyond nature was the redeeming power of God who, by his fathomless love drew men to him as the magnet draws iron. God was not only like a magnet; for Kircher he was the great archetypical magnet of the universe, and God could not be explained by a swarm of particles.[58] We see a comparable sentiment in Newton's later rejection of Descartes's mechanical theory of gravity.

It is significant that three of the characters in this story, Gassendi, Mersenne, and Descartes, were the founders of the mechanical philosophy in France. Kircher had some contact with each of them, and yet he repudiated the mechanical philosophy just as he repudiated the moving earth. For a classical scholar like Kircher, atomism was forever associated with the atheistic philosophies of Democritus, Epicurus, and Lucretius. Gassendi's daring forays into atomism would have frightened him. As a true son of the Church, he used his instruments in the way that tradition and his order had found most valuable.

CHAPTER THREE

The Magic Lantern and the Art of Demonstration

PHYSICS DEPARTMENTS possess two kinds of instruments—demonstration instruments that are normally kept in a storage area next to the major physics lecture hall, and research instruments that are used in the department's research laboratories.[1] There are two major differences between experiments using "demonstration" instruments and those using "research" instruments. First, a "demonstration" experiment "shows" or "exhibits" the phenomena so that students may better understand what is being presented in words. The "demonstration" always presents the phenomena directly to the senses. The data from a "research" experiment, on the other hand, are presented in a written, graphical, or digitized form, but seldom are they "demonstrated" directly to an audience. Second, a "demonstration" experiment teaches what is already known to the lecturer, while the "research" experiment seeks to obtain new knowledge.

When we project these terms back into the seventeenth and eighteenth centuries, however, we are at risk of committing an anachronism. Although Robert Hooke is often called the "demonstrator" for the Royal Society, he was, in fact, one of the "curators" of experiments and had under his direction assistants who were called "operators," "mechanicks," or "laborants," but not "demonstrators." A "demonstration," in its common seventeenth-century meaning, was a rigorous proof, as in the *quod erat demonstrandum* ("that which was to be proved") that traditionally appeared at the end of a Euclidean proof. How "demonstration" moved from rigorous proof to a "showing" of the phenomena is a problem of considerable importance, because, as John Heilbron argues, "the chief agent in changing the scope of physics was the demonstration experiment."[2] Our purpose here is not to explore all the byways of this linguistic journey, but to show how the magic lantern, one of the standard instruments of the demonstration lecture, made its way into experimental science.

DEMONSTRATION IN EXPERIMENTAL PHYSICS

The Greek word *apodeixis* has several meanings. It can be a proof, as in logic or mathematics; it can be an explanation; or it can be an instruction, as in teaching.[3] The scholastic philosophers regularly translated *apodeixis* as *demonstratio*, which also has several meanings, both common and technical, that are not entirely congruent with *apodeixis*.[4] Our word *demonstra-*

tion has a comparable variety of meanings. All of these meanings, however, carry the notion of "showing" or "pointing out" something.

Because of the variety of meanings these words bore in ancient and modern languages, one would not be surprised to find *demonstration* used in different ways in scientific literature. But Aristotle had gone out of his way to define very narrowly what he meant by demonstration or *apodeixis* in science. In the *Posterior Analytics* he argued that a demonstration must be a syllogism, of the first figure, with indubitable but undemonstrable premises that are prior to the conclusion. The conclusion must itself be dubitable, that is, not obvious; otherwise there would be no need for the demonstration.[5] There could be other kinds of demonstration, but they would not constitute a science. Considering the narrowness of Aristotle's definition, it is not surprising that his examples were almost all from geometry, and one could question whether true demonstrations would be possible in natural philosophy. But some of Aristotle's own examples indicate that he did not interpret his definition in this strictest sense. For instance, he said it could be "demonstrated" that the lunar eclipse is caused by the earth's coming between the sun and the moon, and that it could be "demonstrated" that the planets are near because they do not twinkle. These demonstrations were arguments that obviously depended upon observation.[6]

Medieval philosophers accepted Aristotle's definition, almost without exception, but because of the subject's importance they debated at length over how his words should be interpreted. What is perhaps more surprising is that philosophical lexicons and dictionaries retained Aristotle's definition into the nineteenth century. For example, Abraham Rees's *Cyclopaedia* of 1819 adheres to Aristotle's definition of demonstration as a syllogism (against those who would say that a mathematical demonstration is not syllogistic) while calling on the support of such moderns as Leibniz, Wallis, and Huygens.[7] The demonstration lecture as we understand it does not provide a rigorous proof in the Aristotelian sense, which leads us to wonder why other meanings of "demonstration" besides Aristotle's did not find their way into the philosophical dictionaries.[8]

The introduction of philosophical instruments into natural philosophy in the seventeenth century further confused the meaning of demonstration, because instruments did not provide the kind of common, repeatable, direct sense experience from which philosophers usually drew their premises. The new instruments of natural philosophy went beyond common experience. They extended the senses or altered nature in such a way that new things were observed. The telescope and the microscope distorted the vision. Likewise the barometer and air pump produced "unnatural" spaces in their receivers. From these new observations, conclusions might be drawn by "demonstration," whatever that might mean, but it was not obvious that the new instrument-aided observations were sufficiently certain, or that demonstrations using the new instruments could proceed in the traditional manner. The use of philosophical instruments in demonstration had both a positive and a negative side. By going beyond common experience, they allowed the natural philosophers to demonstrate something new; simultaneously, however, they exceeded the ability of the senses to validate the results. Whatever mode of inference the philosopher employed—deduction, induction, or retroduction, or any combination of these—the use of instruments for "showing" the phenomena altered the nature of the demonstration.[9]

The problems raised by the philosophical instruments of the seventeenth century had already been anticipated to some extent in the science of medicine. Physicians wished to make their science demonstrative. Aristotle gave them some hope in the *Posterior Analytics* by claiming that it is for the doctor to know the fact that circular wounds heal more slowly than elongated ones, and for the geometer to know the reason why, implying that if physicians were also geometers they might arrive at causes in medicine demonstratively.[10] But demonstration came to mean something quite different in medicine than in Aristotelian physics. In the preface to *De motu cordis* Harvey writes that he is presenting his theory of the circulation of the blood, "having now for nine years and more confirmed these views [about the motion and function of the heart] by multiplied ocular demonstrations in your presence [*multis ocularibus demonstrationibus in conspectu vestro confirmatam*], illustrated them by arguments, and freed them from the objections of the most learned and skilful anatomists."[11] "Demonstrations" here means direct observations through anatomical dissection, the demonstrations being followed by "arguments" as to the cause of what is observed. Harvey calls these demonstrations "ocular" because they are validated by direct observation rather than by logical argument. He dedicates the book to his friend Doctor Argent, who has been a "faithful witness" to his experiments and has borne out with his testimony Harvey's "ocular demonstrations [*ocularibus demonstrationibus eorum*]."[12] Chapter 14, which contains the conclusions of Harvey's argument and carries the title "Conclusion of the Demonstration of the Circulation of the Blood [*Conclusio demonstrationis de sanguinis circuiter*]" contains the famous statement that "both argument and ocular experiment [*ocularibus experimentis*] show that the blood passes through the lungs and heart by the force of the ventricles."[13] In this case Harvey uses the word "demonstration" to refer to the entire argument, both logical and experimental, and refers to the actual dissection as "ocular experiment."

Medicine also contributed the "demonstrator" to natural philosophy. In the medieval schools anatomy was taught by the physician's reading aloud from the text while the barber surgeon performed the dissection, but there was also present a "demonstrator" or "ostensor" whose task it was to point out the organs as the physician read about them (see fig. 3.1). In the seventeenth century at the Jardin du Roi in Paris the title of *démonstrateur* began to be used more widely. From its founding in 1635 until 1718 the purpose of the Jardin was largely medical. The professors lectured on botany, anatomy, and chemistry, but always as these subjects applied to medicine and

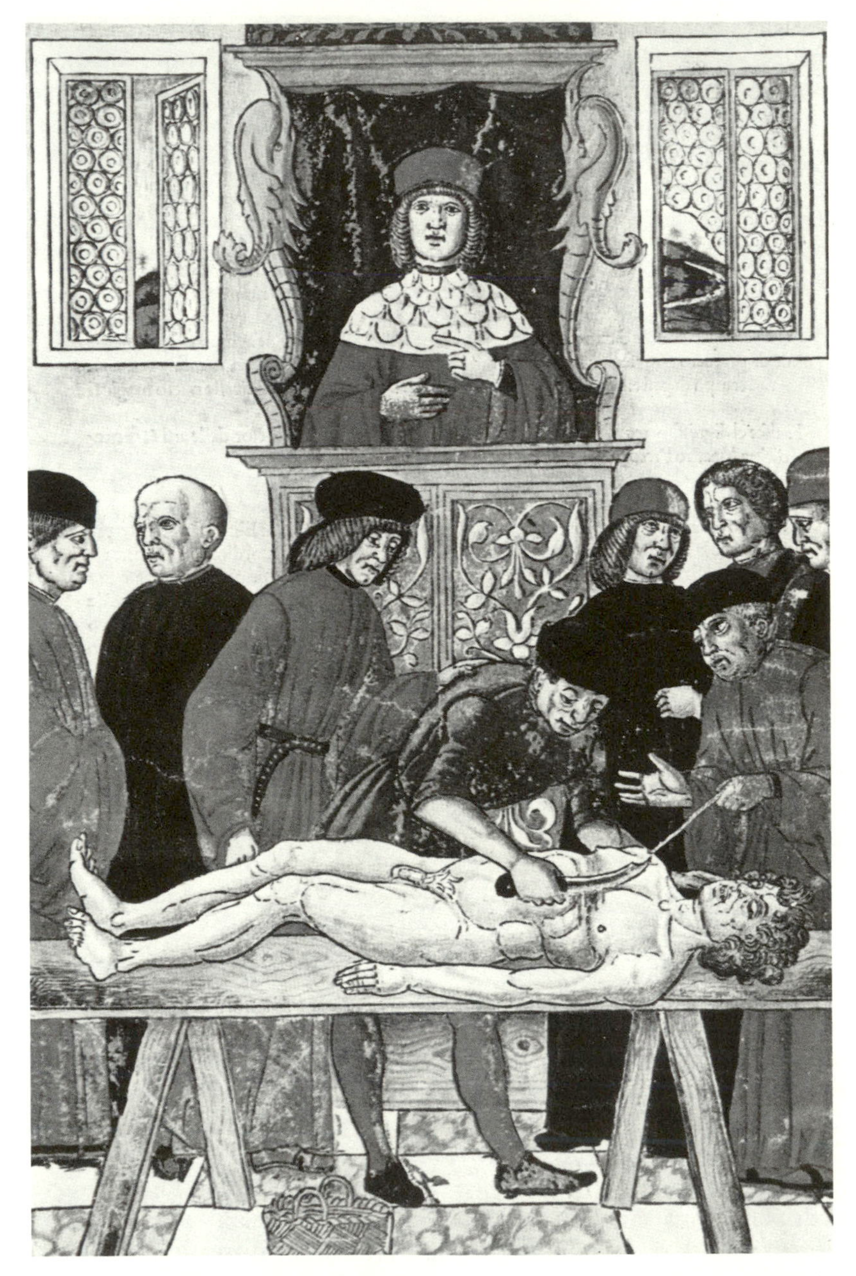

Fig. 3.1. A medieval anatomy lecture. Note the "demonstrator" or "ostensor" pointing out the organs as the surgeon dissects and the professor reads. Courtesy of Yale University, Cushing/Whitney Medical Library, Historical Library.

pharmacology. In 1635 the botany course was taught by the *intendant* assisted by an "under-demonstrator" (*sous-démonstrateur*). Medical botany was divided among three "demonstrators"; the courses were always practical and included the showing of specimens. An ordinance of 1635, for instance, stipulated that one of the three demonstrators should devote himself to "presenting ocular and manual demonstrations of all and each of the operations of surgery," emphasizing again the significance of "ocular" demonstrations.[14] In 1718 the Jardin royal des plantes médicales became the Jardin du Roi with a concomitant reduction in the emphasis on medicine. The "demonstrator," whether in botany, anatomy, or chemistry, exhibited specimens, usually from the king's *cabinet d'histoire naturelle*, to an audience not limited to medical students. Guillaume-François Rouelle, for instance, who held the title of *démonstrateur en chimie* from 1743 to 1768, delivered a famous series of lectures on chemistry that effectively separated French chemistry from its roots in pharmacy. The anatomical "demonstrator" became the "demonstrator" in chemistry and experimental physics—the one who used instruments to point out or "show" new natural phenomena.

As the new philosophical instruments of the seventeenth century made their way into acceptable science, they began to do more than provide premises for, or confirm the conclusion of, a "demonstration"—they became the demonstration itself. The purpose of the experiment was to establish a "matter of fact," an event that indubitably occurred and would occur again under the same conditions.[15] Experimenters called them "ocular demonstrations," because they established matters of fact but were not demonstrations in the traditional sense. Valerio Magni's *Demonstratio ocularis. Loci sine locato: Corporis successive moti in vacuo: Luminis nulli corpori inhaerentis* (1647) is an early example applied to the Torricellian debate. In his "ocular demonstration" Magni claims to give a "historical"—that is, descriptive—account of his experiments.[16]

More significantly, Robert Hooke uses the term "ocular demonstration" at several points in his *Micrographia* of 1665. At the point where he conjectures that the forms of things may be explained by the packing of globules in different regular ways, he writes, "And this I have *ad oculum* demonstrated with a company of bullets."[17] In this case Hooke is giving an experimental solution to a geometrical problem, and it is natural for him to call it a "demonstration," albeit a demonstration *ad oculum*. His plan is to investigate all kinds of geometrical figured bodies and then "demonstrate" which form of geometrical packing is the most likely to produce the observed form.[18] In creating these regular figures nature "plays the Geometrician," says Hooke, and the experimenter meets with nothing less than the "Mathematicks of nature, having every day a new Figure to contemplate."

At this point Hooke argues that investigation requires a new method, "a *novum organum*, some new engine and contrivance, some new kind of Algebra, or Analytick Art before it can surmount" the high, difficult sides of the pyramid of natural knowledge.[19] The precise nature of this "philosophical algebra" has long been debated by historians, but it is clear that Hooke wants it to be demonstrative in the Aristotelian sense, that is, indubitable, like mathematics. He hopes, as had Bacon, to create an inductive method that produces certitude—that with his "philosophical algebra" some day "even physical and natural enquiries as well as mathematical and geometrical will be capable of demonstration."[20] Again in his discussion of optics, Hooke argues that he has "given proof sufficient (*viz. ocular demonstration*) to evince, that there is such a modulation" of light as he has claimed.[21] For Hooke, an "ocular demonstration" is a proof of a physical phenomenon by direct observation, but because he hopes to provide absolute certainty it also often has the character of a geometrical proof; thus his idea of an "ocular demonstration" is closer to the traditional Aristotelian sense of demonstration than was Harvey's. Note that the above examples where he uses the term "ocular demonstration" also involve geometrical forms, geometrical analysis, and geometrical optics.[22]

According to Hooke a demonstration, at least as he commonly used the word, was still a proof that established an indubitable truth. Hooke's quarrel with Newton in 1672 stemmed partly from the fact that Newton claimed certainty in his theory of colors, and Hooke could not accept that Newton had made a real demonstration in physics.[23]

Hooke's colleagues at the Royal Society did not all agree that demonstrations were possible or even desirable in natural philosophy. Thomas Sprat in his *History of the Royal Society* rejected demonstrations understood as rigorous proofs. He claimed that "Whatever they [the fellows] have resolv'd upon they have not reported, as *unalterable Demonstrations*, but as *present appearances*: delivering down to future Ages, with the good success of the Experiment."[24] In describing the activities of the Royal Society he says the fellows "made" or "performed" experiments. They "operated" the instruments. In no case did they "demonstrate" experiments or instruments. Experiments were "labor," not logic, according to Sprat.[25]

Boyle was even more critical than Sprat of supposed demonstrations in experimental philosophy. In the quarrel over the air pump that Steven Shapin and Simon Schaffer have described so skillfully for us, Boyle claimed specifically that it was Hobbes's "demonstrative way of philosophy" that had led him and his followers into error and irreligion.[26] And Hobbes, indeed, argued in his *Dialogus physicus* that natural philosophy was demonstrative from the observation of common phenomena and not from artificial experiments that distorted nature and our perception of it. Attempts to explain the properties of air from experiments with the air pump were bound to be circular, because the experimenters "demonstrate without a principle of demonstration."[27]

A "demonstration" could obviously mean different things in the seventeenth century. As with the original *apodeixis* it could be a proof as in mathematics, an "ocular inspection" as in anatomy, or a "showing" of an instrument as in the popular demonstration lecture of the eighteenth century.

As the eighteenth century progressed, it more commonly took on this last meaning—that is, the showing of an instrument in a formal lecture. All these different uses had one sense in common, however. A demonstration, whether geometrical or ocular, was not, primarily, a method of discovery—that was the function of invention. Hooke said in the *Micrographia* that an inquiry in natural philosophy begins with "a noble *Inventum* that promises to crown the successfull endeavour."[28] In 1673 Huygens pointed out that the proof of the isochronism of the cycloid by Lord Brouncker, president of the Royal Society, was merely a "demonstration of a proposition which had already been discovered," to which Brouncker replied, "As to what he is pleased to say concerning my Demonstration, I doe acknowledge that to Invent is much more than to Demonstrate, and that likely in this case I had never thought of or done the latter [demonstrate] if Mr Huygens had not done and made known the former [invention], nor did I offer it but for my own satisfaction untill he should be pleased to publish his."[29] Demonstration, however one understood the word, was a method of proof or a method of establishing a fact and did not necessarily lead to any new knowledge on the part of the demonstrator.[30] In that sense it was like our modern demonstration lecture that teaches what is already known to the lecturer.

The Magic Lantern

The magic lantern was an *inventum* that, in the eighteenth century, became a staple in popular scientific lectures. Its role has changed from its first appearance around 1659 as an instrument of magic to its present manifestation as the slide projector, the overhead projector, and (with a significant amount of added technology) the movie projector and the cathode-ray tube or television screen. The projected image has become ubiquitous today and challenges the written word as the major vehicle for communication. We take it so much for granted that we no longer consider it a part of "science," any more than writing or talking on the telephone. In the seventeenth century, however, the projected image could be seen only in the instruments of natural magic such as the camera obscura, mirror writing, and the magic lantern.[31] What the magic lantern "showed" or demonstrated were devils, ghosts, and illustrations from fairy tales. A century passed before it was used to show scientific illustrations. Like other "magical" instruments, the magic lantern was regarded with ambivalence by experimental philosophers in the seventeenth century, and it is instructive to retrace its incorporation into the "demonstration" lecture.

The invention of the magic lantern has long been attributed to Athanasius Kircher, but that appears to be a mistake. Kircher claimed the invention (as he did most of the gadgets that came his way) and his students reinforced his claim.[32] Because of his fame and the wide circulation of his books, it is not surprising that he received the credit. In the first edition of *Ars magna lucis et umbrae* (1646) Kircher described a variety of mirrors that projected writing on the wall as well as a lantern that used a concave mirror to create a beam of light, but there was no indication in the text or illustration that this focusing lantern formed an image. In the second edition of 1671 Kircher described what was undoubtedly a magic lantern, but his claim for priority depends on the first edition where the magic lantern does not appear (see fig. 3.2). And what is more, the magic lantern that Kircher described in the 1671 edition would not work! The text describing his lantern was vague in the extreme and the accompanying illustrations were no better. Kircher placed the objective lens *between* the light source and the slide. He also described an upright image, while the magic lantern produces an inverted image, the same as in the camera obscura.[33] It is possible that the engravings added to the second edition of the *Ars magna lucis et umbrae* were done in Holland without Kircher's ever having seen them, but even so, he could hardly claim priority. By 1671 magic lanterns were showing up everywhere.[34]

In describing how he used his magic lantern Kircher employed the verb *demonstrare* only once and then only to say that by means of slides one can "demonstrate" anything at all. He usually used the verb *exhibere*; thus Kircher saw himself "exhibiting" a natural wonder with his lantern, not "demonstrating" a scientific principle.[35]

The earliest reference to the magic lantern is not in the works of any natural magician, but in the correspondence of Christiaan Huygens, a physicist of absolutely sterling reputation. Sometime in 1659 Huygens sketched a group of skeletons that he projected optically by means of convex lenses and a lamp (see fig. 3.3).[36] When his father asked him in 1662 to send a magic lantern, Huygens complained bitterly that such bagatelles wasted his time.[37] As an excuse, he pretended that he could not remember the proper focal lengths of the lenses and instructed his brother Lodewijk on how to remove a lens from the new instrument when it arrived in order to render it inoperable.[38] All this because he did not want his father to bring ridicule on the family by showing off the lantern at the French court.

Huygens called it *laterna magica*, a term that he probably coined (at least his was the first use of the term that we know of). His drawing shows the first workable lantern, complete with parabolic reflector, light source, condensing lens, slide stage, and adjustable objective composed of two biconvex lenses. Since the optics of Huygens's lantern was essentially identical to that of the modern slide projector, we can credit its invention to him. But Huygens regarded it solely as entertainment. The only uses he saw for it were to produce ghosts and to satisfy his father.

Let us compare Christiaan Huygens's scornful letter of 1662 about the magic lantern with one that his father, Constantijn, had written to *his* father, Christiaan senior, forty years earlier. The Huygens family served the House of Orange with great distinction through several generations. In 1624 Constantijn succeeded his father as secretary to the stadtholder, Frederick Hen-

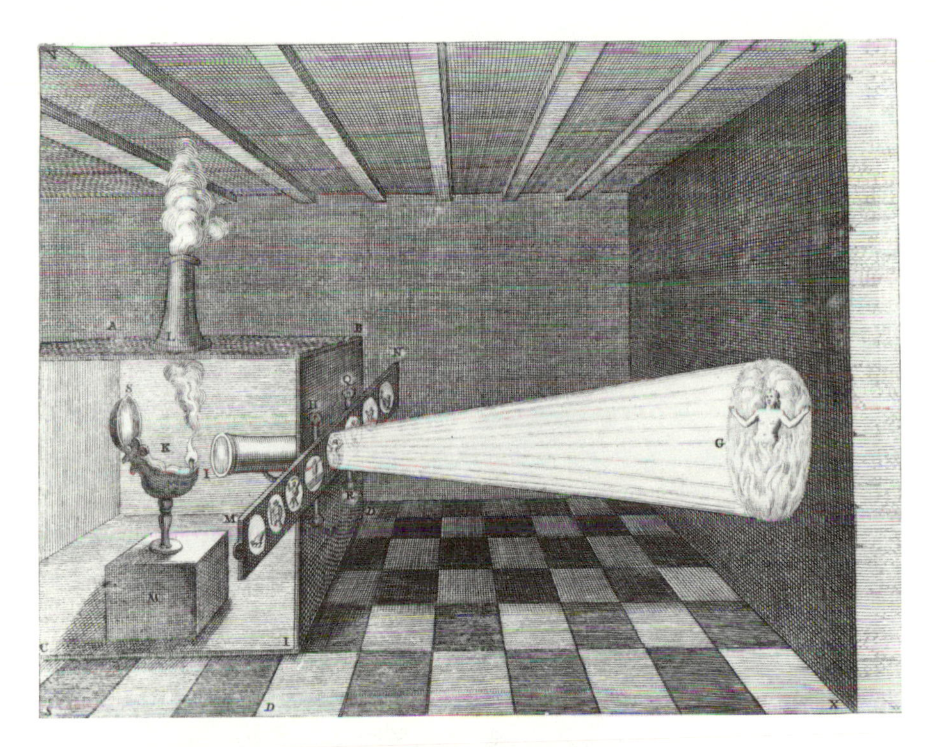

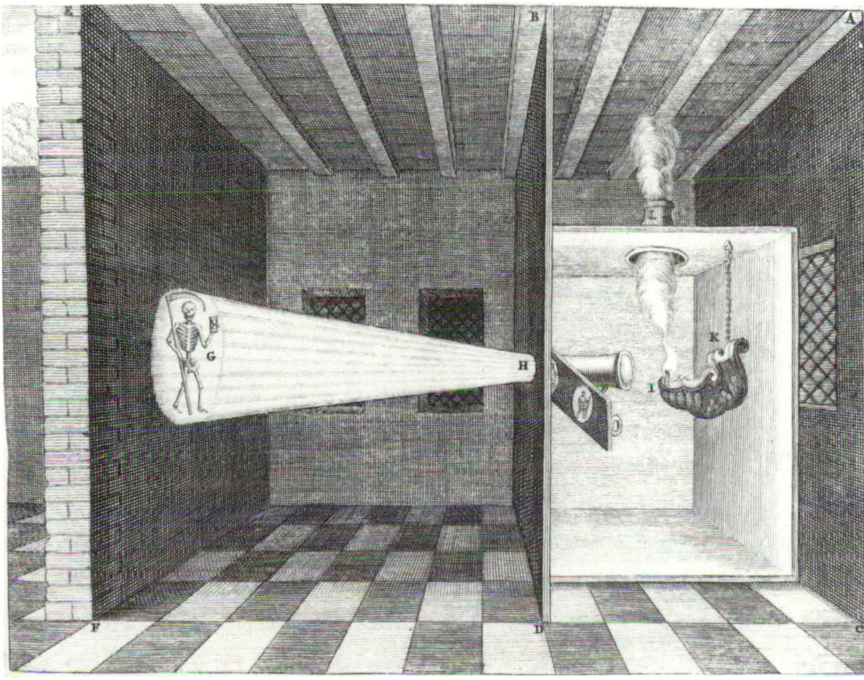

Fig. 3.2. The magic lantern of Athanasius Kircher. Note the erect image and the position of the slide. From Kircher, *Ars magna lucis et umbrae*, 2d ed., pp. 768, 770. Department of Rare Books and Special Collections, Princeton University Libraries.

Fig. 3.3. The image and optics of Christiaan Huygens's *laterna magica*. From Huygens, *Oeuvres*, vol. 13, pt. 2, p. 786, and vol. 22, p. 197. Courtesy of the University of Washington Libraries.

drik, prince of Orange. He had made his first trip to England in 1618 in the company of Dudley Carleton, English ambassador to the Hague, and had returned to England many times afterward in a diplomatic capacity. In 1621 he met Cornelis Drebbel, a Dutch engineer, architect, and natural magician in the service of James I. Constantijn was captivated by Drebbel, who "looks like a Dutch farmer, but whose speech reminds one more of the philosophers of Samos and Sicily."[39] "For a whole year," wrote Constantijn, "I had Drebbel to myself. Me he possessed, who possessed his time, if I mistake not: this he abundantly proved by the many hours of discussion he had with me, favoring me above most of his friends." Among Drebbel's inventions must have been something resembling a magic lantern, because in a letter of 1608 Drebbel had described how he could change the appearance of his clothes, or appear as a lion, bear, horse, or cow, all the while standing in the middle of a room. He could also make ghosts appear in a cloud from the earth and giants twenty or thirty feet high.[40] All of these apparitions point to some kind of projection apparatus, although it need not have been a true magic lantern.

Constantijn's father was less than pleased with his son's newfound friend and warned him that Drebbel's magic might come from the devil, to which Constantijn replied: "I laughed at your letter where you chose to warn me against the magic of Drebbel, and reproached him for being a sorcerer. But rest assured that finding nothing beyond the natural in what he does, it won't be necessary to bridle me." Constantijn came home from England with a microscope from Drebbel and a camera obscura. The camera obscura created an image whose beauty was "indescribable in words," and the microscope was "a passage to a new world by a new manifestation of nature."[41] Constantijn considered his father's fear of sorcery laughable, and yet he reveled in Drebbel's natural magic.

By the 1670s the experimental philosophy was gaining respectability in England. Natural magic no longer carried the threat of sorcery (although Kircher retained a cautious respect for the devil). Just as Constantijn ridiculed his father's fear of sorcery, so Christiaan, the physicist, ridiculed his father's enthusiasm for natural magic. Christiaan's idea of a demonstration in physics was not a magic lantern show. In the preface to his *Treatise on Light*, which he read to the Paris Academy in 1678 and published in 1690, Huygens gave his idea of a demonstration in physics:

> There will be seen in it [the *Treatise on Light*] demonstrations [*démonstrations*] of those kinds which do not produce as great a certitude as those of Geometry, and which even differ much therefrom, since whereas the Geometers prove their Propositions by fixed and incontestable Principles, here the Principles are verified by the conclusions to be drawn from them; the nature of these things not allowing of this being done otherwise. It is always possible to attain thereby to a degree of probability which very often is scarcely less than complete proof [*une évidence entière*]. To wit, when things which have been demonstrated [*démonstrées*] by the

> principles that have been assumed correspond perfectly to the phenomena which experiment has brought under observation; especially when there are a great number of them, and further, principally, when one can imagine and foresee new phenomena which ought to follow from the hypotheses which one employs, and when one finds that therein the fact corresponds to our prevision. But if all these proofs [*preuves*] of probability are met with in that which I propose to discuss . . . this ought to be very strong confirmation of the success of my inquiry.[42]

Huygens had a clear idea of the hypothetico-deductive method. If a hypothesis allowed one to predict a hitherto unobserved phenomenon, and if by experiment that predicted phenomenon did indeed occur, then one could be certain to a high degree of probability that the hypothesis was correct. Here was the kind of "demonstration" that physicists have employed ever since Huygens stated it.

In his method Huygens had obviously gone far beyond his father and Cornelis Drebbel, replacing the production of wonders by a recognizably modern experimental procedure, but it did not necessarily follow that the *instruments* employed in experimental physics had achieved the same degree of separation from the instruments of natural magic. Physical "cabinets" and popular scientific lectures continued to present nature in her oddest and most spectacular form, and it was not yet clear which instruments would best serve the new experimental physics and which ones should be discarded.

Demonstrating the Magic Lantern

In the seventeenth century the primary use of the magic lantern had been entertainment. In 1662 Huygens's friend Pierre Petit asked him for the dimensions of his "lantern of fear" and the focal lengths of the lenses, because he was having trouble making one that projected a proper "species" (*espèce*) or image (see fig. 3.4).[43] In 1664 their correspondence mentioned the lantern of Thomas Rasmussen Walgenstein (1627–1681) who was giving lantern shows and selling instruments at substantial profit.[44] Walgenstein had studied at Leyden at the same time as Huygens and may well have learned about the lantern from him.[45] After Walgenstein's performances there were numerous references to the magic lantern in seventeenth-century literature, but no other regular shows that we know of. The magic lantern appeared in most of the compendia of instruments published in the last three decades of the seventeenth century. These were books in the natural magic, or "mathematical magic," tradition, most notably Francesco Eschinardi, *Centuriae opticae* (1664), Claude François Milliet Déchales, *Cursus seu mundus mathematicus* (1674), Johann Christoph Sturm, *Collegium experimentale, sive curiosum* (1676–1685), Johann Zahn, *Oculus artificialis teledioptricus sive telescopium* (1685–1686), William Molyneux, *Dioptrica nova* (1692),

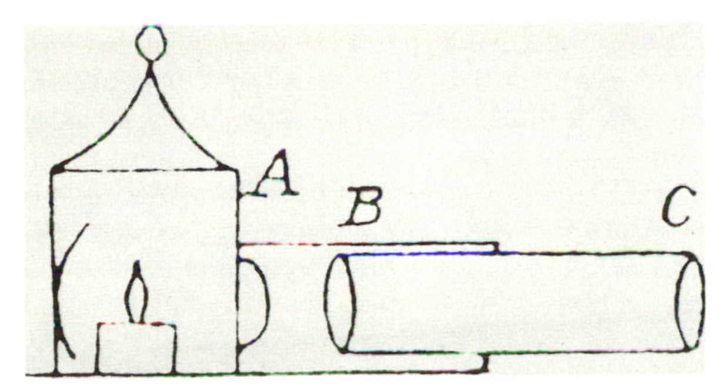

Fig. 3.4. Pierre Petit's letter to Christiaan Huygens contains the first sketch of a magic lantern. From Huygens, *Oeuvres*, 4:269. Courtesy of the University of Washington Libraries.

and Jacques Ozanam, *Recréations mathématiques et physiques* (1694).[46] We know that the magic lantern spread quickly to England, because Samuel Pepys recorded in his diary for August 22, 1666, that he purchased "a lanthorn with pictures in glasse," from the optician Richard Reeves, "to make strange things to appear on a wall, very pretty."[47]

In the eighteenth century the lantern was taken up by itinerant showmen and performances were common, but the instrument also began to receive the attention of natural philosophers. The book that established the "demonstration lecture" as the proper mode for teaching experimental physics was the *Physices elementa mathematica, experimentis confirmata; sive introductio ad philosophiam newtonianam* (1721) of Willem Jacob 'sGravesande. John Keill had already begun to use experiments in his lectures at Oxford in 1704, at the same time that Newton and his curator, Francis Hauksbee, reestablished the tradition of regular experiments at the meetings of the Royal Society, but 'sGravesande's later work had much greater influence.[48] 'SGravesande began teaching at Leyden in 1717 after a visit to England as secretary to the Dutch ambassador; in England he had met Newton, attended Jean Théophile Desaguliers's lectures, and became a convinced Newtonian.[49] His *Physices elementa* was the first significant defense and exposition of Newtonian natural philosophy on the Continent. Moreover, its importance was immediately recognized by Newton's English supporters. Both John Keill and his successor at Oxford, Desaguliers, translated it in competing English editions. The book attained two more Latin editions, a French edition, and six English editions by 1747, and Voltaire made a special journey to Leyden to obtain help from 'sGravesande for his own Newtonian *Lettres philosophiques* (1735) and *Élémens de la philosophie de Newton* (1738).[50] Its novelty is perhaps best exhibited by Louis-Bertrand Castel's negative review of it in the *Journal de Trévoux*. Castel found the book "full of experiments that are rare, curious and ingenious . . . without any of those simple, naive easy observations that nature affords abundantly to all countries and to all minds."[51] Castel did not like 'sGravesande's Newtonian philosophy exhibited by instruments any better than Newton's own version. He still believed that demonstration should be based on common experience, not on particular experiments done with instruments that distorted the senses.

The table of contents to 'sGravesande's *Mathematical Elements* reveals what "experimental physics" meant to him. The subjects covered were motion and mechanics including simple machines, fluids under pressure and in motion, the air as an elastic fluid (including sound), fire (including heat and electricity), geometrical optics and color, the system of the world (largely planetary motion), and gravitation as a cause of celestial motion. Mechanics, optics, hydrostatics, astronomy, and music (sound) were traditional subjects that had been treated mathematically in antiquity. "Gravity" was, of course, a Newtonian addition to the repertoire of physics. "Fire" and "air" were brief but important additions that came not so much from Newton as from 'sGravesande's predecessor at Leyden, Hermann Boerhaave. Medicine, botany, and physiology were noticeably absent.

'SGravesande not only set the style for experimental physics, he also designed and described a magic lantern that became the model for all lanterns during the subsequent century and was surpassed only by Philip Carpenter's Phantasmagoria Lantern in 1820 (see fig. 3.5). 'SGravesande's lantern employed a four-wick oil lamp to increase the illumination along with an adjustable concave parabolic mirror, a condenser, and an objective consisting of two biconvex lenses with a diaphragm stop between them. It could be used as much as thirty feet from the screen. An instrument like 'sGravesande's could only have been the result of much practice and tinkering.

Because of the importance of 'sGravesande's book for experimental physics, it is appropriate to ask what his magic lantern "demonstrated." In the illustration accompanying 'sGravesande's text, it demonstrated only a particularly horrible devil—not very enlightening as a subject of experimental physics, but it is clear that for 'sGravesande the "demonstration" of the magic lantern was not in the image it projected, nor in the instrument itself, but in the argument from geometrical optics that "demonstrated" how the image was formed. The magic lantern was, for 'sGravesande, the most interesting of "several machines made by the combination of mirrors and lenses which afford useful and pleasant appearances whose explanation may be easily deduc'd from what has been said," and what had been said were the laws of geometrical optics. He included the magic lantern in his experimental physics because it illustrated the laws of optics, not because of any "show" that it was able to make. 'SGravesande was consistent in his use of the term "demonstration." For instance, he wrote an essay describing two camera obscuras of his own design. In the essay, which he presented in the form of definition and theorem, the only thing that he called a demonstration was a calculation of the correct angle for a mirror that reflected the desired scene onto the screen inside the camera. No other aspect of the instrument or its use was a demonstration.[52]

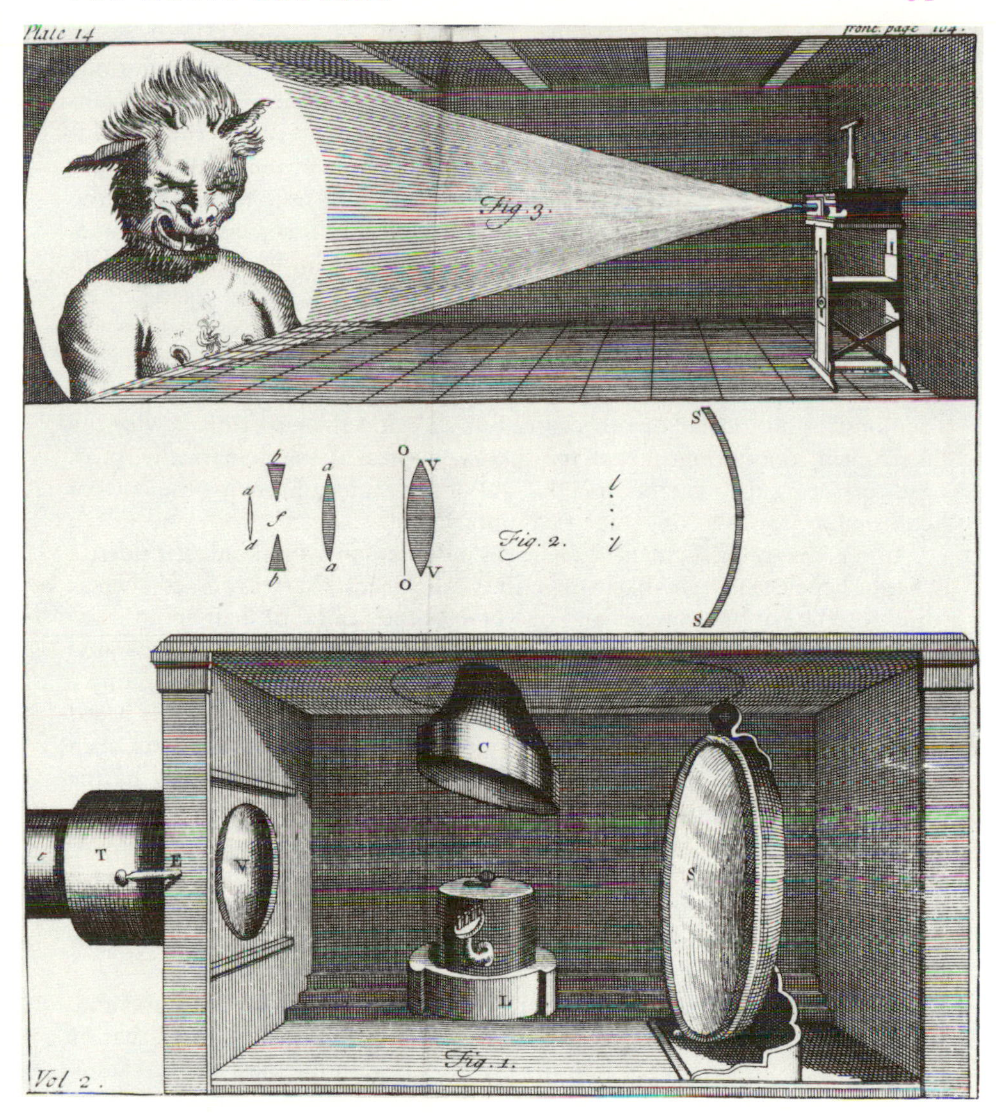

Fig. 3.5. Willem 'sGravesande's magic lantern in use and cutaway. From 'sGravesande, *Mathematical Elements*, vol. 2. Department of Rare Books and Special Collections, Princeton University Libraries.

'SGravesande's intent was to expound Newtonian philosophy, but in his constant use of instruments to illustrate Newton's arguments, his approach was quite different from Newton's. Newton had wished to instruct, but not necessarily to please. 'SGravesande explained that "in order to render the Study of Natural Philosophy as easy and agreeable as possible, I have thought fit to illustrate every Thing by Experiments, and to set the very Mathematical Conclusions before the Reader's Eyes by this Method."[53]

'SGravesande was defensive about his method, because often "Mathematicians think Experiments superfluous where Mathematical Demonstrations will take Place: But as all Mathematical Demonstrations are abstracted, I do not question their becoming easier when Experiments set forth the Conclusions before our Eyes; following therein the Example of the English, whose way of teaching Natural Philosophy gave me occasion to think of the Method I have followed in this Work."[54] In both of these passages 'sGravesande states that his purpose is to set forth "before our eyes [*sub oculos*]" the conclusions of mathematical demonstrations by experiments. The first advantage of using instruments is that it makes the demonstrations less abstract and easier to grasp. The second advantage is that it confirms the conclusions by direct observation. Showing the instrument is not, in itself, a demonstration, ocular or otherwise. But showing the experiment with the instrument does confirm that the theory, expressed mathematically, gives the correct results, whether it is the law of the center of gravity, the laws of collision, or the laws of geometrical optics.

In an "Essay on Evidence" 'sGravesande explains the kind of evidence that he believes is attainable in experimental physics. Matters of fact (such as the existence of the Romans and of Rome as the capital of their empire) are truths that we accept as indubitable even though they are not demonstrated mathematically. Their evidence is historical and is based on "testimony." We also believe that the sun will rise tomorrow. This is a physical fact based on "analogy," meaning the accepted uniformity of nature. Other facts, such as the phenomena observed in experiments, are perceived directly by the senses. All three kinds of facts are known with a "moral" certitude, which, though not deductive, can bring the same kind of conviction as that of mathematics. Thus sense, testimony, and analogy are the basis for moral evidence.[55] In mixed mathematics, "the demonstrations are . . . grounded on a hypothetical foundation," and if the foundation is morally certain, so is the conclusion of the demonstration.[56]

Of course 'sGravesande was attempting to follow Newton's method, which was mathematical. But Newton would never have admitted that he employed hypotheses, even morally certain ones. In his most famous statement on method, Newton wrote, "The whole burden of philosophy seems to consist in this—from the phenomena of motions to investigate the forces of nature, and then from these forces to *demonstrate* the other phenomena."[57] Note that the "demonstration," as Newton described his method, came after the "forces of nature" had been discovered from the phenomena. From the forces of nature one could then deduce further phenomena by mathematical demonstration. 'SGravesande took yet another step to confirm and illustrate these "further phenomena" by experiments performed *sub oculos*.

The instrument maker Benjamin Martin used an argument similar to that of 'sGravesande, although he called it an "ocular demonstration," reviving the term employed by Harvey and Hooke. In describing the camera obscura he claimed that it was of importance not only for drawing and painting, "but the optician himself is greatly interested therein. By this grand experiment he *demonstrates ocularly* the principles of his art. For by admitting the sun-beams thro' the hole of the window-shut into the darkened chamber, he can actually shew the focus of parallel rays by reflection from concave mirours."[58] Notice that in this "ocular demonstration" one sees not only the image but also actual convergence of the rays. In the experiment the geometrical laws of optics are visibly displayed.

By the middle of the eighteenth century, the meaning of "demonstration" had undergone several changes as a result of the new instruments of natural philosophy and the experimental method that employed them. Galileo used the term ambiguously—sometimes it meant a mathematical or logical argument based on premises drawn from common experience; sometimes it meant an observation with an instrument like the telescope that led to a mathematical argument. Some, like Boyle and Sprat, rejected the idea of a demonstrative natural philosophy altogether. Others, like Harvey, Hooke, and Benjamin Martin, distinguished a new kind of demonstration that they called "ocular," although they disagreed as to how it should be defined. For Harvey, ocular demonstration was a literal "showing" or "pointing out" of bodily parts and their motions. For Hooke and Benjamin Martin it was a visual observation that led to a geometrical argument. Newton reinforced the importance in natural philosophy of mathematical demonstrations based on premises that had been established by experiment. 'SGravesande continued Newton's mathematical emphasis, but with experiments that were meant to illustrate and confirm rather than to initiate inquiry. It was with 'sGravesande that the "demonstration" lecture using instruments became a standard method of instruction in natural philosophy. There is a paradox in the fact that 'sGravesande's instruments designed to illustrate Newton's mathematics led to a form of "demonstration" that was entirely nonmathematical. 'SGravesande's *Physices elementa* established the "demonstration lecture" as the proper way to teach experimental physics, but it was also one of the last physics textbooks to employ the geometrical format of axiom, corollary, and theorem.[59] As the "demonstration" part of the physics lecture passed from geometrical proof to the exhibition of phenomena, the rhetorical geometrical form also disappeared from the textbooks.[60]

The Image as Demonstration

During the eighteenth century the magic lantern appeared regularly in books on optics and experimental physics. Pieter van Musschenbroek's *Essai de physique* (Leyden, 1739) described a magic lantern modeled after 'sGravesande's, as did Jean Antoine Nollet's *Leçons de physique expérimentale*.[61] The van Musschenbroek brothers may well have been the source of these lanterns, because 'sGravesande, in the preface to his *Physices elementa mathematica* identified the maker of his machines as the "very ingenious

Artist of this Town, and no unskilful Philosopher, whose Name is John van Musschenbroek and who has a perfect knowledge of every Thing that is here explained."[62] Van Musschenbroek added mechanical slides that showed motion, including a windmill with rotating sails, a man drinking from a goblet, a tightrope walker, and a girl curtsying (see fig. 3.6). Pieter tells us that these slides are available from his brother Jan, who is the leading instrument maker of Leiden. Nollet saw the Musschenbroek slides on a visit to Holland in 1736, and so it is likely that the van Musschenbroeks were responsible for the improved magic lantern.[63]

Lecturers during the eighteenth century were sensitive to criticism that the magic lantern was entertainment, not science. Nollet, for instance, had been criticized for performing experiments that were the "plaything of childhood and the instrument of charlatanism" under the "perfidious name of experimental physics"; he introduced his discussion of the magic lantern by saying that it was one of those instruments that had been "rendered ridiculous in the eyes of many people by its too great popularity," the problem being that three-quarters of those who saw it had no idea how it worked. But, asked Nollet, was this any reason for not explaining it? After all, Newton studied soap bubbles, which proves that nothing is too puerile for the philosopher.[64] Benjamin Martin implied that the magic lantern had been subverted from its original philosophical purpose, which had been to magnify small objects in a dark room, and was now used "rather to surprize and amuse ignorant people, and for the sake of lucre, than for any other purpose." Yet Martin believed that "it might be applied to more useful purposes, in magnifying the transparent parts of animal and vegetable substances, as wings of flies, membranes, etc. especially if enlightened by the sun-beams in a darkened chamber as I have many times experimented" (see fig. 3.7).[65]

In the above passage Martin described a new version of the magic lantern called a "solar microscope" that promised to project more than just ghosts and goblins. The solar microscope brought a shaft of sunlight into a dark room by means of a mirror. The light illuminated a drop of pond water, a flea, a butterfly wing, or some other natural specimen that was at least partially transparent. Lenses of short focal length produced a greatly magnified image. Nathaniel Lieberkuhn of Berlin exhibited the solar microscope to the Royal Society in 1739 and he is usually considered its inventor, although various kinds of magnifying lanterns using solar illumination had been described before Lieberkuhn.[66] The solar microscope allowed lecturers and teachers to project something of value to natural philosophy. The projected image itself, rather than the instrument or its internal optics, could become the subject of scientific "demonstration." Nollet, who obtained his solar microscope from London, probably from John Cuff, wrote that it was, "properly speaking, only a magic lantern illuminated by the sun," but "much more curious and interesting" (see fig. 3.8).[67] Nollet designed his own improved version, which he offered for sale.[68] His most spectacular

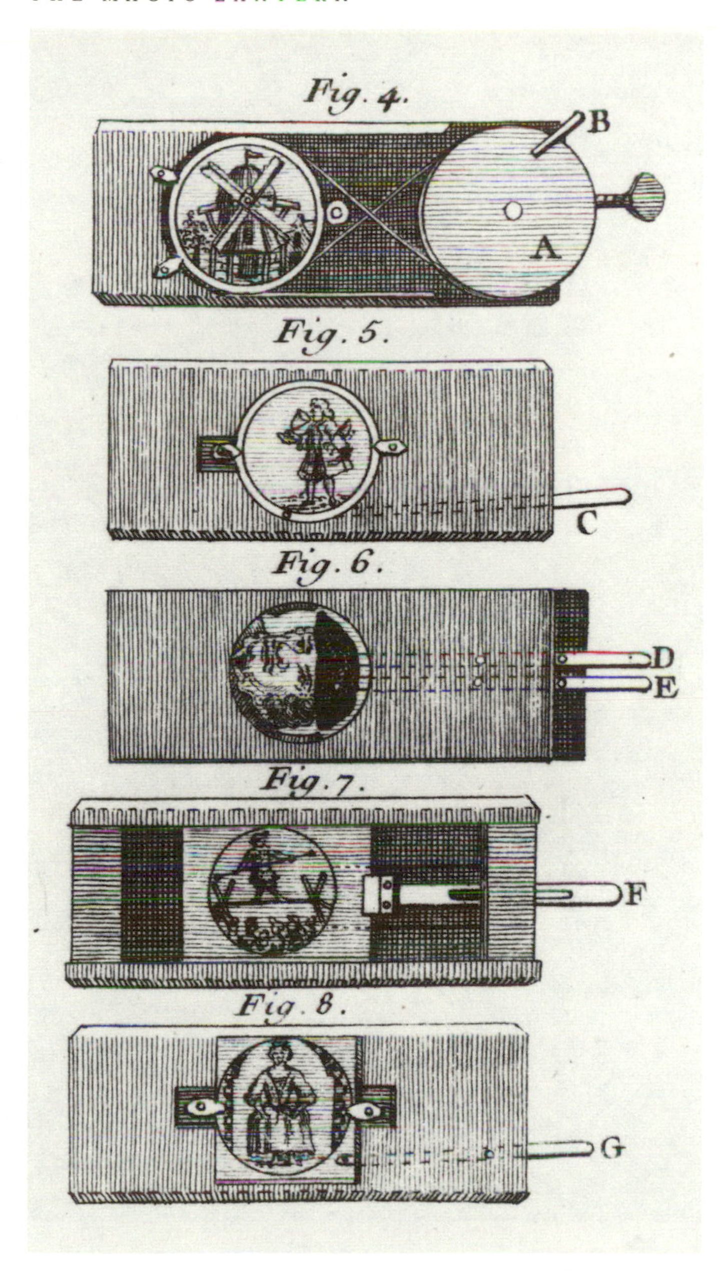

Fig. 3.6. Pieter van Musschenbroek's mechanical slides. From Musschenbroek, *Essai de physique* (1751), 2:628, table 21. Department of Rare Books and Special Collections, Princeton University Libraries.

Fig. 3.7. The magic lantern show. Attributed to Rowlandson. In the eighteenth century, lanternists were depicted as lower-class. Courtesy of the Science Museum/ Science & Society Picture Library.

On opposite page: Fig. 3.8. The solar microscope. From Liesegang, *Dates and Sources*, p. 15. Courtesy of the *New Magic Lantern Journal*: Magic Lantern Society of Great Britain.

John Cuff's trade card. Courtesy of the Science Museum/Science & Society Picture Library.

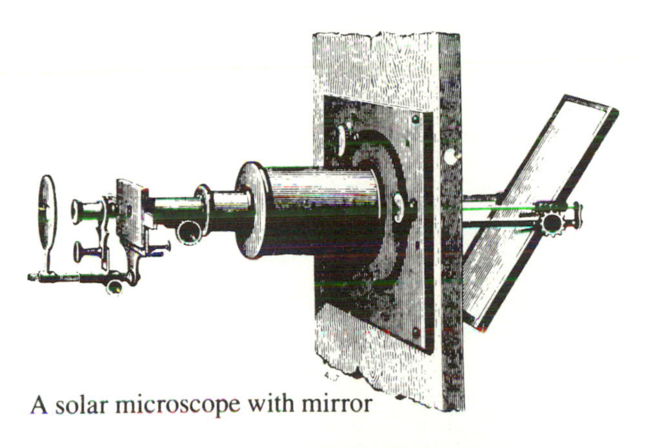

A solar microscope with mirror

JOHN CUFF,

Optician, Spectacle, *and* Microſcope *Maker*,

At the Sign of the REFLECTING MICROSCOPE and SPECTACLES, againſt *Serjeant's-Inn* Gate in *Fleet-ſtreet*,

MAKES and Sells all Sorts of the moſt curious Optical Inſtruments, ſuch as SPECTACLES and READING GLASSES, ground on Braſs Tools in the Method approved by the Royal Society, of *Brazil* Pebbles, Rock Chryſtal, or the fineſt Flint Glaſs, and ſet either in Silver, Mother of Pearl, Tortoiſe-ſhell, *&c*, in the moſt convenient Manner for Uſe.

REFRACTING TELESCOPES, proper for celeſtial or terreſtrial Obſervations, either by Land or Sea; and alſo REFLECTING TELESCOPES, as invented by Sir *Iſaac Newton* and Mr *Gregory*.

MICROSCOPES of ſeveral Kinds, both Single and Double, (either for the Study or the Pocket) particularly a new-conſtructed DOUBLE MICROSCOPE, invented by the ſaid *John Cuff*, to remedy the Inconveniencies complained of in all the former Sorts; the SOLAR MICROSCOPE, as improv'd by him with a moſt convenient and manageable Apparatus for viewing Objects, and a Contrivance for drawing the Pictures of them with great Eaſe and Exactneſs, even by thoſe that have never drawn before; the AQUATIC MICROSCOPE, invented alſo by him for the Examination of Water Animals; the OPAKE MICROSCOPE, which by a Light reflected upon Objects that have no Tranſparency, renders them diſtinctly viſible; *Culpepper*'s MICROSCOPE, *Wilſon*'s Pocket MICROSCOPE, *&c*.

The CAMERA OBSCURA for exhibiting Proſpects in their natural Proportions and Colours, together with the Motions of living Subjects: MAGIC LANTHORNS; Convex and Concave SPECULUMS or LOOKING GLASSES, MULTIPLYING GLASSES, BAROMETERS, THERMOMETERS, OPERA GLASSES, SPECTACLES of true *Venetian* Green Glaſs, SPEAKING TRUMPETS, ſeveral Mathematical Inſtruments, and many other Curioſities, are ſold by the ſaid Operator at reaſonable Prices, either Wholeſale or Retale.

experiment was showing the circulation of the blood in the mesentery of a frog.

In 1750 Leonhard Euler designed a projecting microscope and an opaque projector for larger objects, such as miniature paintings.[69] Euler was willing to apply his extraordinary mathematical skill to almost any subject, but it is perhaps surprising that he should write about the magic lantern. The solar microscope and his opaque projector, however, could project images of objects of art and natural history and not just the ghosts and goblins of entertainment. In his *Letters to a German Princess* Euler described these inventions again. The magic lantern was called magic, he explained, because "the first inventors wished to persuade people that it involved magic or sorcery."[70] His lantern, on the other hand, had the purpose of education.

Instruments for projecting opaque objects were continually being reinvented throughout the following century and given various names: in England "opaque lantern," in Germany "Wunderkamera" or "Episcope," and in France "megascope." Along with the solar microscope they became standard items in every *cabinet de physique*.[71]

The "Demonstrator" and the Demonstration Lecture

The demonstration lecture appeared to lose some of its appeal after the middle of the eighteenth century, but it recovered with a vengeance around 1780 and became ever more popular during the century's final decades. This renewed enthusiasm for the demonstration lecture coincided with a striking new turn to quantification in what we would now call the "research" part of physics.[72] Since the popular demonstration lecture carefully eschewed all mathematics and quantitative measure, there began to be, for the first time, a significant difference between "demonstration" instruments and "research" instruments. Antoine Lavoisier's gasometers, for instance, with which he claimed "proofs of the demonstrative order" were very expensive high-precision instruments. Lavoisier used them in 1783–1785 to prove that water was a compound of oxygen and hydrogen.[73] Equally precise was Coulomb's electrostatic torsion balance (also in 1785) with which he measured the force of electrical attraction. Neither of these instruments was, or could be, part of the popular physics lecture because they were designed to obtain precise numerical data, not dramatic, visible effects. This trend to greater quantification opened an increasing divide between popular and professional science. It also distinguished further the different meanings of "demonstration."

It was at this same time that one began to see the terms "demonstration" and "demonstrator" applied to the popular lecturer. In the work of 'sGravesande, experimental physics had lost all connection with medicine and the life sciences, but the medieval "demonstrator" migrated from anatomy to experimental physics, nevertheless, when it became necessary to specify a person who performed experiments for an audience. In the fourth volume of the *Encyclopédie* (1754) under the entry "Démonstrateur" one reads, "One gives this name particularly to those who give anatomy lessons on the cadaver in a public or private amphitheater."[74] There is no mention of a physics "demonstrator," even though lecturers at the Jardin du Roi had long been given that title.[75] By 1781 the demonstrator had migrated into experimental physics. Sigaud de la Fond, "maître de mathématiques," was able to announce in his *Précis historique et expérimental des phénomènes électriques* that anyone wishing to acquire instruments like his could obtain them from his nephew, M. Rouland, who assisted him as "démonstrateur" and was also "démonstrateur de physique de l'Université," a position that the "maître" had previously held.[76] In 1760 Sigaud de la Fond had succeeded to the abbé Nollet's chair at the Collège Louis-le-Grand after attending Nollet's famous lecture course on experimental physics. But he also taught anatomy and physiology and practiced medicine throughout his career. Between 1770 and 1782 he held a chair of surgery. So it was natural for him to refer to "demonstrators" both in medicine and in experimental physics.[77] It is certain, however, that this "demonstrator," whether anatomical or physical, was not presenting mathematical proofs. "Demonstration" had now become the manipulation of apparatus to instruct and edify an audience.

Physics lecturers were caught between the enthusiasm of the public for drama and the scorn of the academic physicists, who did not hesitate to express their distaste for the entertainers. Jean-Paul Marat (1743–1793), after a generally successful career as a physician, decided in 1776 to devote his attention to the study of physics. He used the solar microscope for the first time as a research instrument and soon made what he believed to be a revolutionary discovery. After removing one of the lenses, he found that he could project the image of a flame on the wall of a darkened room. About the flame and streaming above it he saw an aura of some sort (see fig. 3.9). The same could be observed around any hot object such as a burning coal or a red-hot iron ball. Marat concluded that the aura was igneous fluid flowing out of the hot object and made visible by its interaction with the light coming from the sun. From these and similar experiments with the solar microscope Marat explained the nature of heat, light, and electricity. He regretted having to disprove the theories of all previous physicists on these subjects, especially Newton, but his honor would not allow him to conceal the truths that he had discovered.

Marat's "igneous fluid" was actually the heated air around the object, which, because of its lower density, refracted the light from the solar microscope and left a shadow. Marat insisted that the shadow was not caused by the air, because he persuaded himself that he still saw the same shadows when his experiments were performed in a vacuum.[78]

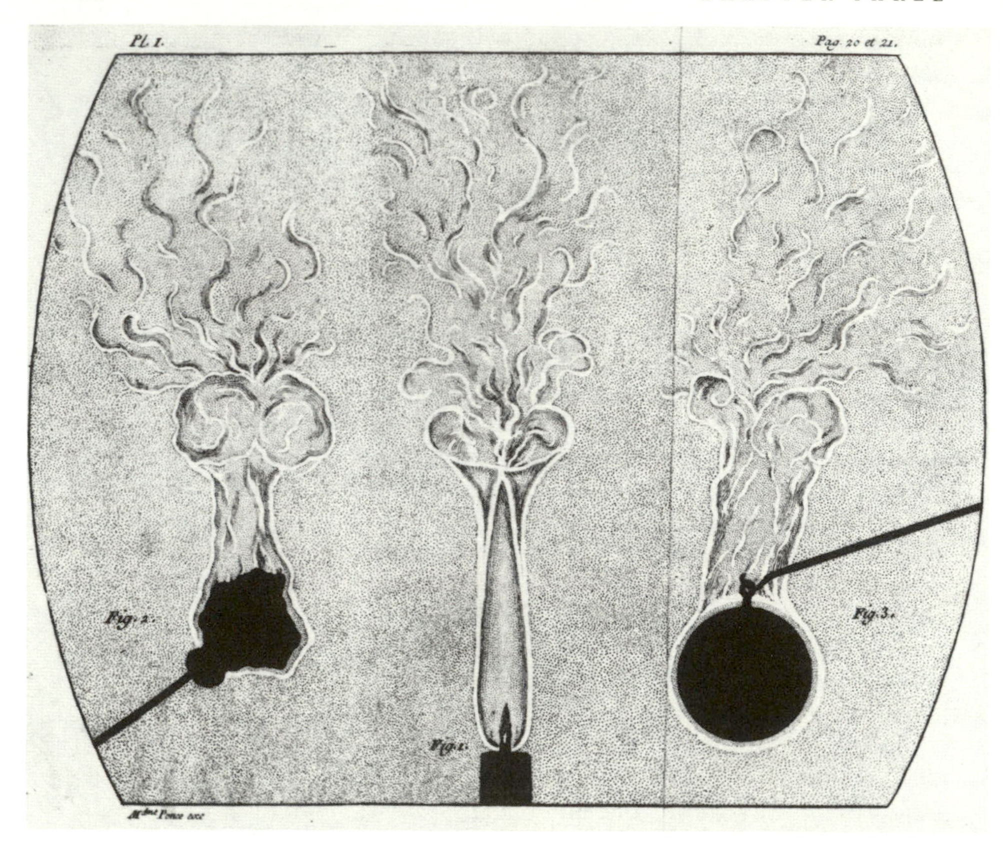

Fig. 3.9. Jean-Paul Marat observed "igneous fluid" flowing from a hot coal, a candle flame, and a red-hot iron ball. From Marat, *Recherches physiques sur le feu*, p. 20. Department of Rare Books and Special Collections, Princeton University Libraries.

Marat presented his discoveries to the Académie des sciences, expecting extravagant praise and immediate election to membership, neither of which happened. The committee appointed to review his discoveries found Marat's use of the solar microscope "ingenious," but when the commissioners were asked to review his further findings on light, which contradicted Newton, they became cautious. They concluded that as the many experiments "appeared not to prove what the author imagines that they prove, and because they are in general contrary to the most familiar parts of optics, we believe it would be useless to enter into the great detail that would be necessary to explain them," and the committee refused to give its approbation.[79] Marat, whose megalomania and poison pen were to serve him better in revolutionary politics than they had in natural philosophy, claimed that it was the geometers of the academy who forced the commission to reject his experiments, and he, the honest experimenter, had been destroyed by the mathematical elite of an elitist institution.[80] However feeble Marat's claims were, it was an argument that succeeded in bringing down the academy in 1793.

Marat also had a *cabinet* and appealed to the public through a course of physics lectures that he offered at the Hôtel d'Aligre, rue Saint-Honoré, and which featured his experiments with the solar microscope. Because Marat himself was not a very successful lecturer, the actual instruction was carried out by his "disciple" the abbé Filassier. The most successful physics lecturer at the time was Jacques-Alexandre-César Charles (1746–1823), who possessed the most extensive *cabinet de physique* in France. In 1792, when he gave his instruments to the nation, he had more than 330 pieces, most of them very fine. In 1780 Charles invented the "megascope," an optical instrument for projecting the magnified image of any object.[81] He was particularly proud of this instrument because it allowed him to present experiments on a grand scale, grander than anything his competitors could stage. Charles also used the solar microscope to advantage in his optical lectures, which he gave in the summer when there was reliable sunlight.[82]

On March 15, 1783, Marat went to the home of Charles, accused Charles of making fun of him in his lectures, and attacked him with a sword. Charles, who was not armed, apparently got the better of his assailant and, with some help, threw Marat out into the street. Marat complained to the police, who managed to prevent a duel (much to Marat's relief, we might suppose).[83] In describing the events to friends, Charles defended himself by saying that "if one is to be beaten [for criticizing Marat], all Europe must arm."[84]

While this contretemps is largely attributable to Marat's pugnacity, it is worth trying to discover what was at the bottom of the quarrel. Both men were competing for the attention of Parisian society, both used the magic lantern and the solar microscope prominently in their lectures, and both had an interest in the nature of heat. Charles had begun his lectures in 1771 at the place des Victoires and was sufficiently successful to be elected to the Académie des sciences. In 1783 he built the first hydrogen balloon, and in 1787 he established "Charles's Law" that the densities of all gases decrease in proportion to the increase in temperature. We can assume that Marat's "igneous fluid," observed through the solar microscope, would have been of interest to him.

Marat told the police that Charles had compared him to "Sieur Comus" and therefore had contemned his scientific abilities. Comus (the god of feasts, lovers, and debauchery) was the stage name of Nicolas-Philippe Ledru, who presented his show at the boulevard du Temple. Comus had made his reputation as a magician. His repertoire included prestidigitation, mind reading, and fortune-telling, along with optics and electricity.[85] But he had also learned about the construction of scientific instruments in England, and in 1781 Louis XVI commissioned him to build instruments and prepare meteorological maps for the navy. Comus gained further respectability when he began to use electrotherapy to treat epileptics at his *cabinet*. In 1784 Louis conferred on him the title "Physician to the King" and helped him set up an expanded clinic.

Comus was remarkably successful at obtaining patronage using his own particular blend of entertainment and serious science.[86] In 1781 Marat and Comus had had an argument over the priority of certain electrical experiments. When Charles compared Marat to Comus, it is not surprising that Marat took offense. During the 1780s physics lecturers in Paris were competing for a limited fund of support and patronage. Membership in the Académie des sciences was the most desirable prize, but there was also much to be gained from the public and from the nobility, whose interests tended to be on the side of entertainment. Natural magic and experimental physics blended completely in this love of spectacle. The Académie des sciences and the Faculté de médecine attempted to draw a line between the two, but it was not always easy to distinguish between, say, Comus's electrotherapy and Mesmer's animal magnetism. Nowhere was this combination of science and spectacle more attractive than in ballooning. Human flight had always been a dream of the natural magicians. Now with the discovery of new chemical "airs" and a better understanding of heat it became a real possibility.

The first Montgolfier balloon ascension took place in Annonay on June 4, 1783, three months after Marat's attack. Charles sent up his first balloon on August 27 from the Champ-de-Mars in Paris, and on December 1 Charles himself made his one and only balloon ascent from the courtyard of the Tuileries palace.[87] The Montgolfier balloon was a hot-air balloon and Charles's balloon was inflated with hydrogen, but the difference between these two substances was not obvious. In 1783 heat was as much a chemical as a physical process. This is evident in the words of Charles's promoter, Barthélémi Faujas de Saint-Fond (1741–1819), who referred to the "gas or rarified air" that the hydrogen balloon contained; similarly, Jean-François Pilâtre de Rozier (1756–1785), the first human aeronaut, referred to hot air as "igneous gas" and to hydrogen as "inflammable gas," which indicates that he believed some principle of fire or heat was responsible in both cases for the lower density of the gas.[88] Next to electrical experiments using large electrostatic generators and banks of Leyden jars to produce prodigious sparks, the most spectacular popular demonstrations employed the newly discovered gases such as imflammable air (hydrogen) and eminently respirable air (oxygen). The facts that hydrogen burns explosively and that all manner of substances, such as iron, will burn in pure oxygen provided numerous opportunities for impressive demonstrations.

Pilâtre de Rozier also gave a course of demonstration lectures beginning in 1781 at the "musée de Monsieur," later known as the "Lycée." He had as many as seven hundred subscribers to his lectures, many of them women from distinguished families.[89] It was the same clientele that frequented the lectures of Marat and Charles. The association of the physics lecturers with ballooning emphasizes the romantic nature of the subject in the eyes of the populace, who swarmed to witness the ascensions. Ballooning joined the magic lantern, the solar microscope, and the electrostatic generator as the chief scientific spectacles of the late eighteenth century.

Another physicist turned balloonist, Étienne Gaspard Robertson, produced the most spectacular magic lantern show of the eighteenth century. He called it the "phantasmagoria." Robertson, whose original name was Robert, was born in Liège in 1763 and moved to Paris shortly before the outbreak of the Revolution. In 1792 he enrolled in the physics course given by Charles, who inspired him to work with the solar microscope, the megascope, the magic lantern, and the mirror of Archimedes. During the Terror Robertson found it convenient to return to Liège for his health, where he devised a mechanism for directing the Archimedean mirror, and in 1796 the Convention granted him a *laissez-passer* to return to Paris and report on the military potential of his giant mirror.[90] While waiting for action from the Institut, he refined his lantern, which he later called the "phantascope," and staged his first performances in the *cabinet de physique* of a M. de Beer. When these arrangements proved unsatisfactory, he moved his show to the Pavillon de l'Echiquier, and finally to the abandoned Couvent des Capuchins, which had the proper gothic atmosphere.[91] The phantasmagoria was mostly magic with just enough physics to make it respectable (see fig. 3.10). The audience entered through a room containing Robertson's *cabinet* and then wound its way through "a series of dark passages, decorated with weird and mysterious paintings . . . the very door was covered with hieroglyphics. The chapel itself was hung with black and was feebly illuminated by a single sepulchral lamp."[92] In the pitch darkness the audience saw dim ghosts that seemed to rush toward them to the sound of Benjamin Franklin's glass harmonica. The phantasmagoria was a huge success and drew large

Fig. 3.10. Depiction of a performance from Robertson's phantasmagoria. From *New Magic Lantern Journal* 4 (1986): 4. Courtesy of the *New Magic Lantern Journal*: Magic Lantern Society of Great Britain.

crowds. It was emulated first in London and then in other major cities around Europe.

Robertson was able to present his show to a relatively large audience, because instead of projecting his images onto a wall, he back-projected them onto a translucent screen so that his lantern was directed toward the audience. This greatly increased the illumination, as did the new Argand lamp that he employed in his projector. And finally, he mounted his "phantascope" on wheels, which allowed him to roll the lantern toward or away from the screen. The image would increase dramatically in size and give the impression that it was charging the audience. In order to keep the image in focus and the illumination constant, the "physicist" (Robertson's name for the projectionist) continually adjusted the focus and the light stop. After the phantasmagoria reached England, Thomas Young devised a linkage to control the focus automatically. The program contained such presentations as *The Death of Lord Littleton*, *The Pilgrimage of Saint Nicholas*, *Preparations for the Witches' Sabbath*, *Diogenes with His Barrel*, *The Birth of Rustic Love*, *The Temptation of Saint Anthony*, and the like. These were not physics lectures, obviously, but Robertson did add physics experiments, especially galvanic experiments, to his show. Robertson knew Volta and was a member of the Paris Galvanic Society. Like Charles and Pilâtre de Rozier he recognized the opportunities, both theatrical and scientific, of ballooning; he made fifty-nine ascents, the most famous at Hamburg on July 18, 1803, where he set a new altitude record. Robertson also employed Charles's megascope to project enlarged images of animated figures. His *Apotheosis of Héloïse* was almost certainly the projected image of a living person.[93]

The Magic Lantern in Education

The divide in experimental physics between quantification and spectacle continued from 1780 through the century, with the magic lantern obviously on the side of spectacle. In the hands of 'sGravesande it had been an instrument to be explained geometrically; in the hands of Robertson it was primarily entertainment. But in both capacities the magic lantern had shown ghosts and fairy-tale figures. It is surprising that during a century so interested in education there were not more efforts to use the lantern for instruction. Benjamin Martin argued in 1781 that the magic lantern should be used for education as well as for entertainment, but with little success.[94] The comte de Paroy, who claimed to have inspired Robertson's phantasmagoria and Charles's megascope, persuaded Marie Antoinette to educate the dauphin with the magic lantern. The dauphin, aged six, was too easily distracted to learn from books, and Paroy argued that the magic lantern would make an impression where the written word had failed. In fact he saw a great future for the magic lantern in educating the entire world, not just the *enfant de France*.[95] The queen reacted with skepticism at first but was won over, according to Paroy, and the dauphin would have been the first pupil to be instructed via the magic lantern if the events of the Revolution had not intervened.

In France the lantern was first used in education in 1839 by Jules Duboscq, François Soleil, and the abbé Moigno. Duboscq and Soleil were the first to use electrical arc light and incandescent lamps as illuminants and designed many different kinds of projecting lanterns and microscopes.[96] The abbé François-Napoléon-Marie Moigno (1804–1884) campaigned vigorously for the educative use of the lantern. In 1852 he proposed the introduction of audiovisual education to the ministère de l'instruction publique, and in his *L'art des projections* of 1872 he advocated free evening lectures for workers. He finally succeeded in 1880 in persuading the Maison de la Bonne Presse to circulate lanterns and slides for religious instruction in order to curb the trade in licentious literature that was corrupting the country. The Ligue française de l'enseignement offered an anticlerical version to compete with Moigno's program.[97]

In England the lantern was first used in a scientific capacity to project astronomical diagrams.[98] In 1849 Henry Mayhew began publication of a series of letters in the *Morning Chronicle* describing various trades, including the manufacture of magic lanterns. Mayhew's toy maker told him:

> I have known the business of magic lantern making thirty-five years. It was then no better than the common galantee shows in the streets, Punch and Judy, or any peepshow common thing. There was no science and no art about it. It went on so for some time. . . . About thirty years ago [1820] the diagrams for astronomy were introduced. These were made to show eclipses of the sun and moon, the different constellations, the planets with their satellites, the phases of the moon, the rotundity of the earth, and the comets with good long tails. . . . This I consider an important step in the improvement of my art. Next, moving diagrams were introduced. I really forget, or never knew, who first made these improvements.[99]

The most important "improver" was Philip Carpenter. Carpenter had experience with optical instruments and had contracted with Sir David Brewster in 1819 to manufacture the kaleidoscope, which Brewster had just invented. In 1820 he began to market his Phantasmagoria Lantern, which was superior to all previous models and permitted a much higher quality of projection in the home. Carpenter also developed a "copper plate" process for mass-producing lantern slides of high quality. In 1823 he published *Elements of Zoology* accompanied by 56 slides covering 256 natural history subjects. In addition to the slides of animals, Carpenter also sold slide sets of astronomical diagrams along with more conventional series, such as "Portraits of the Kings and Queens of England," "Costumes of the Ancients," and the like. Moreover, Carpenter offered his slides at a cut rate if they were to be used for "educational purposes."[100]

One reason why the magic lantern was only gradually introduced into education was the difficulty of getting adequate illumination. As long as the only source of light was a candle or smoky oil lamp, dim ghostly figures were about all that one could see, and even then the lantern had to be close to the screen. Carpenter's Phantasmagoria Lantern and the Sciopticon that L. Marcy of Philadelphia marketed in 1872 greatly improved the illumination and made the lantern more usable in the home. The introduction of limelight made the lantern suitable for large lectures. Lewis Wright used an oxyhydrogen mixed jet apparatus to reach an illumination of one thousand candles and in 1884 successfully responded to the Microscopical Society's challenge to make a microscope that would project "the tongue of a blow-fly six feet long."[101]

Even more important than illumination for the success of the magic lantern was photography, which made possible faithful reproductions at greatly reduced cost. The first public show using photographic slides was staged by the Langenheim brothers in Philadelphia in 1849. Called "Hyalotypes," these slides quickly became known in Europe after they were shown at the Great Exhibition in London in 1851.[102]

The lantern as spectacle reached its apogee at the Royal Polytechnic Institution in London, founded in 1838 to present science to a popular audience. The Polytechnic employed large lanterns that took glass slides as large as 8½ by 7 inches (see fig. 3.11). This allowed the slide painters to include much greater detail in their paintings, and the show used as many as six lanterns at a time, superimposing the images to create dissolving views, motion, and other special effects.[103] The Polytechnic advertised "Lectures, Experiments, and Scientific Productions," while London guidebooks urged that it "be visited by all when in town, who will leave it with remembrances of electricity, oxygen, hydrogen, and the diving bell." Another guidebook stated that it "partakes of the quadruple character of a Lecture Room, a Concert Hall, a Museum, and a Temple of Magic." This was not high science—the most popular attraction for many years was "Professor Pepper's ghost"—but it was the kind of setting in which the magic lantern was most effective.

The magic lantern also quickly made its way into more serious physics demonstrations. We can now talk about "demonstration" in its modern sense without anachronism. The subtitle of Lewis Wright's *Optical Projection* (first edition 1890) was *A Treatise on the Use of the Lantern in Exhibition and Scientific Demonstration.* Chapter titles are "Apparatus for Scientific Demonstrations," "Demonstrations of Apparatus in Mechanical and Molecular Physics," and "Physiological Demonstrations." In Germany, where the use of special scientific lanterns was first systematized in the schools, the most important book was Adolf Weinhold's *Physikalische Demonstrationen.*[104] It seemed that everything could be projected in the nineteenth-century demonstration lecture—Newton's rings, Airy's spirals, Marey's pulse mirror, Ludwig's kymograph, Lissajous figures, Wheatstone's kaleidophone, König's manometric flames, Tyndall's sensitive smoke-jets, Taylor's phoneidoscope, Lippman's capillary electrometer, Fresnel's prism,

Fig. 3.11. The Great Hall and the "Optical Box" at the Royal Polytechnic Institution. From *New Magic Lantern Journal* 4 (1986): 48, 51. Courtesy of David Henry and the *New Magic Lantern Journal*: Magic Lantern Society of Great Britain.

Fig. 3.12. Scientific lanterns from the nineteenth century. From Wright, *Optical Projection*, pp. 119, 160, 169, and 209. Courtesy of the University of Washington Libraries.

Atwood's machine, ad infinitum. These lanterns projected not just pictures on glass slides, but actual physical phenomena, especially phenomena involving very small forces (see fig. 3.12). The apparatus in such experiments was necessarily small, and a light beam could magnify the effect enormously without interfering with the phenomenon being demonstrated.

The greatest physics demonstrator of all time was John Tyndall, lecturer at the Royal Institution. Twenty-five prominent American men of science, including Joseph Henry, Louis Agassiz, and Ralph Waldo Emerson, persuaded Tyndall to undertake a speaking tour of the United States during the winter of 1872–1873. In each city he gave a series of six lectures on optics assisted by two demonstrators using instruments from the Royal Institution. Central to each demonstration was a magic lantern using a carbon arc and powered by twenty voltaic cells. Tyndall opened his lectures by describing how energy from the "burning" of zinc in the voltaic cells was transferred electrically to the blinding light of the arc. Anyone reading the *Lectures on Light Delivered in the United States in 1872–73* cannot help but be impressed by Tyndall's spectacular demonstrations.[105] Attendance at a single lecture sometimes reached 1,500, and the *New York Daily Tribune* printed each lecture as it was given. A special edition of all six lectures sold over 300,000 copies.[106] The press referred to Tyndall's instrument as a "magic lantern," but Tyndall himself called it a "camera." Presumably he did not want his demonstrations to be thought of as "magic."

Conclusion

The magic lantern was an instrument of natural magic that kept its "magical" character longer than almost any other. It began to lose its name only in 1872 when Marcy advertised his Sciopticon as an "optical lantern" rather than as a magic lantern. In 1874 Edward L. Wilson of Philadelphia founded the *Magic Lantern*, the first journal devoted exclusively to the instrument, and in 1877 Paul Liesegang founded a second entitled *Laterna Magica*. In 1889 J. Taylor compromised on the instrument's name when he entitled his new journal the *Optical Magic Lantern Journal*, with the word "magic" dwarfing the word "optical" in the masthead. But when the second editor took over in 1902, he staged a competition which resulted in a new masthead that magnified "optical" and shrunk "magic" into insignificance (see fig. 3.13).

The magic of the lantern never disappeared, however. When it successfully incorporated motion and became the cinema, its first achievement was not to produce art, but to put stage magic out of business. Georges Méliès's first film in 1896 was a version of the vanishing lady trick, and he proceeded to the "Man with the Rubber Head," "The Terrible Turkish Executioner," and "The Over Incubated Baby." When the great Houdini visited the Théâtre Robert-Houdin in 1901, he discovered that stage magic had been

Fig. 3.13. Masthead of the *New Magic Lantern Journal*. The magic lantern becomes the "optical lantern" with just a bit of magic left. Courtesy of the *New Magic Lantern Journal*: Magic Lantern Society of Great Britain.

completely replaced by the cinema, which continues today to work its magic through special effects.[107]

The history of the magic lantern reveals once more the great complexity of the rise of modern science. New instruments and new methods were an essential part of the Scientific Revolution, but they did not always go hand in hand. Instruments could remain the same while methods changed and vice versa. Natural magic, experimental philosophy, and mathematics overlapped and were woven together in a complex, constantly changing structure. Words such as "demonstration," "fact," and "experiment" changed their meanings or acquired multiple meanings. The demonstration experiment and the demonstration lecture became institutionalized after a long process of development, but they were in no way an inevitable consequence of the rise of modern science.

The magic lantern is an especially interesting example, because it retained its association with natural magic longer than most instruments from the seventeenth century. The purpose of the instruments of natural magic had been to display the wonders of nature. The purpose of the demonstration instruments of experimental physics was the same, with the added assumption that the causes of the phenomena being exhibited could be explained by some coherent physical theory, and that the instruments assisted in "demonstrating" that theory.[108] What it meant to "demonstrate" with instruments remained a subject of debate throughout the Scientific Revolution. We have observed a transition in experimental philosophy from "demonstration" as a logical argument to "demonstration" as an exhibition of phenomena. The institution of the demonstration lecture cemented this transition, although the word has always retained its useful ambiguity.

In the modern physics demonstration lecture, the lecturer displays the phenomena to the audience—this is one kind of "demonstration"—but he also claims that his experiments confirm a theory that is being "demonstrated" in a different sense. The ambiguity is convenient, because it suggests that one kind of demonstration or "showing" includes the other. To some extent, this has always been the purpose of the demonstration lecture, and the changing meaning of "demonstration" is, therefore, largely a shift in emphasis.

We use slide projectors and overhead projectors now without thinking of them as scientific instruments and without a thought about magic. We regard them as practical instruments for presenting images and imparting information. But we should remember when we show a slide, watch television, or go to the movies that we are experiencing the fulfillment of Kircher's dreams. The attempts by Kircher and his fellow natural magicians to produce wondrous effects appear ludicrous alongside the achievements of a Galileo or a Newton, but unlike the experimental philosophy (which was still philosophy in spite of Francis Bacon), natural magic was essentially practical. Many of the instruments that we take for granted, not only in scientific research, but in our daily lives, find their origin there.[109] The magic lantern reminds us that we are connected to Kircher's world more closely than we realize.

CHAPTER FOUR

The Ocular Harpsichord of Louis-Bertrand Castel; or, The Instrument That Wasn't

IN HIS ACCOUNT of the Great Cat Massacre Robert Darnton brings to history a lesson learned from anthropology, that one can enter an unfamiliar culture most easily by studying those aspects that are most incomprehensible. From a bizarre massacre of cats by printer's apprentices in Paris during the 1730s Darnton explains the apprentices' life, their ceremonies, their behavior, their hatred for their master, and the peculiar significance of cats in their rituals. The apprentices found the torture of cats hilariously funny, while we, reading about it in the twentieth century, "don't get the joke." Precisely the fact that we don't get the joke means that we have something to learn.[1]

Historians of science have traditionally ignored that which they do not "get." If an idea, book, organization, or instrument does not make sense from the perspective of twentieth-century science, it is ignored, and if it is found in the writings of someone we have learned to revere, it is regarded as downright embarrassing. The last fifteen years have seen a great change in this regard, and historians of science have learned that they cannot study what used to be called the "progressive element" of science in isolation without doing violence to history as a whole.

One problem with studying the unfamiliar in science is that we dissolve the disciplinary boundaries of our subject. We have no objective criterion by which we can say whether an instrument or idea is "scientific." This is not altogether bad. By dissolving our own disciplinary boundaries, we can then ask the more important historical question of how the instrument or idea was regarded by its creator and by those who used it, and how it fit *their* disciplinary boundaries.

"Philosophical" instruments like the telescope, microscope, and air pump were new in the seventeenth century and still carried the flavor of natural magic. As a result they were suspect and their value had to be demonstrated. The process of determining what was acceptable practice in natural philosophy also required a decision about what were acceptable instruments. And since the new instruments were radically different from the old ones and so important for the new experimental philosophy, the choice of instruments helped to define the philosophy.

Not all instruments were accepted, of course. If they had been we would be hard pressed to say what we mean by "natural science." The telescope, microscope, and barometer were big winners. The speaking tubes, magic glasses, and hydraulic fountains were losers. Of most interest to us as historians are those instruments that were, so to speak, "on the margin"—those instruments that caused confusion as to whether they were truly philosophical.

From Cat Piano to Ocular Harpsichord

In keeping with Darnton's methodology and subject matter we might want to look at the cat piano. Athanasius Kircher first wrote about it in his great *Musurgia universalis* of 1650, and it has reappeared occasionally since (see fig. 4.1). In order to raise the spirits of an Italian prince burdened by the cares of his position, a musician created for him a cat piano. The musician selected cats whose natural voices were at different pitches and arranged them in cages side by side, so that when a key on the piano was depressed, a mechanism drove a sharp spike into the appropriate cat's tail. The result was a melody of meows that became more vigorous as the cats became more desperate. Who could not help but laugh at such music? Thus was the prince raised from his melancholy.[2] The cat piano confirms Darnton's discovery that most early modern Europeans found the torture of cats funny. It also illustrates Kircher's fascination with the relationship between the art of music and the natural production of animal sounds. But for us it is an instrument that has mercifully been forgotten.

However, the cat piano did appear once during the eighteenth century in a place prominent enough to attract notice. Louis-Bertrand Castel described it in 1725 in an article announcing his famous *clavecin oculaire* or ocular harpsichord. The ocular harpsichord was like a standard harpsichord

Fig. 4.1. The cat piano. From *La Nature*, pt. 2 (1883): 519–520. Courtesy of the University of Washington Libraries.

except that it played colors instead of sounds. The possibility of such an instrument depended on the analogy between the seven spectral colors and the seven tones of the musical scale. He used the example of the cat piano to show that sound was not beautiful by itself and that the beauty of music lay only in the sequence and harmony of the notes. The cat piano might conceivably have produced a recognizable tune, but the effect would certainly not have been one of harmony. It was only a joke to illustrate Castel's important discovery. The ocular harpsichord was a different matter. It would produce beautiful harmonies for the eye. According to Castel, it would be the "universal instrument of the senses."[3]

Whether the ocular harpsichord was a scientific instrument or not depends on one's point of view. Castel claimed in his announcement that his harpsichord would not merely give a simple impressionistic idea of sound in color but would really paint sounds by a precise and natural correspondence between color and pitch, so that a deaf listener could enjoy music that was originally written for the ear. He would demonstrate this correspondence following reasons of fact and geometrical analysis. He would accept only that which was proven.[4]

Reaction to Castel's announcement of the ocular harpsichord was not generally favorable, but it did cause considerable excitement—enough that Castel could reasonably ask why his opponents were willing to spend so much time combating what they claimed was a worthless idea.[5] Part of the problem was Castel's independence of mind, which led him to argue with everyone. Voltaire called him the "Dom-Guichotte des mathématiques" because of his tendency to attack the giants, including Newton, Leibniz, Réaumur, and Maupertuis.[6] Voltaire could have included Rameau, Rousseau, Dortous de Mairan, and Voltaire himself. That Castel should have warranted the attention of such illustrious foes is in itself remarkable.

Castel had joined the Jesuits as a novice in 1703 at age fifteen. In 1720 he came to the notice of Fontenelle, who was instrumental in having him transferred from Toulouse, where he had been teaching rhetoric, to Paris, where his teaching expanded to include physics, infinitesimal calculus, mechanics, pyrotechnics, and architecture. In Paris he became the unofficial science editor for the Jesuit *Journal de Trévoux* and in this capacity wrote on every conceivable subject from the Northwest Passage to the squaring of the circle. In this he followed the tradition of the great Jesuit polymaths like Kircher, who admitted no limits to their breadth of knowledge.

He announced his ocular harpsichord in 1725 at the urging of the composer Jean-Philippe Rameau, who had been organist at Clermont when Castel taught there. The analogy between color and musical tone was by no means original with Castel. Newton had stated it very prominently, as had Kircher. Newton had studied musical harmony in 1664–1666 and throughout his life retained a belief in the *musica mundana*, or universal harmony of the world. His attention was called to the analogy between color and tone by Robert Hooke, who mentioned it in his criticism of Newton's first optical

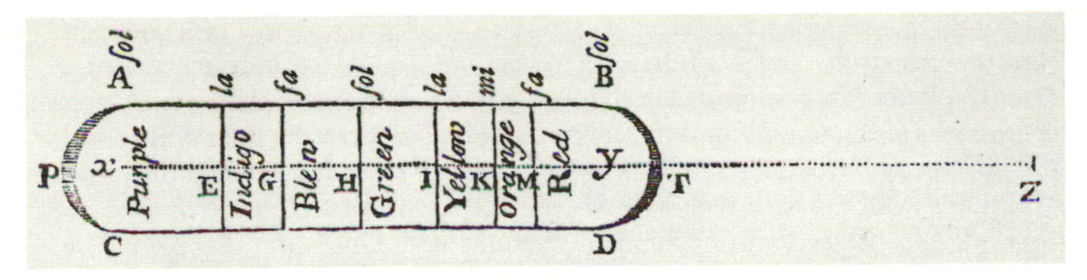

Fig. 4.2. Isaac Newton's illustration of the color-tone analogy. From Birch, *The History of the Royal Society of London*, 3:263. Courtesy of Special Collections, the University of Washington Libraries.

paper of 1672; Newton, in his second optical paper of 1675, went Hooke one better by showing that the seven bands of color in the spectrum have widths in the same harmonic ratios as the string lengths on the monochord that produced the musical scale (see fig. 4.2).[7] Because Newton also read Kircher it is possible that Kircher was the source for Newton's analogy, as Voltaire claimed, but it is also certain that Newton's supposed discovery of a new harmonic relation between the colors in the spectrum brought the color-tone analogy into prominence. Newton wrote: "As the harmony and discord of sounds proceed from the properties of the aerial vibrations, so may the harmony of certain colours . . . and the discord of others . . . proceed from the properties of the aetherial. And possibly color may be distinguished into its principal degrees, Red, Orange, Green, Blew, Indigo and deep Violet on the same ground, that sound within an eighth is graduated into tones."[8]

The most immediate stimulus for Castel was probably Nicolas Malebranche, who in the sixteenth elucidation to his *Recherche de la Verité* referred specifically to the analogy between light and sound. Malebranche used Newton's experiments as evidence for his theory that both light and sound were caused by vibrations propagated in media composed of small vortices, and Castel adopted the same analogy of similar vibrations, although he repudiated Malebranche's little vortices.[9]

Castel's most important patron was Charles de Secondat Montesquieu, with whom he began correspondence soon after his arrival in Paris. For a while he had Montesquieu's son as a pupil at the Collège Louis le Grand and hoped through that contact to persuade Montesquieu to publish in the *Journal de Trévoux*. In 1735 he wrote an extremely long and verbose account of "new experiments on optics and acoustics" in the form of letters addressed to Montesquieu and published in the *Journal de Trévoux*.[10] Montesquieu's friendship was valuable to Castel, but it did not include any great enthusiasm for the ocular harpsichord.

The greatest boost for the ocular harpsichord came from Voltaire, who devoted chapter 14 of his *Eléments de la philosophie de Newton* (1738) to the color-tone analogy and to Castel's instrument. Voltaire wrote that he believed Kircher to be the source for Newton's analogy between light and sound, and he praised Kircher as "one of the greatest mathematicians and most learned men of his times." Kircher had argued entirely by analogy, and Voltaire favored instead Newton's experimental method. Yet even Voltaire was willing to admit that "this secret analogy between light and sound leads one to suspect that all things in nature have hidden connections, that perhaps will be discovered some day."[11] In spite of his sympathy (limited, to be sure) for Castel's ideas, Voltaire quarreled with him and took revenge by attacking him in the public "Letter to Rameau," in which he also ridiculed the ocular harpsichord.[12] What disturbed Voltaire was not the idea of an ocular harpsichord (after all, Newton had given serious attention to the color-tone analogy) so much as Castel's style of inquiry, which employed analogy in place of experiment and was, therefore, very different from that of Newton. And even though he may not have been able to follow all of Newton's mathematical arguments, Voltaire understood Newton's style and method as well as anyone in France. He concluded that Castel's style did not sufficiently grasp "the spirit of this century."[13] Castel was certainly not a child of the Enlightenment.

Others examined directly the analogy between color and tone. Jean-Jacques Dortous de Mairan criticized Castel's ideas in 1737. In 1739 the composer Georg Philipp Telemann wrote *Beschreibung der Augen-orgel, oder des Augen-clavicimbels*, based on his observations of the instrument during his visit to Paris in 1737–1738. In 1742 the Saint Petersburg Academy also devoted a séance to the ocular harpsichord, at which Georg Krafft expressed his doubts about the usefulness of the analogy.[14] Even Jean-Jacques Rousseau, who befriended Castel in 1741, had no use for the instrument.[15] Thus one can conclude that Castel's ocular harpsichord received plenty of attention, but only limited acceptance.

The *philosophe* most willing to give serious consideration to Castel's invention was Denis Diderot, who found in it a natural theme for his *Lettre sur les sourds et muets* (1751). When Diderot's imagined deaf-mute sees Castel's machine, he thinks the colors are a form of speech and concludes that the inventor must have been a deaf-mute too. Diderot's interest in the formation of the senses meant that he would take the color-tone analogy seriously, but in the *Encyclopédie* he joined the chorus of those urging Castel to make the instrument and demonstrate the harmony of colors directly rather than talking about it interminably.[16]

One would expect that having conceived of an instrument to exploit the analogy between color and tone, Castel would have been eager to make the instrument or have it made. This was not the case, however, and there is reason to doubt that a working ocular harpsichord was ever made during Castel's lifetime—by him or by anyone else.

Part of the problem was the technical difficulty of making such an instrument in the eighteenth century. In 1730 Castel had exhibited some kind of device, but apparently all it did was raise colored slips of paper into view.[17] Supposedly this modest instrument created so much excitement in Paris that

Castel was obliged to close his rooms to visitors and postpone his efforts. On December 21, 1734, with much fanfare, he demonstrated a more advanced instrument but admitted that it was "only a model and therefore very imperfect."[18] His anonymous English assistant later made an instrument that he demonstrated in London after Castel's death. This harpsichord contained five hundred lamps (probably candles) and must have given off a prodigious quantity of heat. That is probably why a manuscript note attached to the description of the English ocular harpsichord says that it was there to be observed in Soho, but was never played.[19] All descriptions of the instrument during Castel's lifetime are distressingly vague. It was not that he had any problem obtaining support for his invention. The prince de Conti offered his support, and Castel actually accepted two thousand livres from Comte Maillebois and a thousand crowns from the duke of Huescar, the Spanish ambassador.[20] With this money Castel was able to employ workmen to help with the construction, but their efforts came to naught.

Yet even aside from the technical difficulty of building an ocular harpsichord, Castel seems to have had no desire to build the instrument in the first place. His response to critics after he announced his harpsichord in 1725 was, "I am a mathematician, a philosopher . . . and I have no desire to make myself into a bricklayer in order to create examples of architecture."[21] For Castel the idea and not the artifact was what counted. It apparently did not occur to him that one might construct an instrument for the purpose of testing a theory. We are confronted here with a thoroughly unfamiliar approach to the natural world, one that we could easily dismiss as unfruitful and therefore unimportant. But Castel's disinclination to make the ocular harpsichord demands an explanation, and it is our task to try to understand it.

The Harpsichord as Thought Experiment

The ocular harpsichord was a kind of "thought experiment," a realization of an idea in an imagined instrument. Castel claimed that even if he did actually construct an instrument, it would not and could not decide whether there was a real analogy between light and sound. As he explained to Montesquieu, the public clamor to see the ocular harpsichord was misguided. Montesquieu would understand that it was nobler and more scientific to approach the problem through the mind than through the senses. And besides, it would not be possible to judge color harmony immediately from the ocular harpsichord in any case. Castel insisted that one had to become accustomed to any kind of music to appreciate it. "One has to learn to appreciate even Homer."[22]

The cat piano can assist us again in understanding Castel's argument. In his letters to Montesquieu he claims that animals cannot create or appreciate music; the sounds they make are only cries. Therefore the cat piano is a product of human art, not cat art, and it produces music only to the extent that the sounds are controlled by the human playing it. Animals cannot make music at all. Music can be created and appreciated only by the human *mind*; it will not "make sense" to the senses alone.[23] Nor will just any mind appreciate just any kind of music. Only a mind prepared by previous experience can respond to a new kind of music or instrument. Castel uses the quarrel over the relative superiority of French and Italian music to illustrate this last argument, asserting that French music portrayed the French character in a unique way, and that a Frenchman could not immediately appreciate Italian music.[24]

This characteristic of music leads Castel to argue that the ocular harpsichord is artificial, even though it is based on a real analogy in nature. In fact, he argues that just as the best fruits and flowers are the product of the art of agriculture, so is the best music the most artificial. The less natural the ocular harpsichord is the better: "All of which leads me to say: 1. That the more color-music is refined, artificial, scientific even, that is, nonhabitual, the more beautiful and agreeable it will be, not at first, but *col balsamo di costume*; and thus 2. I must attempt to make it known to the taste, to the mind, to the reason, to the internal sense in order to make it felt by the external sense, the eye."[25] Of course the reality of the color-tone analogy must not be denied. It exists in nature, but it must be revealed to the mind before it can be appreciated by the senses.

Moreover, the purpose of an instrument like the ocular harpsichord is not to test a theory or to produce a new idea. Physics is the subject of our everyday experience: "Everyone is a bit of a physicist to the extent that he has an attentive mind capable of natural reasoning." Castel bases his physics, "not on arbitrary hypothesis or particular and personal experience, but uniquely on history and on the general observation of nature and art."[26] Therefore, an instrument in physics has the purpose of confirming what we already know to be true from reason applied to our general experience. It cannot by itself be the basis for constructing a theory that generalizes beyond the single phenomenon that it produces.[27]

This conception of the role of an instrument explains in part Castel's hostility to Newton. Newton's prism experiments are entirely different from Castel's ocular harpsichord. They rest on an *experimentum crucis*, a single test of a single idea. The ocular harpsichord, on the other hand, illustrates an analogy understood from general experience. It is not surprising, then, that Castel dislikes Newton's prism: "I distrust the prism and its fantastic spectrum. I regard it as an art of enchantment, as an unfaithful mirror of nature, more proper by its brilliance to create flights of imagination and to serve error than to nourish minds solidly and to draw obscure truth from deep wells." The prism is the apparatus of the imposter and the instrument of the "spectre magique."[28]

Castel asks what right the prism has to credence. Does its geometrical shape prove that the colors coming from it are primitive? Why are the colors produced by the prism any more fundamental than the colors of the tricolor flower? Besides, the prism provided Newton with only a single unique fact, that is, the dispersion of the colored rays, from which Newton constructed his entire theory. But a unique fact is a monstrosity, a single event, from which no general conclusion or universal theory can be drawn. "My philosophy . . . considers only facts, but facts that are natural, daily occurring, constant, and a thousand times repeated, habitual facts rather than facts of the moment, facts of humanity rather than facts of one man. A unique fact is a monstrous fact."[29] Castel uses the word *fact* here partly in its original Latin meaning of something made or done. Thus the validity of a "fact" depends on the testimony of observers and on the veracity of the person claiming the fact. Because he depends on facts, Newton makes the error of turning effects into causes, phenomena into principles, and experiments into explications. While Castel does not doubt the experiments that Newton describes, he dislikes his tendency to claim as "fact"—that is, as a deed—what is only an interpretation of a phenomenon. Moreover, Newton's jargon is meaningless. His notion of a "ray" of light makes no more sense than his notions of "attraction" and "gravitation."[30]

Newton is imperious and his followers are far more dogmatic than the Cartesians. They have to accept his arguments without question, because Newton's arguments demand complete assent or complete denial. At least Descartes was modest enough to realize that his system of the world was a hypothesis that could be modified by subsequent reasoning and experience. Descartes's hypotheses have flexibility. They are intelligible and his followers can reason with him.[31] But not the Newtonians. They are not allowed to question. Newton transforms his readers into spectators, not participants. The Newtonians claim that they present "facts," not hypotheses, and argue that the facts cannot be denied. This makes them totally unyielding. Reasoning does not force consent, but facts do, and only God can claim facts. The method of facts is emphatic and disdainful. It leads only to occult qualities and error.[32] It is a mistake to claim that a system contains absolute truth.

Newton's system also is difficult and inaccessible. His experiments require that he remove himself from the natural world and enter an artificial world of prisms and rays. There is no need to shut oneself up in a camera obscura in order to understand light. Nature is everywhere and reveals itself constantly to our senses. The rainbow appears in the presence of the sun.[33]

In these criticisms we recognize an attitude toward instruments that preceded what we call the Scientific Revolution. Experiment had value only to the extent that it confirmed experience, and reason naturally preceded experiment, so that the necessity of an experimental test could be regarded only as a sign of defeat. There could be no crucial experiment, because a crucial experiment was only a single instance, a monstrous event.[34] In fact, experiment should be the last resort of the natural philosopher, not the first step of an investigation, as Newton had argued.

The Harpsichord as Rhetoric

The rhetorical form of argument was also important for Castel. He followed his 1725 announcement of the ocular harpsichord by a "geometrical demonstration" of it. This geometrical demonstration, however, did not contain any geometry as such. It was a set of propositions, followed by demonstrations with an occasional scholium thrown in for good measure, and it could be called geometrical only because it discussed musical harmony, a subject that was traditionally part of mathematics. As Peter Dear has shown, it was characteristic of Jesuit scientists after Christoph Clavius to use the form of a geometrical proof in order to give universality to experiential statements, and this was obviously Castel's purpose.[35] He wrote that in his first publication he merely stated the question and proposed the possibility of an ocular harpsichord. In his second article he wanted to extend his demonstration to all the senses, because "a discovery that is fecund ought always to move forward into a new order." An important discovery cannot exist alone, because it will always lead to more discoveries "as one harvest provides the seeds for a new harvest." In fact Castel claimed that the ocular harpsichord, or at least the *idea* of the ocular harpsichord, would become the universal instrument of the senses and poets would discover in it a complete *musurgie* that would account a priori for "all sounds, tones, accords, dissonances and, what has never yet been attempted, for the pleasure of all things." By casting his argument in geometrical form, Castel generalized it and extended it to all the senses (see fig. 4.3).[36]

Castel's use of geometry was obviously different from Newton's use of it. Castel called Newton an excellent geometer, but a poor physicist (a criticism that he probably borrowed from Malebranche), and while he praised the geometrical method, he criticized Newton for overreliance on mathematics in his optics. In applying mathematics to physics, Castel ascribed the greatest value not to theorems and calculations, but to the logical form and the generalizing power of geometry.

Castel claims that analogy is the basis for discovery in natural philosophy and that analogy reveals important connections between science, art, and literature. While there may be many arts and sciences, there is only one truth, which the arts and sciences express from different points of view. In particular, philosophy and poetry have the same object, the same nature, and the same truth—a sublime thought in poetry is equivalent to a discovery in natural philosophy. Therefore analogy is crucial for making the transition from one expression of truth to another: "Now it is analogy that renders these poetic flashes fecund in discoveries. Because what one calls among the poets and orators *metaphor, similitude, allegory, figure*; a philosopher, a geometer will call *analogy, proportion, ratio*. All our discoveries, all our scientific truths, are only truths of ratio. And from there often the figurative sense degenerates into the proper sense and the figure into reality" (emphasis

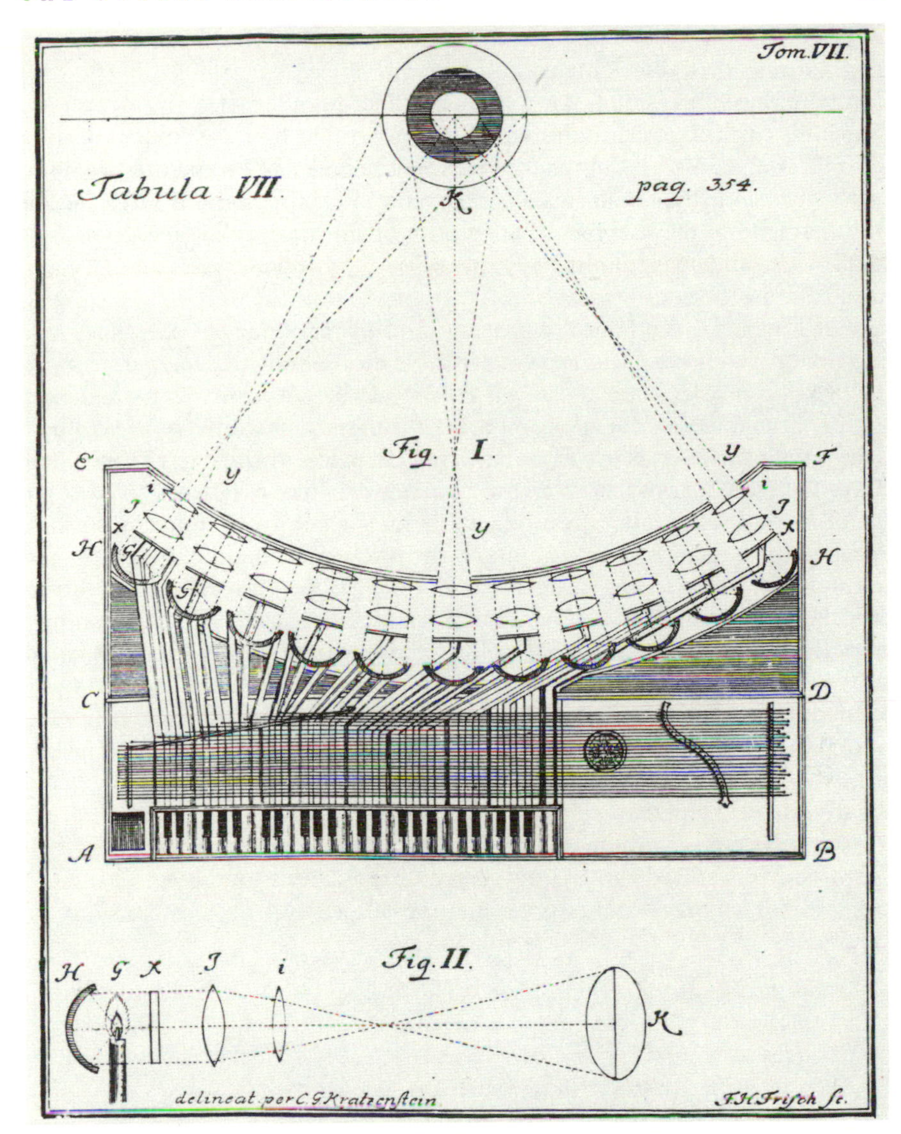

Fig. 4.3. Castel left no illustration of his ocular harpsichord. The instrument depicted here is a variant proposed by Kruger in his "De novo musices quo oculi delectantur genere," p. 354. Courtesy of Special Collections, the University of Washington Libraries, neg. no. 14219.

added). Castel gives as his rule the following. When he encounters a poetic or other literary statement about nature that is especially beautiful and sublime, he applies the method of geometrical analysis; that is, he assumes it to be true and sees what consequences he can derive from it. From the truth of the consequences he verifies the original statement, and if he is persuaded of its truth he then attempts to demonstrate it to others.[37]

For example, Virgil "paints the night" when he writes, "Rebus nox abstulit atra colores [Black night took the color away from things]." The sublimity of this expression lies not in tropes, figures, allegories, or metaphors, but in its truth. It is Descartes who has shown that because colors are only modifications of light, they cannot exist in the dark, and therefore when the night chases the light it also chases the colors. "This thought of Virgil has all the character of the sublime, of the grand, of the beautiful, being in the first place true, and in addition new, marvelous, profound, paradoxical even, and contrary to our presumption."[38]

For Castel the aesthetically pleasing and the rational are the same. It is also the basis of his disagreement with Newton about the color-tone analogy. Newton associated the seven colors of the spectrum with the seven notes of the musical scale by comparing the measured widths of the colored bands with the lengths of vibrating strings that sound consonant tones, but Castel's argument is very different: "Among the colors, violet is a sad color and one that takes much from black, being the color of mourning for our kings and for the Church. . . . violet is the passage from affliction to joy; the rainbow is a sign of joy, but of a joy which follows an affliction, and to which the affliction serves as a contrast and as a base."[39] Therefore violet should serve as the base for the color scale. Later, however, Castel decided that blue was the "fundamental bass" for color harmony, because the study of dyes and pigments convinced him that there were three primary colors—red, yellow, and blue—that corresponded to the major triad in music. Beginning with his color triad, he filled in the rest of the colors to create the twelve-note chromatic scale. Of course he did not hesitate to point out that these twelve tones had long been called "chromatic," indicating that musicians had recognized the analogy between color and tone long before the seventeenth century—and that the analogy could not be purely verbal:

> But why is this scientific system of half-tones called chromatic and colored? It is doubtless a metaphor, a comparison, an analogy of discourse, and consequently, it seems to me, of thought, of reasoning, of science. Because in the arts above all, and in the sciences, there is no affected term, [no conceit] that does not express an idea, and is often the result of several truths and an implied theory.[40]

Not only does analogy serve as a means of scientific discovery; it is also a valuable rhetorical tool. This is because any new truth, and especially a scientific discovery, is shocking and revolts the reader. It should be enclosed in a rhetorical "envelope" that conceals the full harshness of the new truth, piques the curiosity of the reader, and provides only analogies to the new idea. This was Descartes's error. He should have presented his ideas in poetry and allowed the commentators to reveal his principles in full light.[41] While Descartes's style was too direct, Newton's was even worse, and the plain declarative style of writing so favored by the British philosophers was, for Castel, a detriment to the proper pursuit of science.

We can now understand why it was a Castel and not a Newton who came up with the idea of an ocular harpsichord, why Athanasius Kircher was Castel's hero, why Voltaire changed his mind so completely when he learned what was behind the instrument, and why the ocular harpsichord was destined to remain a "marginal" scientific instrument. It was an instrument perfectly suited to Castel's way of studying the natural world. It was based on analogy, the analogy between color and tone, and it connected the aesthetic with the rational. Castel argued that man inhabited an artificial world intermediate between the supernatural and the natural and that as an artificer he was an intermediary between God and nature.[42] Instruments like the ocular harpsichord are one means man has of illustrating the hidden analogies that rule nature.

The Harpsichord after Castel

Castel's ocular harpsichord had much in common with other instruments in the natural magic tradition that combined aesthetics, entertainment, and natural philosophy in a single apparatus. Electrical instruments before 1780 had much the same character. They did not measure anything and were designed to elicit wonder in the spectator. During the last quarter of the eighteenth century, when instruments became much more quantitative, the ocular harpsichord became increasingly irrelevant to most natural philosophers.

One natural philosopher, however, did advance arguments on color similar to those of Castel: Johann Wolfgang Goethe. Most striking is Goethe's attack on Newton's color theory. Both he and Castel insist that color is a modification of white light caused by the interaction of light and dark. Both argue that the spectrum observed by Newton does not occur at all distances from the prism and therefore that Newton was looking at a special case. Both claim that Newton's prismatic colors were produced by modification of the edges of a beam of white light, and that green is not a primary color, but a mixture of blue and yellow rays coming from the edges of the white beam. Castel compares the white light "shattered" by the action of the prism—splintered into colors when bent by it—to a wooden rod that splinters when it is bent.[43]

Even more striking is the similarity in Castel and Goethe's criticisms of Newton's method. Both locate the error of Newton's method in his rhetorical style. Both argue against the authority of fact as Newton uses it, Goethe accusing Newton of "insufferable arrogance," and both claim that Newton's arguments assume what they set out to prove. Both deny the validity of a single experiment and both argue that only a collection of observations will lead to an understanding of the phenomena.[44]

Both Castel and Goethe insist on the subjective nature of experiment, Goethe going so far as to argue that "insofar as he makes use of his healthy senses, man himself is the best and most exact scientific instrument possible" and that artificial instruments which set nature apart from man are a great misfortune for physics. Not surprisingly, both Goethe and Castel criticize Newton's abstract concept of a "ray," and both approach the phenomenon of color through the study of pigments and dyes, not the "adventitious" colors produced by the prism. Goethe does not like the prism any more than Castel does, and he insists that man can never come to understand nature by subjecting her to torture.[45]

But Goethe did not share Castel's enthusiasm for the ocular harpsichord. One might expect that Castel's desire to find a truth which transcends both science and poetry and gives validity to both would appeal to Goethe as well.[46] While he sympathized with much of Castel's theory of color, Goethe criticized Castel's excessive use of analogy, and since analogy was at the root of the entire concept of an ocular harpsichord, Goethe could not accept it.[47] Because he employed analogy willingly in his own natural philosophy, we must conclude that he was not opposed to analogy as such, but only to the kind of analogies employed by Castel. Of course Goethe had read the criticisms of the color-tone analogy by Voltaire, Dortous de Mairan, and Krafft, and knew that it had few supporters. It is likely, however, that Goethe's criticism came not from the opinions of others, but from a feeling that Castel's method represented an outdated, naive, and undisciplined search for cosmic harmony which ignored any close study of natural phenomena.[48]

Castel had frankly admitted that he did not like bothering with details and that he preferred to grasp the truth by generalizing from daily experience. Goethe, on the other hand, described his own method in natural philosophy as "concrete thinking" (*gegenstandliches Denken*). "My thinking does not separate itself from concrete objects; . . . the elements of the objects or rather my perception of them, enter into my thinking and are most intimately penetrated by it; and . . . my perception itself is thinking, my thinking perception."[49] Goethe was a close observer who worried very much about the details. In fact he described subjective color phenomena like "colored shadows" and afterimages better than anyone before him.[50] He did not use complex apparatus in his experiments, and he denied the possibility of an *experimentum crucis*, but his hostility to Newton did not mean that he neglected experiment. Castel's method of grasping at analogies without worrying about "the details" could only have exasperated Goethe. The ocular harpsichord was a product of this unsatisfactory method. It was also an artifice, an artificial way to create an analogy, which, if it truly existed in nature, should be evident without a complex mechanism.

Although Goethe repudiated the color-tone analogy, it became an important theme in romanticism as an example of synesthesia, the substitution of one sense for another. Poetry, music, and painting all employed the analogy during the nineteenth century, but in a very different way from that used by Newton and Castel. Castel's ocular harpsichord depended on a precise correspondence between color and tone. A particular color corresponded to a particular musical pitch, not to a mood or emotion. While the precise correspondence claimed by Castel still held for those individuals who were "synesthetic," that is, who could find a given pitch by associating it with a given color, the color-tone analogy as it was used by the romantics usually associated color with the mood of the music and not its pitch.[51]

The ocular harpsichord did continue to suggest itself to inventors after Castel, most of whom reinvented the instrument and discovered afterward that it had been suggested long before. On June 6, 1895, Alexander Wallace Rimington performed on his great color organ for the first time at St. James Hall in London, and Alexander Scriabin's symphony *Prometheus* (1911), which has a part written especially for a color organ, continues to be performed.[52] Thomas Wilfred toured the United States and Europe in the 1920s with his clavilux, a modern ocular harpsichord. Performances of this sort led Albert Michelson to exclaim in *Light Waves and Their Uses*:

> Indeed, so strongly do these color phenomena appeal to me that I venture to predict that in the not very distant future there may be a color art analogous to the art of sound—a "color-music"—in which the performer seated before a literally chromatic scale, can play the colors of the spectrum in any succession or combination, flashing on a screen all possible gradations of color, simultaneously or in any desired succession, producing at will the most delicate and subtle modulations of light and color, or the most gorgeous and startling contrast and color chords![53]

As this quotation shows, the ocular harpsichord is too attractive an idea to disappear completely, and we can expect it to reappear in one form or another, although perhaps not as the instrument that Castel envisioned. So far painting and photography appear to have been the most important media for exploiting the color-tone analogy.[54]

The ocular harpsichord was one of those marginal instruments that served science for a while and then disappeared, only to pop up again occasionally in subsequent history. One cannot really say that the analogy upon which it was based was proven false, just that it did not lead anywhere in the form that Castel proposed, nor did it point in the direction that natural science subsequently took. From the way that Castel looked at the world, it made perfect sense. From the way that we look at the world, it belongs in the same category as the cat piano. In the eighteenth century it was not obvious where it belonged.

CHAPTER FIVE

The Aeolian Harp and the Romantic Quest of Nature

IN HIS *Edge of Objectivity* Charles Gillispie ends his discussion of romanticism with the statement that although "deep interests have been bound up with the romantic view of nature, deep interests and deep feelings . . . it is the wrong view for science."[1] Any categorical statement like this one is bound to raise our historiographical hackles. We immediately want to know for whose "science" the romantic view is the wrong one, and why it is not permissible to approach nature from any methodological direction. We quickly point out the importance of the *Naturphilosophen*—Julius Robert von Mayer, Hans Christian Oersted, Lorenz Oken—for breaking the stranglehold that Laplacian mechanism held over scientific explanation in the early nineteenth century. *Naturphilosophie* allowed important new concepts like energy, the magnetic field, archetypal morphology, cell theory, and evolution to emerge, ideas that were inconceivable in the unromantic Enlightenment. We point out that in England and Ireland Sir Humphry Davy, Michael Faraday, and William Rowan Hamilton, great scientists all, were sympathetic to these views.

Nevertheless, we know what Gillispie means. The romantics wanted to create a speculative physics. They were not really interested in science as we know it, but in metascience, that is, in the fundamental relationship between man and nature that makes science possible. They wanted to study the whole of nature rather than its parts, because they believed that any part could be known only after the whole was understood. They resisted the analysis of natural phenomena. Wordsworth condemned those who would "murder to dissect," and Coleridge hated Locke and the other "little-ists" who would pick the world to pieces without any concern for its unity or for man's place in it.[2] The little-ists, however, have enjoyed great success, and we are forced to agree with Gillispie that the romantic view is not the direction that modern science has taken.

On the other hand, the word "science" has meant different things to different people at different times, and who is to say what is the "correct" meaning? We historians of science need to study romanticism because it is the most important alternative in the West to the "scientific" mode of thought engendered by the Scientific Revolution. We have learned at considerable cost that these alternative modes of thought are seldom exclusive and that we make mistakes when we do exclude them from what we might wish to regard as "real" science. The natural philosophy of the Scientific Revolution may have had to "overcome" the alternatives of Aristotle and the occult sciences, but it did so only by incorporating large parts of those philosophies that it "overcame." Likewise natural science "overcame" romanticism but did not remain unaffected by it.[3]

We also need to reconsider the approach to romanticism which considers it as a literary movement that later "influenced" science. Romanticism included the study of nature at its very heart, and therefore the study of nature should be more a guiding principle for romanticism than a subject to be "influenced."[4]

Of course it is dangerous to generalize about any group of thinkers as diverse as those we denote as romantics. Some, like Shelley, welcomed the accomplishments of the Scientific Revolution; others, like Keats and Blake, condemned them. Still others, like Goethe and Coleridge, attempted to create new philosophical pathways for science to follow. As a result, defining "romantic science" is as difficult as defining romanticism itself. But all the romantic philosophers struggled with certain paradoxes that defined a common approach to nature. The most important of these for science was their attention to the particular among the universal, the single flower that spoke for all of organic nature, the church bell that resonated to the harmony of the heavens. They avoided abstractions like mathematics and they detested analysis, which meant that for them, the universal had to be perceived almost intuitively in the particular object. Moreover, they refused to stand apart from nature and insisted that the objects perceived could not be separated from the subjects perceiving them. For these reasons, they allowed little room for any mediators between sense and the transcendent.

Because the romantics wanted to move directly from sense to universal truth, they left little room for instruments. Goethe, as we have seen, condemned Newton's use of the prism to analyze light. His own experiments, of which he did many, studied the direct perception of color without any elaborate intervening instruments. The instruments that the romantics did appreciate were those that revealed the unpredictability and complexity of nature, the "grotesques and arabesques of nature" as Novalis called them.[5] Thus it is not surprising that natural magic, which operated by analogy and used instruments to illustrate the wonders of nature, provided the most important instrument of romantic poetry.

The one instrument that was ubiquitous in romanticism and the one that best served the needs and purposes of the romantic quest for the harmony of nature was the Aeolian harp. Because it disappeared almost completely at the end of the nineteenth century, it requires an explanation. The Aeolian harp is a stringed instrument played by the wind. The most common form is a rectangular closed box about three feet long, six inches wide and three inches deep. Three to twelve strings, tuned in unison, are stretched the length of the box between two bridges, and one or more sounding holes are cut in the top of the box below the strings. The harp sits on a windowsill with the sash drawn down just above the strings. When there is a draft through the window, the harp will sound one or more notes, the pitch depending on the strength of the wind. The music has an eerie quality and is difficult to locate. Very slight changes in the draft will bring on different notes, at first harmonious and indolent, but as the wind strengthens, marked dissonances occur until in a strong wind the music becomes more like a scream. Nothing could better match the sentiment of the romantic soul.

Marjorie Nicolson and M. H. Abrams have taught us the importance of scientific themes in the literature of the eighteenth and nineteenth centuries.[6] They document the use of instruments as analogies to nature. In particular Abrams uses the figures of the mirror and the lamp to depict the shift from neoclassic to romantic criticism, illustrating a "radical alteration in the typical metaphors of critical discourse."[7] The mirror is an image of the mind reflecting nature through poetry; this is the *neoclassical* goal of clear picturing. The lamp, on the other hand, portrays the mind as a radiant projector illuminating the objects perceived and actively operating on the world that the poet inhabits; this is the *romantic* goal of man *in* nature, not man observing nature. Abrams has not chosen these metaphors arbitrarily; they are constant themes in the poetry and criticism of the age. They are, however, metaphors of description and therefore apply more to criticism than to the production of poetry. The Aeolian harp, on the other hand, was an instrument of inspiration.

The Aeolian harp was superior to other instruments because its music was unpredictable and because it was played, not by man, but by the breath of Nature herself. Shelley perhaps best illustrated the role of the Aeolian harp, because he used it to define poetry: "Man is an instrument over which a series of external and internal impressions are driven, like the alternations of an ever-changing wind over an Aeolian lyre, which move it by their motion to ever-changing melody."[8] The harp corresponds to the poet's soul waiting to be touched by the wind of inspiration.

The Aeolian harp became such a common analogy for romantic creativity that one wonders if it might have been more than an analogy. William Jones, who popularized the Aeolian harp in 1781, certainly believed that it was. He argued that "its harmony is more like to what we might imagine the aerial sounds of magic and enchantment to be, than to artificial music. We may call it, *without a metaphor*, the music of inspiration" (emphasis added).[9] We wonder if perhaps Coleridge was *actually listening* to his harp in the window at his cottage in Clevedon when he expressed his love to Sara Fricker, if Wordsworth was *actually listening* for the sound of the harp while writing *The Prelude* and was frustrated when it failed to inspire, and if Melville *actually heard* the harp scream like the sound of rigging in a gale when he recalled a storm at sea.[10] Abrams seems to agree when he states, "It is possible to speculate that, without this play-thing of the eighteenth century, the romantic poets would have lacked a conceptual model for the way the mind and the imagination respond to the wind, so that some of their most characteristic passages might have been, in a literal sense, inconceivable."[11] The actual harps sitting in the windows and gardens of poets around the globe inspired some of the most characteristic poetry of the romantic era. Abrams calls the Aeolian harp a "play-thing," dismissing the instrument itself, and pursuing its imagery in romantic poetry. But our subject is instruments, and therefore while Abrams pursues the image, we will pursue the instrument.

The Origin of the Aeolian Harp

As with many of the instruments described in this volume, the Aeolian harp goes back to Athanasius Kircher and his famous museum of curiosities at Rome. In this particular case we can be pretty certain that he was its inventor. There were, of course, numerous references in mythology to instruments playing by themselves. The supposed discovery of the earliest musical instrument, the lyre, occurred when Hermes heard the wind playing music on dried sinews stretched across a tortoise shell. King David's harp sang in the wind when he hung it before his tent at night. And Saint Dunstan's harp miraculously played an anthem by itself around A.D. 1000. But none of these stories suggests an instrument crafted for the purpose. Della Porta noted in his *Natural Magick* (1540) that when the winds are "very tempestuous" stringed or wind instruments will play by themselves if turned to the wind.[12] Kircher's student Gaspar Schott tells us that Kircher made his harp on the lines indicated by Della Porta, but this was giving Porta more credit than was his due, because, again, Porta had not proposed a new instrument.[13]

Kircher described his harp in his *Musurgia universalis* (1650) and again in his *Phonurgia nova* (1673).[14] For Kircher it was another instrument of natural magic along with the speaking heads and echoing mirrors. "In my Museum," he wrote, "it is listened to with very great amazement. . . . No one will ever suspect what kind of instrument it is, or by what hand or pump or artifice it creates its melodious sound. This instrument will be so much the more recherché and worthy of wonder to the extent that it is more hidden and concealed."[15] Schott tells with pleasure how Kircher tricked the minister of the abbey, who sought in vain for what he thought must be an organ in Kircher's rooms.[16] And yet Kircher's harp was more than just a practical joke. He discovered that it would play complicated melodies and chords even when its fifteen strings were tuned in unison. Moreover, he discovered that "one and the same string is able to emit infinite, diverse sounds," often at the same time.[17] He equipped his harp with wooden doors that funneled the air directly over the strings, and found that he obtained more sound if the air struck the strings slightly obliquely (see fig. 5.1). He noted that the harp did not sound the fundamental of the string, that is, the pitch produced by plucking the strings. Rather "the string will give forth now the third, now the fifth, now the fifteenth or the twenty-second," a striking variety of tones from a single string. The music was a kind of "warbling" sound (*tremulum*), sometimes like a bird, sometimes like an organ or some other instrument. It

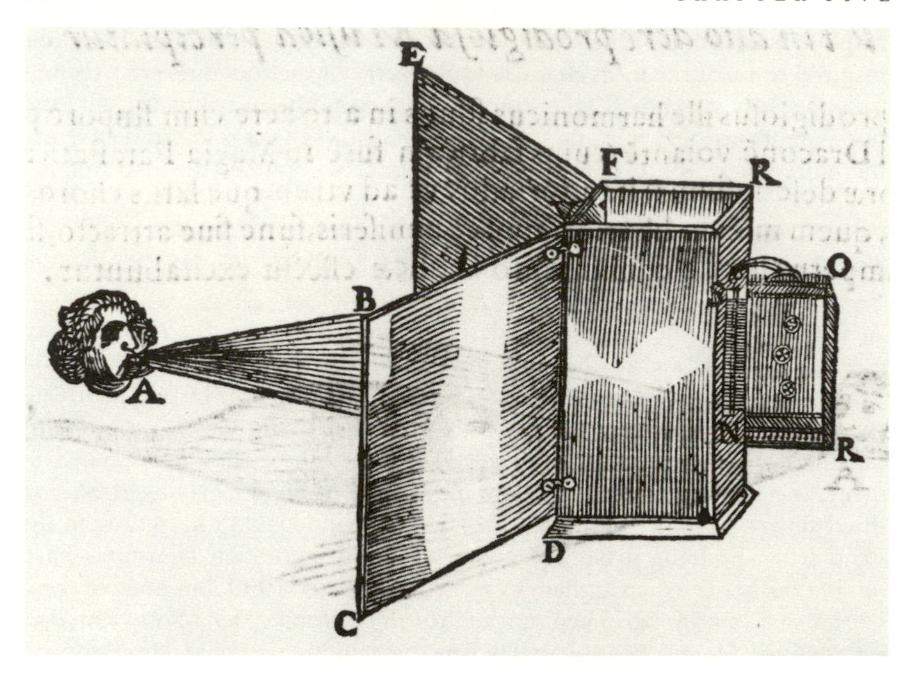

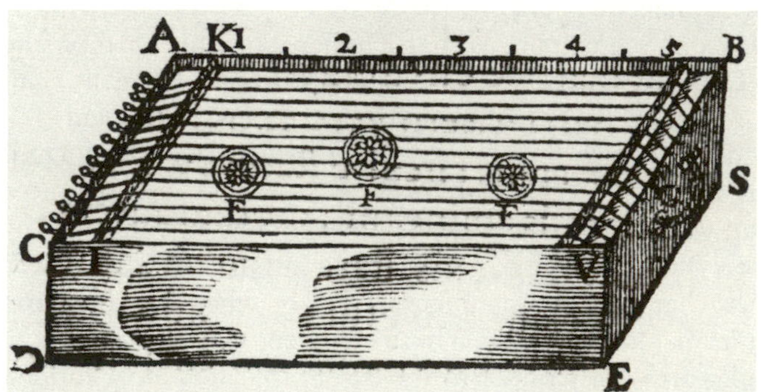

Fig. 5.1. The first Aeolian harp. From Kircher, *Musurgia*, 2:352–353. Graphic Arts Collection, Department of Rare Books and Special Collections, Princeton University Libraries.

was very sensitive to any change in the wind. Just opening a door in another part of the house would cause it to sing or fall silent.[18] All this suggests that Kircher spent many hours perfecting and listening to his instrument.

To Kircher it was "Machinamentum X" or a "Machinam harmonicam automatam," a "self-operating harmonic device." The first person to call it an Aeolian harp was Johann Jacob Hofmann, who in his *Lexicon universale* (1698) quoted Kircher on all the details but referred to the instrument as "Æolium instrumentum."[19] Other mentions of the Aeolian harp were infrequent in the seventeenth century and the instrument dropped from view.

The Aeolian Harp in Britain

When the Aeolian harp does reemerge, it is in a most unlikely place—among a group of Scottish poets and musicians in London. It appears, suddenly in 1748, in James Thomson's *Castle of Indolence* and sets the theme for all future poetic uses of the harp:

> Each Sound too here to Languishment inclin'd,
> Lull'd the weak Bosom, and induced Ease.
> Aereal Music in the warbling Wind,
> At Distance rising oft, by small Degrees,
> Nearer and nearer came, till o'er the Trees
> It hung, and breath'd such Soul-dissolving Airs,
> As did, alas! with soft Perdition please:
> Entangled deep in its enchanging Snares,
> The listening Heart forgot all Duties and all Cares.
>
> A certain Music, never known before,
> Here sooth'd the pensive melancholy Mind;
> Full easily obtain'd. Behoves no more,
> But sidelong, to the gently-waving Wind,
> To lay the well-tun'd Instrument reclin'd;
> From which, with airy flying Fingers light,
> Beyond each mortal touch the most refin'd,
> The god of Winds drew Sounds of deep Delight:
> Whence, with just Cause, *The Harp of Aeolus* it hight.
>
> Ah me! what Hand can touch the Strings so fine?
> Who up the lofty Diapasan roll
> Such sweet, such sad, such solemn Airs divine,
> Then let them down again into the Soul?
> Now rising Love they fan'd; now pleasing Dole
> They breath'd, in tender Musings, through the Heart;
> And now a graver sacred Strain they stole,
> As when Seraphic Hands an Hymn impart:
> Wild warbling Nature all, above the Reach of Art![20]

In a footnote Thomson adds: "This is not an Imagination of the Author; there being in fact such an Instrument, called Æolus's Harp, which, when placed against a little Rushing or Current of Air, produces the Effect here described."[21] The same year Thomson also produced an "Ode on Æolus's Harp" in which he gave more information, again in a footnote: "Æolus's Harp is a musical instrument, which plays with the wind, invented by Mr. Oswald; its properties are fully described in the Castle of Indolence."[22] The ode appeared in Dodsley's *Collection of Poems* in 1748 after the *Castle of*

Idolence appeared, but was probably written earlier, because Dr. Charles Burney claimed in his memoirs that he set the ode to music in 1747; he added that "it was performed one morning at Lady Townshend's to whom I had the honour of being introduced by . . . Mr. Hume."[23] By underscoring the harp twice and by ascribing it to "Mr. Oswald" Thomson seemed to be doing Oswald a favor, and sure enough in October 1751 the *General Advertiser* carried the announcement: "By Authority This Day is Published Aeolus's Harp—A new-invented musical instrument, which is played by the wind, as described by Mr. Thomson, in his *Castle of Indolence*. Sold only by the Inventor. J. Oswald, at his music-shop in St. Martin's church-yard."[24]

James Oswald was a Scottish composer, music publisher, and dancing master who moved to London in 1741, first working as a hack composer and then setting up his own music store and publishing firm in 1747.[25] His shop became the center for a group of Scots in London including Thomson, Charles Burney, Tobias Smollett, and Christopher Smart. Thomson died in August 1748, so he survived his poetic invention by only a few months, but Oswald lived on till 1769, gaining reputation and influence. He was appointed chamber composer to George III in 1761 shortly after George's accession to the throne. Burney moved to London from Scotland in 1744 and was apprenticed to Thomas Augustine Arne at the Drury Lane Theatre. There he wrote parts for music, most notably for the masque *Alfred*, performed in 1745, which contained Thomson's "Rule Britannia." Burney also wrote parts for Arne's setting of "God Save the King" and Tobias Smollett's "Tears of Scotland," both of which acquired popularity after the collapse of the Jacobite cause at the Battle of Culloden on April 16, 1746.

Burney became a regular visitor at Oswald's shop in 1746 and the two joined forces to form the Society of the Temple of Apollo, which produced music for David Garrick at Drury Lane. Scholars have speculated on the membership of this society, proposing Thomson, David Mallett, Burney, and other Scots in the Oswald circle, but it appears that it was a creation of Oswald alone. Burney wrote the music, Oswald obtained a sole patent for all the music composed by the society, and Garrick was persuaded that the members were "gentlemen of taste and talent," whose work was worthy of performance.[26] Garrick probably did not care who the gentlemen were, because the pantomimes *Queen Mab* (performed 1750) and a revised *Alfred* (1751), for both of which the society provided the music, were huge successes.[27]

Considering Oswald's skill as a promoter and his willingness to stretch the truth, it is perhaps not surprising that he claimed the Aeolian harp for himself. His pupil William Jones gave a long account of this supposed discovery in his *Physiological Disquisitions* of 1781. According to Jones, Oswald heard that when Alexander Pope was translating Homer, he consulted the Greek commentary of Eustathius, where he found a passage suggesting that the blowing of the wind against musical strings would produce harmonious sounds. Oswald tried to make the sounds with a lute but was unsuccessful and concluded that the story was fabulous. He then heard of a harper whose instrument sounded in the wind by accident. Oswald then persevered and finally obtained the aeolian music.[28] There may be some truth to the story, but Burney later gave what surely is a more accurate account. After referring to Kircher's account of the harp, Burney writes: "It was thence that Thomson the poet took it, who wrote an ode on this aerial instrument, which was set to music . . . Oswald, the celebrated player of old Scots tunes on the violoncello, and composer of many new, passed for the inventor of the Aeolian harp; but as he was unable to read the account of it in the Musurgia, written in Latin, Thomson gave him the description of it in English, and let it pass for his invention, in order to give him a better title to the sale of the instrument at his music-shop in St. Martin's Church-yard."[29] The only problem with this account is that Kircher did not call his instrument an Aeolian harp, so Thomson or some associate must have read the description in Hofmann's *Lexicon universale* and been led from there to Kircher.

The Scots continued to incorporate the harp in their poetry. Christopher Smart composed *Inscriptions on an Aeolian Harp* (1750) and included it in his *Jubilate Agno* (1756–1763) composed during his madness. Tobias Smollett used it as an instrument of seduction in his *Adventures of Ferdinand Count Fathom* (1751), but the harp does not appear to have caught on in any major way. A description of it appeared in *Gentleman's Magazine* for February 1754 with an explanation that the author ("A. Z.") believed it "not to be thoroughly known."[30] Another London instrument maker began advertising the harp in 1763, but references to it were still uncommon.[31]

William Jones on the Aeolian Harp

The period of great popularity for the Aeolian harp appears to have begun with the *Physiological Disquisitions* of William Jones in 1781.[32] Jones's book was not especially popular, but it gave a long description of the harp, an account of its history, and a theory of how it operated. While the harp had already emerged during the eighteenth century in poetry and song, Jones placed it in a book on physics and used it to explain the nature of sound. From this beginning, the harp found its way, on one hand, into the emerging science of acoustics, and, on the other hand (with Jones's peculiar ideas about sound), into romantic poetry. These two different treatments of the Aeolian harp represent two different versions of natural "science" and two different ways in which instruments can be used to comprehend the physical world.

William Jones, commonly known as "Jones of Nayland" (to distinguish him from his contemporary Sir William Jones, the prominent Orientalist), had matriculated at University College, Oxford, in 1745, was ordained in 1751, and became a member of the Royal Society in 1775. He wrote on both natural philosophy and theology, and joined the two to the extent that their juncture was possible. His greatest strength in natural philosophy was music, which he studied with Oswald. On the subject of music, at least, he wrote from considerable knowledge. Jones also designed a harp of his own, which was sold in London by Longman and Broderip.[33] Subsequent writers on the Aeolian harp either quoted Jones directly or drew much of their information from him.[34]

To a musician the Aeolian harp presented two problems: the first was the notes that one heard coming from the harp, and the second was the cause of the music in the first place. Because the harp's strings were not fretted and were tuned in unison, one would expect to hear only the fundamental note plus the harmonics, which are the notes produced by equal divisions of the string. Thus when the string vibrates as a whole, one hears the fundamental; when it divides in two parts, one hears the octave, in three the fifth, and so forth. Without fretting the string one hears no other notes. A trained ear can detect these harmonics in the sound from any stringed instrument. But the strings of the Aeolian harp emitted notes other than the harmonics. "When it plays, the unison itself is plainly heard as the lowest tone, and the combinations of concords, though consisting chiefly of the harmonic notes, are by no means confined to them, but change, as the wind is more or less intense, with a variety and sweetness which is past description."[35] It is very difficult to see how the harp string could vibrate in any other way than in an integral number of parts, and yet Jones heard other notes as well.[36]

It is also very difficult to see how the wind caused the harp strings to vibrate. Kircher surmised that the wind came in "rays" that plucked the string and that only the part of the string struck by the ray of wind would vibrate, the rest of the string remaining still.[37] How part of a string could remain still while the rest of the string vibrated he did not explain, and his theory was met with derision by those who, unfortunately, had no better answer.[38] Jones's explanation was at first glance equally far-fetched. He claimed that just as a string created music when it struck the air, so the air created music when it struck the string. He then drew on the analogy between light and sound to explain how the harp, acting like a "sound prism," refracted the wind to produce music. "When any body inflects the rays of light or refracts them, it does not give the colours that are seen, but it makes the light give them: so a sonorous body does not give musical sounds, but makes the air give them." And just as light is composed of different colored rays, so is air composed of different parts, each carrying a different pitch. "There is no reason to suppose that air is homogeneous in its parts, any more than light: and if air consists of heterogeneous parts, they will be differently refrangible according to their magnitudes, and excite different sounds, as they are accommodated to different vibrations and capable of different velocities."[39] Just as white light shows no color until its heterogeneous parts are separated by the prism, "so the air yields no particular musical tone without the assistance of some sonorous body to separate its parts and put them into a vibratory motion." Like Castel earlier in the century, Jones argues that "the analogy between sounds and colours is very strict and may be carried very far. . . . Upon the whole, the Eolian harp may be considered as an air-prism, for the physical separation of musical sounds."[40] Most marvelous of all is the fact that a single string on the harp will sound seven or eight different notes at the same time, suggesting that the string merely separates the vibrations that are already present in the air.

The Aeolian Harp and the Science of Acoustics

Jones's theory of the Aeolian harp came from a variety of concerns, both mathematical and musical, that were important in the eighteenth century. While the mathematical science of harmonics was very ancient, having been founded by Pythagoras in Greek antiquity, there was a new application of mathematics to music during the Enlightenment in the description of the vibrating string. The science of harmonics related pitch and consonance to the length of the string by ratio and proportion; the new differential calculus could describe the string's actual motion. In 1746 Jean d'Alembert derived and found a solution for the wave equation, which gave the motion of the string. This was one of the very first uses of partial differential equations, and d'Alembert soon became engaged in a three-way debate with Leonhard Euler and Daniel Bernoulli over the proper way to mathematize the motion of the string. All the leading mathematicians of the eighteenth century attacked this problem because more hinged on it than just a question of music.[41]

The debate over the vibrating string was part of the gradual divorce of acoustics from music that took place in the eighteenth century. Music, as one of the four sciences of the medieval quadrivium, was traditionally part of mathematics. The tradition still held in the eighteenth century to some extent. The beginning of acoustics as a branch of physics is often dated from Ernst Chladni's *Entdeckungen über die Theorie des Klanges* (Leipzig, 1787), but throughout the century natural philosophers raised questions about the production and propagation of sound that were not properly part of harmonics. Thus Jones's theory of the air-prism was more natural philosophy than music.

A major debate of the eighteenth century was the proper way to temper the musical scale. It is impossible to tune a stringed instrument so that all the intervals are true in every key. When musical practice required key changes within a piece, it was necessary to compromise the tuning so that some of the intervals, even if not perfectly true, were close enough not to offend the ear. Many different systems of temperament were suggested, one of which was favored by Jones, but none could be shown to be "scientifically" superior, because as Jones frankly admitted, "after all the researches I have been able to make, I am still at a loss for the *physical* principle of musical consonance."[42]

Also musicians discovered in the middle of the century that two loud sustained tones relatively close together in pitch would produce a third note an octave lower than the lower of the primary tones. Jones knew it as the "Tartini tone" after the violinist Giuseppe Tartini, who gave the most important, but by no means the first, announcement of the phenomenon.[43] The best explanation of these "combination tones," as they were later called, appeared to be that they were notes at the beat frequency. The beats created by the two primary tones were rapid enough to be perceived as a third tone rather than as separate beats.[44] It was possible that the nonharmonic tones coming from the Aeolian harp were Tartini tones. Jones recognized, however, that they did not seem to be the proper pitches and therefore called them "secondary harmonics," but he had no physical explanation for them.[45]

Three years after Jones's investigation of the Aeolian harp, Matthew Young at Trinity College, Dublin, concluded from his experiments that a strong wind sounded the higher harmonics rather than the fundamental tone, because it exerted enough force on the string to prevent it from vibrating as a whole and to force it to vibrate in smaller parts.[46]

In 1830 Charles-Émile Pellisov again took up the problem of the Aeolian harp and poured scorn on Young's explanation, but still he could not explain how the string could vibrate in frequencies other than harmonic.[47] Especially when the wind was dying, the harp would slide from one harmonic to another in a glissando, totally inexplicable in the theory of harmonics (see fig. 5.2). Pellisov concluded that the vibrations producing the sound were longitudinal in the wire and that "the tone that a string gives or is able to give is, in general, completely independent of the transverse vibrations in the string."[48] Longitudinal vibrations could not transmit energy into the air directly, but acting through the bridge they could cause the sounding board to vibrate, which in turn would create the compression waves in the air that we hear as sound.

A new solution came only in 1878 when V. Strouhal proposed a third way in which a string could make a tone in addition to transverse and longitudinal vibrations. He called these tones "frictional sounds" and said they were made whenever an air current passed over a thin wire or a sharp edge.[49] It had been known since Galileo that the pitch of a vibrating string is a function of its length, tension, and mass density, but the pitch of Strouhal's frictional sound was dependent only on the velocity of the air and the diameter of the string, factors totally unrelated to the pitch of a string that is plucked or bowed.[50] Strouhal guessed that the cause of his frictional tones was the production of a turbulent wake behind the string, a guess that proved to be correct.

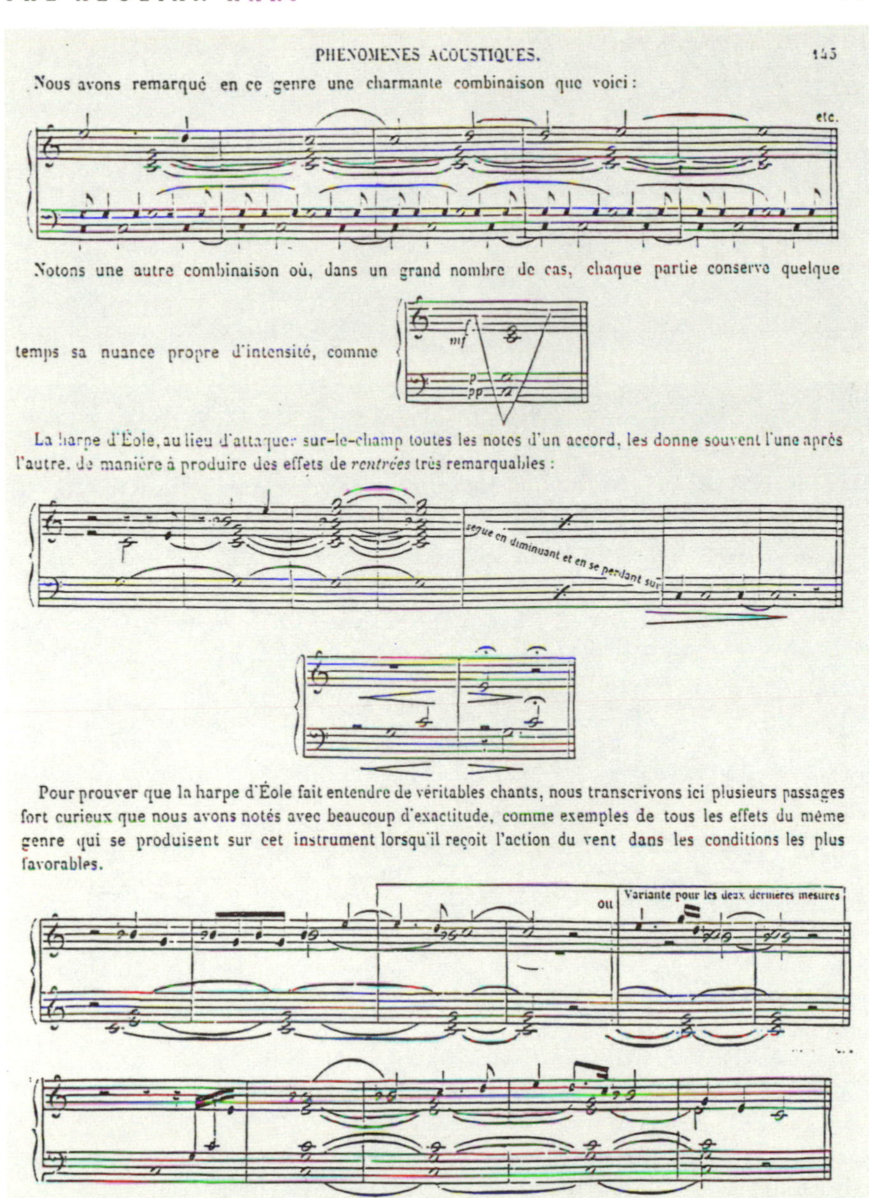
PHENOMENES ACOUSTIQUES. 145

Nous avons remarqué en ce genre une charmante combinaison que voici:

Notons une autre combinaison où, dans un grand nombre de cas, chaque partie conserve quelque temps sa nuance propre d'intensité, comme

La harpe d'Éole, au lieu d'attaquer sur-le-champ toutes les notes d'un accord, les donne souvent l'une après l'autre, de manière à produire des effets de *rentrées* très remarquables:

Pour prouver que la harpe d'Éole fait entendre de véritables chants, nous transcrivons ici plusieurs passages fort curieux que nous avons notés avec beaucoup d'exactitude, comme exemples de tous les effets du même genre qui se produisent sur cet instrument lorsqu'il reçoit l'action du vent dans les conditions les plus favorables.

Fig. 5.2. Georges Kastner attempted to capture the sounds from the Aeolian harp and reproduce them with an orchestra. From Kastner, *La harpe d'Éole*, p. 145.

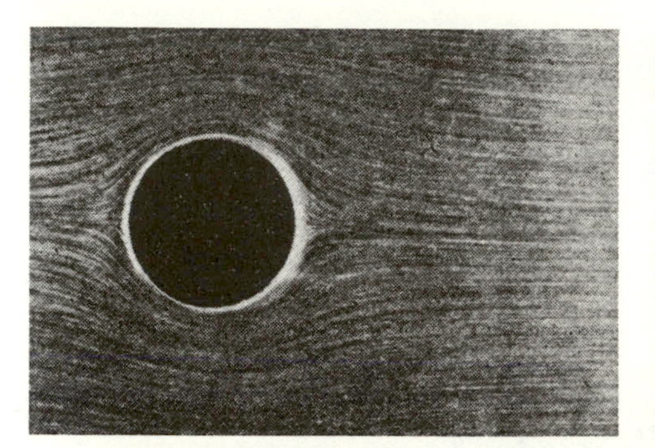

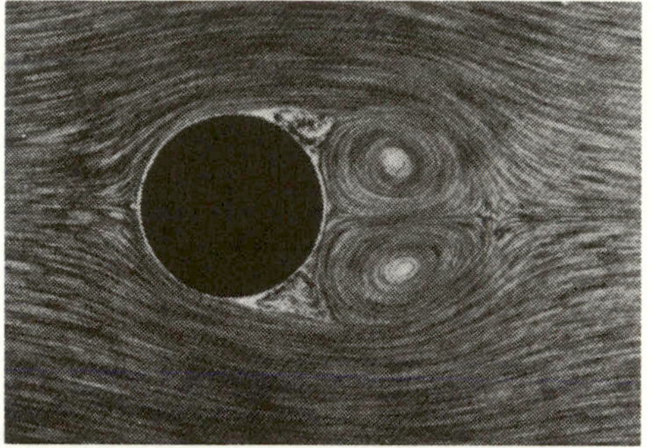

Fig. 5.3. Air flowing past the cylindrical string produces eddies that drive the string from side to side. From Tietjens and Prandtl, *Applied Hydro- and Aerodynamics*, pp. 279–303. Courtesy of the University of Washington Libraries.

At low velocities the air passes over the string in streamlines and does not cause it to vibrate. As the air velocity increases, two symmetrical eddies form behind the string (characteristic of fluid flow around any cylindrical body). At higher velocities the eddies break away, first on one side and then on the other, forming what is called a Von Kármán trail (see fig. 5.3). As each eddy breaks away, it causes lift on that side of the string. The string is thus driven from side to side at the frequency that the eddies break away. Even if the string were rigid and could not vibrate, the eddies breaking away would produce a tone, but it would be soft. If, however, that tone corresponds to one of the harmonics of the string, the string will also begin to vibrate strongly and the sound will be greatly increased. This is why Strouhal's frictional tone became louder when it approached one of the natural harmonics of the string and died away as the wind velocity increased. According to this explanation, the Aeolian harp can produce tones other than the harmonics, even glissandos from one harmonic to the next.[51]

The science of acoustics finally explained why the Aeolian harp is unlike any other stringed instrument. The fact that it is played by the wind rather than by being plucked, hammered, or bowed is more than symbolical. The tone production is unique, and therefore the sound is unique, and while the romantic poets did not have, and probably did not *want* to have, a mechanical explanation of what they heard, they knew that the Aeolian music affected them deeply. This is how it affected Hector Berlioz:

> On one of those sombre days which sadden the close of the year, read Ossian and listen to the fantastic harmony of an Aeolian Harp hung at the top of a tree stripped of its leaves, and I defy you not to experience a deep feeling of sadness, of surrender, a vague and boundless yearning for another existence, an immense loathing for this one; in a word, a sharp attack of spleen linked to a temptation toward suicide.[52]

Metaphysics and the Harp

William Jones's description of the Aeolian harp lent itself to two different kinds of "natural science." One employed experiment and mathematical analysis to describe the motion of a string vibrating in the wind; the other was more subjective and explored the color-sound analogy suggested by Jones's theory of the "air-prism." It was this second kind of question that attracted the romantic poets to the Aeolian harp.

To understand Jones's theory of the air-prism we must look at his broader philosophical position, for Jones was a "Hutchinsonian," a follower of John Hutchinson (1674–1737), author of *Moses Principia* (1724). Hutchinson argued that the Bible gave information about the natural world through analogy, and that without the aid of the Bible, natural philosophy was impossible. These analogies between man and nature had been covered up by corrupt translations from the original Hebrew Scriptures, especially the introduction of "points" assigning vowel sounds to the text. Without points, many of the original Hebrew words in the Bible had two interpretations, one spiritual, the other natural. Thus *khoved* could mean both "glory" and "gravity." Light, which Hutchinson believed was the cause of gravitational attraction, was "emblematically" (in Hutchinson's words) also the glory of Christ.[53] More important, *shem* (a name) and *shamaim* (the heavens) in their unpointed form are very similar and therefore must have the same root meaning. The materials composing the heavens should, therefore, properly be called "the names" and so Hutchinson designated them. The names were capable of three modifications, fire, light, and spirit (or air), which were related by analogy to the three parts of the Trinity. Just as the Trinity was unity, so the heavens were a single substance appearing in three different manifestations. The "names" were distinct from the gross matter composing the earth and the seas, and because our senses can give us information only about the gross matter, we must learn about the names through revelation.[54]

Hutchinson's system was mechanical, because he rejected any powers, principles, or souls inherent in natural objects. Any such powers would take away from the single omnipotent power of God. The names did act to move and alter the gross bodies, but not as independent agencies, only as agents of the Creator. Hutchinson's entire system had a strong Cartesian flavor, but with a very un-Cartesian justification. Hutchinson was critical of Newton, whose philosophy he believed led to materialism and pantheism. Newton's greatest methodological error was his dependence on mathematics, which Hutchinson abhorred, and his ignorance of Scripture.

Hutchinsonianism should have died an early death by all rights of reason, but it had remarkable permanence. While the natural philosophy was absurd, the biblical support for it appealed to those High Church Anglicans who opposed the latitudinarian leanings of the Church of England in the wake of the Glorious Revolution. Centered at Oxford University, Hutchinsonianism was biblical, Trinitarian, and apostolic. Its followers claimed that the analogical method it employed, linking nature and the Bible, was the method of the early church fathers.[55] Newton's secret Arianism—which was not so secret among his associates, Samuel Clarke, Edmond Halley, William Whiston, and John Toland (who coined the term "pantheism")—raised Hutchinson's suspicions, and he tarred Newton's religion, philosophy, and friends all with the same brush.[56]

William Jones was prepared to adapt the teachings of his master to fit the changing needs of natural philosophy. He recognized that Hutchinson's diatribe against Newton was ill-founded; in the *Physiological Disquisitions*, he asked that his own ideas not be identified with Hutchinson lest they be rejected out of hand.[57] There was no doubt about Jones's religious position, however. From his years at Oxford he was a close associate of George Horne, president of Magdalen College, vice-chancellor of Oxford, and bishop of Norwich. Jones and Horne published numerous tracts defending the High Church position, and the Tractarians Pusey, Keble, and Newman constantly quoted Jones and Horne as authorities on episcopacy and the sacraments. Jones's *Essay on the Church* became a standard reference for the principles of the Oxford Movement.[58] Thus Hutchinsonianism was given credit, in some circles at least, for keeping alive in the Church of England the doctrines and practices of the early church.

Jones may not have followed Hutchinson in all of his natural philosophy, but he did follow him in making the "divine analogy" the foundation of all natural knowledge. He wrote:

> Scripture is found to have a language of its own, which doth not consist of words, but of signs or figures taken from visible things. It could not otherwise treat of God, who is spirit, and of the spirit of man, and of a spiritual world, which no words can describe. Words are the arbitrary signs of natural things; but the language of revelation goes a step farther, and uses some things as signs of other things; in consequence of which, the world which we now see becomes a sort of commentary on the mind of God, and explains the world in which we believe.[59]

While he did not talk about "the names," Jones did ground all of natural philosophy on one law: "the natural agency of the elements." He was particularly interested in the agency of fire and air, the two active elements of nature.[60]

When Jones says that the analogy between sounds and colors is "very strict and may be carried very far," it is because he believes that in some sense they are the same thing and may be expected to act in the same way. The sound is not in the gross bodies of atmospheric air, but in a subtler, more spirituous fluid like electricity, which moves quickly through the pores of solid bodies.[61] And of course this spirituous substance finds profound sympathy in the human spirit. "This effect of music upon the human mind is most elegantly alluded to by the Royal Psalmist, that great musician of the Hebrews: *Awake then lute and harp; I myself will awake right early:* by which it is signified, that the mind of man is excited to devotion by the same art which excites the harp to musical sounds, and that when the one is touched the other will answer it."[62]

William Law, the Anglican mystic and contemporary of Hutchinson, argued in *A Serious Call to a Devout and Holy Life* (1728) that private devotion should always begin with a psalm. "Imagine to yourself that you saw holy David with his hands upon his harp and his eyes fixed upon heaven, calling in transport upon all the creation, sun and moon, light and darkness, day and night, men and angels, to join with his rapturous soul in praising the Lord of Heaven."[63] Among the Oswald circle, Christopher Smart, who studied both Law and Hutchinson, carried this injunction to the extreme and was twice confined to an asylum for "religious mania." When seized by illness he would burst forth in praise in the street, at the dinner table, and at other inappropriate occasions. During his confinement Smart wrote in his *Jubilate Agno*:

> For GOD the father Almighty plays upon the HARP of stupendous
> magnitude and melody.
> For innumerable Angels fly out at every touch and his tune is a work
> of creation.
> For at that time malignity ceases and the devils themselves are at peace.
> For this time is perceptible to man by a remarkable stillness and serenity
> of soul.
> For the Æolian harp is improveable into regularity.[64]

Smart wrote the poem in the form of an antiphon after the model of Hebrew poetry. In "A Song to David" he returned to the Aeolian harp as an instrument of adoration: "For Adoration on the strings / The Western breezes work their wings, / The captive ear to sooth."[65] This passage illustrates how the Aeolian harp combined two powerful images of inspiration: the harp of David, and the wind as the sign of the Holy Spirit.[66]

The association by Jones and Smart of the Aeolian harp with the Psalms coincided with the rediscovery of the oracular poetic voice in the second half of the eighteenth century. It was reasonable to seek to understand the mystery of poetic inspiration by emulating the poetry that was most certainly inspired by God. Also the great antiquity of the Psalms meant that they were closer to the time of Creation and were, therefore, more directly inspired than the later books of the Bible.[67] Nor was the oracular voice limited to the Bible, for the Greek poets as well as the Hebrews made use of the harp. Smart insisted that "the story of Orpheus is of the truth. / For there was such a person a cunning player on the harp. / For he was a believer in the true God and assisted in the spirit. / For he playd upon the harp in the spirit by breathing upon the strings. / For this will affect every thing that is sustained by the spirit even everything in nature."[68] The Aeolian harp told not only of God, but also of his Creation, and it spoke with the voice of the psalmist.

The Harp in Romantic Literature

When the Aeolian harp reached the romantic poets, it arrived (thanks to Jones) with considerable philosophical baggage. The Germans were first introduced to the English version of the harp in 1789 and 1792 by H. Lichtenberg's *Göttingen Taschen-kalendar*, which carried articles drawn from Jones.[69] At the same time that Jones was publishing his observations on the Aeolian harp, it acquired new scientific importance as a meteorological instrument. In 1782 the mathematician Jakob Bernoulli reported on a "barometric harp" that announced changes in the weather, and the following year the abbot Giulio Cesare Gattoni of Milan built a giant "Armonica Meteorologica" consisting of wires stretched from his house to a fifty-two-foot tower 150 paces away.[70] In 1787 another "Gigantic Meteorological Eolian Harp" went up near Basel.[71] The creators of these meteorological harps thought that the music was caused by atmospheric electricity that changed with the weather, although they soon found that the harps were not as effective in predicting the weather as they had at first thought.

These meteorological harps were the obvious source for E.T.A. Hoffmann's giant "Wetterharfe" that appeared in "Automata" and again in "Opinions of Tomcat Murr." He described the harps as "thick cords of wire, which were stretched out at considerable distances apart, in the open country, and gave forth great, powerful chords when the wind smote them." Hoffmann wrote, "I truly feel that some hostile power has forced itself into my deepest inwardness, smiting all my hidden strings, and making them resound at its arbitrary will, even if I should perish as a result."[72]

From the time of Lichtenberg's announcement Aeolian harps more modest than the giant weather harps proliferated throughout Germany. By 1796 Friedrich Schiller was honoring woman by comparing her to the Aeolian harp:

> Alive, as the wind-harp, how lightly soever
> If wooed by the zephyr, to music will quiver,
> Is woman to hope and to fear;
> Ah, tender one! still at the shadow of grieving,
> How quiver the chords—how thy bosom is heaving—
> How trembles thy glance through the tear![73]

The following year Goethe opened *Faust* by identifying his poetic utterance with the Aeolian harp.

> And I am seized by a long forgotten yearning
> For that kingdom of spirits, still and grave;
> To flowing song I see my feelings turning,
> As from aeolian harps, wave upon wave;
> A shudder grips me, tear on tear falls burning,
> Soft grows my heart, once so severe and brave;
> What I possess, seems far away to me,
> And what is gone become reality.[74]

The Aeolian harp's significance in England differed from its significance in Germany. The Germans heard in the harp's music a melancholy longing for another world, a yearning for the transcendent, while the English romantic poets heard the wind of inspiration.[75] They also used their harps in different ways. The Germans placed their harps in gardens, or even better in romantic ruins, while the English placed them in their windows (see fig. 5.4).

The first poetic use of the harp in England after Jones's *Physiological Disquisitions* of 1780 was Coleridge's "The Eolian Harp" of 1795. M. H. Abrams has analyzed the philosophical background to this poem in detail.[76] Coleridge wrote "Effusion 35. Clevedon, August 20th, 1795" at his cottage with Sara Fricker whom he was engaged to marry. His poem followed so closely the themes of Thomson's *Castle of Indolence* that one cannot doubt Thomson's influence. Coleridge was also reading the poetry of Erasmus Darwin at this time, especially the "Loves of the Plants." He spoke favorably of Darwin's poetry until 1817 when he and Wordsworth turned against Darwin's "contrived couplets" in the *Lyrical Ballads*. Darwin was an inveterate inventor and gadgeteer. It is no surprise that he incorporated the Aeolian harp in his poetry, but his use of it was like Thomson's. Five swains waiting on the fair Chondrilla (a flower) sigh when she sighs: "So tuned in unison, Eolian Lyre! / Sounds in sweet symphony thy kindred wire."[77]

When Coleridge published his poem the following year, he greatly ex-

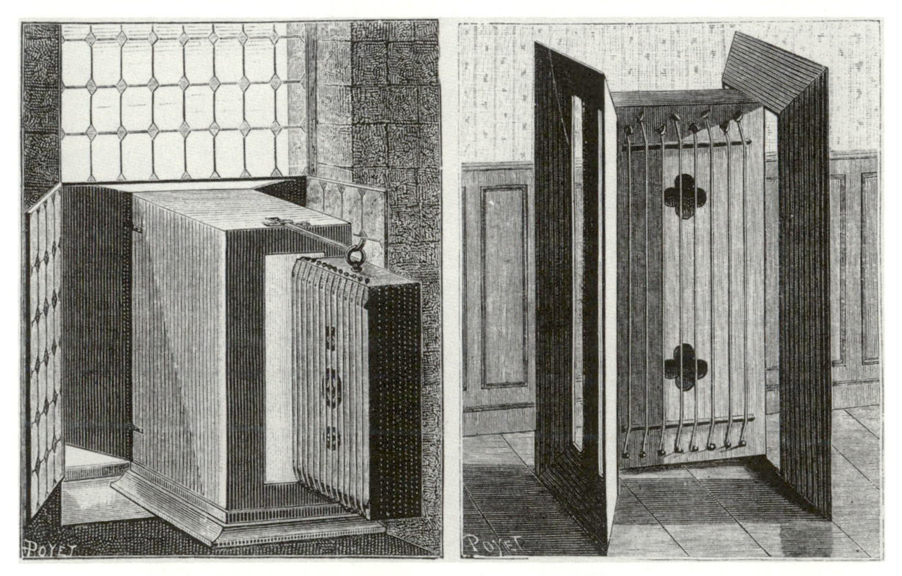

Fig. 5.4. Continental harps. From *La nature*, pt. 1 (1993): 44–45. Courtesy of the University of Washington Libraries.

panded it and added more philosophical verses. In 1803 he changed it again, and again in 1817 when he titled it *The Eolian Harp* and published it in his *Sibylline Leaves*. Abrams uses Coleridge's changes to chart his metaphysical journey from his early stance as a disciple of Hartley and a necessitarian to his later position as a follower of Schelling and the sixteenth-century mystic Jacob Boehme.

The first important philosophical addition to the poem occurred in 1796:

> And what if all of animated nature
> Be but organic Harps diversely fram'd,
> That tremble into thought, as o'er them sweeps
> Plastic and vast, one intellectual breeze,
> At once the Soul of each, and God of all?

Whereupon Sara upbraids him in the poem for succumbing to such "shapings of the unregenerate mind." Coleridge's image of animated nature derives undoubtedly from the "plastic natures" of Ralph Cudworth, but it also reflects Coleridge's recent infatuation with Hartley's associationist psychology, which he was in the process of rejecting.[78] Coleridge's fear of pantheism turned him from his early flirtation with Hartley and brought on the scolding from Sara.

The second important addition came in the *Sibylline Leaves* of 1817:

> O! the one Life within us and abroad,
> Which meets all motion and becomes its soul,
> A light in sound, a sound-like power in light,
> Rhythm in all thought, and joyance every where—
> Methinks, it should have been impossible
> Not to love all things in a world so fill'd;
> Where the breeze warbles, and the mute still air
> Is Music slumbering on her instrument.[79]

Abrams points out that this new quatrain was composed at the same time Coleridge was writing his most important philosophical works—the *Biographia Literaria*, *Theory of Life*, two *Lay Sermons*, the revised *Friend*, and the *Philosophical Lectures*—all of which contain references to the analogy between light and sound. The immediate source was Friedrich Schelling, whom Coleridge began to study in 1808 and whose *Naturphilosophie* he absorbed into his own philosophy. In September 1817 he wrote to C. A. Tulk that "Color is Gravitation under the power of Light . . . while Sound on the other hand is Light under the power or paramountcy of Gravitation. . . . The two Poles of the material Universe are Light and Gravitation."[80] It is the interaction, or conflict, of these two poles that produces the phenomena of color and sound; thus color and sound are generated from the same action and differ only in degree.

Coleridge had earlier run across the idea that phenomena are produced by the tension between polar opposites in Jacob Boehme's *Aurora* (1612), which he says he had "*conjured over*" as a schoolboy.[81] Boehme claimed that the creation of the world was effected by two elements, sound and light—light, because in Genesis that was the first thing God created, and sound, because, according to the Gospel of John, "In the beginning was the Word." In an alchemical allegory light becomes the "Salitter" (niter, or potassium nitrate), a union of all the divine powers. "The *second* Form or Property of Heaven . . . is *Mercurius*, or the Sound, as in the *Salitter* of the Earth there is the Sound, whence there grows Gold, Silver, Copper, Iron, and the like; of which Men make all Manner of *Musical Instruments* for sounding, or for Mirth, as Bells, Organ-Pipes, and other *Things* that make a Sound: There is likewise a Sound in all the Creatures upon Earth." To which Coleridge remarks: "[This] is admirable—the Messenger or Mercury of the Salitter is indeed Sound, which is but Light under the paramouncy of Gravitation. It is the Mass-Light. The Granit-blocks in the vale of Thebais still send forth sweet *Sounds* at the touch of Light—a proof the Granit is a metallic composition."[82] Here Coleridge is referring to the statue of Memnon in the Valley of Thebes that supposedly resonated with song when struck by the first rays of the rising sun. The statue of Memnon had been associated with the Aeolian harp ever since Kircher because both instruments made music without human intervention, and because there was a mistaken assumption among eighteenth- and nineteenth-century English poets, probably owing to Kircher, that the statue held a harp or lyre from which the music issued. Often the harp of Memnon and the harp of Aeolus were used interchangeably.[83]

The sound-light-air Trinity of Boehme had obvious similarities to the "names" of John Hutchinson and may have been one of Hutchinson's sources. He was certainly a source for William Law and Christopher Smart. But Coleridge was less than enthusiastic about Hutchinsonianism. In more charitable moments he recognized it as a serious attempt to obtain the correct literal meaning of the Bible. In less charitable moments it was "the dotage of a few weak-minded individuals."[84] Coleridge approved of the biblical foundation that Hutchinson gave to natural philosophy, but strongly opposed his mechanism.[85] We know, however, that a philosopher or a poet often takes the most from whomever he despises the most, and Coleridge was, if nothing else, an enthusiastic borrower. His reading of Hutchinson may well have reinforced the imagery that he drew from Boehme and Schelling.

I have found no evidence that Coleridge read Jones directly, but by 1817 he could have encountered the Aeolian passages from the *Physiological Disquisitions* in Bloomfield's *Nature's Music* and a number of other sources. Whether or not there was any direct borrowing, we can see common harp themes in both Jones and Coleridge, especially in the lines that Coleridge added in 1817. Coleridge linked the Aeolian harp to the sound-light analogy in the lines "A light in sound, a sound-like power in light, / Rhythm in all thought, and joyance every where—" an analogy that Jones argued could be "taken very far." Also Coleridge echoed Jones's description of the Aeolian harp as a passive instrument that merely actuated the music which lay dormant in the air, just as a prism actuated the colors that lay hidden in white light. For Coleridge, "the mute still air / Is music slumbering on her instrument," music that can be heard only when the wind carries the air through the "sound prism" of the harp.

In that same year of 1817 Shelley also used the sound-light analogy in his poetry and illustrated it with the Aeolian harp. At least one scholar claims to recognize a direct influence of Jones's "air-prism" in Shelley's *Alastor*, where the analogy returns relentlessly, the sounds of the harp blending with the multicolored hues of the rainbow.[86]

> . . . its music long,
> Like woven sounds of streams and breezes, held
> His inmost sense suspended in its web
> Of many-coloured woof and shifting hues. *(Lines 154–157)*

By 1817 the Aeolian harp was everywhere in England, both in poets' windows and in their poetry. So was the sound-light analogy that had attracted so much attention in the eighteenth century. Assigning a specific source for Coleridge or Shelley is probably beside the point, because they undoubtedly encountered the harp in many different ways, in natural philosophy, in poetry, and in their drawing rooms and gardens. Like Castel's ocular harpsichord, it was an obvious emblem of synesthesia. Unlike the ocular harpsichord, it actually existed, producing its own unique harmony and providing inspiration for the listening poet.

There was uncertainty, however, as to how the Aeolian harp inspired the poet. In the *Eolian Harp* it was a passive instrument stirred to life by the wind. In *Dejection: An Ode* (1802) it announced a coming storm, a storm that raged in the poet's soul. Deserted by his "shaping imagination" Coleridge heard in the Aeolian harp the cry of a "little child / Upon a lonesome wild, / Not far from home, but she hath lost her way: / And now moans low in bitter grief and fear, / And now screams loud, and hopes to make her mother hear" (lines 121–125). The harp voiced the poet's agony at being lost in a spiritual desert.

In *Dejection* the Aeolian harp was far from passive, far from the "mute still air . . . slumbering on her instrument" of *The Eolian Harp* (lines 32–33). Instead Coleridge made the poet's creative power a force of nature. He had discovered that nature could not be understood "objectively" apart from the sensing subject; he wrote, "We receive but what we give, / And *in our life alone* does Nature live"(*Dejection*, lines 47–48, emphasis added).[87]

The difference between these two poems illustrates Abrams's distinction between the "mirror" and the "lamp"—between the neoclassical imitation of nature and the romantic voice of the subject in nature. But it also illustrates the ongoing debate over the nature of divine inspiration. The rediscovery of Hebraic poetry raised anew the question of whether the poetry in the

Bible was the word of God speaking through the poet (the so-called dictation theory) or the work of the poet's creative imagination, inspired by God, but not dictated by him. In his posthumous *Confessions of an Inquiring Spirit* Coleridge denied that the Bible could be the work of a "superhuman ventriloquist," and insisted that the prophets and witnesses in the Bible remained human. In fact they were only fully human when God moved them to speak.[88] Coleridge related how he submitted himself to the "royal Harper" as a "*many-stringed instrument* [Coleridge's emphasis], for the fire-tipt fingers to traverse, while every several nerve of emotion, passion, thought, that thrids the flesh-and-blood of our common humanity, responded to the touch." According to Coleridge the poetry of the Bible is inspired, but it is also totally human. When the Aeolian harp speaks for the poet, it can no longer be a passive instrument.[89]

The Aeolian Harp in America

The Aeolian harp inevitably came to Concord and inspired Ralph Waldo Emerson and Henry David Thoreau, both of whom wrote harp poems to the sounds of their own instruments.[90] Emerson was tone deaf and claimed that in place of a musical ear he had "musical eyes," but while he could not appreciate music that was artificially composed, he was not deaf to the sound of the Aeolian harp.[91] As had Coleridge, Emerson described the Aeolian harp as a passive instrument, waiting to be stirred by nature. But when it did speak, it spoke truly, more truly than any music of human creation.[92] In his essay "Education" Emerson wrote, "As every wind draws music out of the Æolian harp, so doth every object in Nature draw music out of [a man's] mind."[93] And in his poem "The Harp" he tells us how this drawing out occurs.[94] The harp is "One musician sure" whose "wisdom will not fail." It is the "chief of song where poets feast," speaking with the voice of Merlin from "the casement by my side." No poet, not even Homer, could match the sounds of nature that Emerson had heard as a boy wandering through the hills in spring. "These syllables that Nature spoke, / And the thoughts that in him woke, / Can adequately utter none / Save to his ear the wind-harp lone." The Aeolian harp opened a window on the panorama of Emerson's youthful memories. In his *Journal* he recalled with nostalgia his college life: "The thought, the meaning, was insignificant; the whole joy was in the melody. . . . What joy I found, and still can find, in the Æolian harp!"[95] Emerson, who demanded direct inspiration and could brook no artificial intermediaries between his soul and that of nature, heard in the voice of the Aeolian harp—which was no more than "a couple of strings across a board and set . . . in your window"—tidings from nature more authentic than any that poets could bring.[96]

Thoreau had a better ear and wrote about music all the time. In fact, he claimed that a man's entire life "should be a stately march to an unheard music."[97] It was music that could be better heard out of doors in cadence with the step of the solitary wanderer. On September 3, 1851, Thoreau took his daily walk along the new railroad track between Boston and Concord and returned to record in his journal: "As I went under the new telegraph-wire I heard it vibrating like a harp high overhead. It was as the sound of a far-off glorious life, a supernal life, which came down to us, and vibrated the lattice-work of this life of ours."[98] The telegraph was a new source of Aeolian sounds to conjure with; as it spread across the continent, it became everybody's Aeolian harp. Here was Hoffmann's *Wetterharfe* restrung a thousand times.

Thoreau described his first encounter with the telegraph harp in his *A Week on the Concord and Merrimack Rivers.*[99] There was a "faint music in the air like an Aeolian harp," which he suspected of coming from the telegraph wire. Applying his ear to one of the posts he confirmed that the sound was indeed coming from the wire. It was a message "sent not by men, but by gods." It recalled the statue of Memnon and "the first lyre or shell heard on the seashore—that vibrating cord high in the air over the shores of earth."[100] Thoreau returned to the telegraph harp over thirty times in his journal. He noticed seasonal change in the telegraph's song and linked it to his own mood:

> The heat to-day (as yesterday) is furnace-like. It produces a thickness almost amounting to vapor in the near horizon. The railroad men cannot work in the Deep cut, but have come out on to the causeway, where there is a circulation of air. They tell with a shudder of the heat reflected from the rails. . . . I have scarcely heard one strain from the telegraph harp this season. Its string is rusted and slackened, relaxed, and now no more it encourages the walker. I miss it much. So is it with all sublunary things. Every poet's lyre loses its tension. It cannot bear the alternate contraction and expansion of the season.

But the cold winds of winter restored tension to the wire, and in January Thoreau wrote:

> The telegraph harp again. Always the same unrememberable revelation it is to me. It is something as enduring as the worm that never dies. . . . I never hear it without thinking of Greece. How the Greeks *harped* upon the words immortal, ambrosial! They are what it says. It stings my ear with everlasting truth. It allies Concord to Athens, and both to Elysium. It always intoxicates me, makes me sane, reverses my views of things. I am pledged to it. I get down the railroad till I hear that which makes all the world a lie. When the zephyr, or west wind, sweeps this wire, I rise to the height of my being. A period—a semicolon, at least—is put to my previous and habitual ways of viewing things. This wire is my redeemer.[101]

The music of the telegraph harp spoke for Thoreau and measured his mood. He did not miss the irony. As electrical messages darted back and forth along the wire transmitting the business of the day, the harp telegraphed to Thoreau a message from God. "Thus I make my own use of the telegraph, without consulting the directors, like the sparrows, which I perceive use it extensively for a perch. Shall I not go to this office to hear if there is any communication for me, as steadily as to the post-office in the village?"[102] And in *A Week on the Concord and Merrimack Rivers* the telegraph harp brought him "fairer news than the journals ever bring." It told of "things worthy to hear, and worthy of the electric fluid to carry the news of, not of the price of cotton and flour, but it hinted at the price of the world itself and of things which are priceless, of absolute truth and beauty."[103]

The Aeolian harp was an instrument of inspiration for Emerson and Thoreau as it had been for Wordsworth and Coleridge, but for the Americans it was more than an "intellectual breeze." It was also a voice from that transcendent world which they so wanted to reach. It seems strange that a man-made instrument could equal or maybe even surpass the direct sensual contact with nature. This instrument, however, was an intermediary of a different kind from most scientific instruments. It was sensitive—listening and responding to nature, rather than invading and dissecting nature. Its appeal was quasi-magical. Its music brought to the senses a wonder or harmony of nature that was not otherwise perceived. Thoreau, quoting from the Neoplatonist Iamblichus, explained that Pythagoras heard the harmony of the spheres, not through any instrument or voice, but by means of "a certain ineffable divinity" that allowed him to fix his intellect "in the sublime symphonies of the world."[104] Thoreau, who despised abstractions, heard the "sublime symphonies" through his ears and was redeemed by a telegraph wire that served *his* purpose, not that of the railroad.

The Aeolian Harp as an Instrument

Instruments could be used to heighten the sensual experience of nature and enhance the poetic vision, whether it was the instrument of opium or the instrument of the Aeolian harp. Nor was this discovery the exclusive property of the romantics. Thomas Gray, on his walking tour through the Lake Country in 1769, constantly used his "Claude Lorraine Glass" to miniaturize and intensify the landscape and to add the hazy patina so admired in the paintings of Claude Lorraine.[105] But optical instruments, which satisfied the poetic needs of the eighteenth century, objectified nature too much for the romantics. Coleridge, whose poems were often visionary in both senses of the word, condemned the "slavery of the eye" that reduced "the conceivable . . . within the bounds of the *picturable*."[106] Poetic visions needed to come from within rather than from without the mind, and however much optical instruments like the telescope, the microscope, and the Claude Lorraine Glass heightened the images of nature, they still produced images of physical objects.

Sound was a more subjective sense, often not localizable in any physical object and closer to the actual vehicle of poetry, which was, after all, spoken words. Poetry was a kind of music, traditionally sung to the lyre. What better instrument to speak nature's poetry than the Aeolian harp?

One might argue that the Aeolian harp in the hands of the romantic poets was more of an antiscientific instrument than a scientific one, but that would misinterpret the romantics' intent. Blake certainly despised science, and Keats, at the famous dinner given by Benjamin Haydon on December 28, 1817, proposed a toast to the confusion of Newton.[107] Wordsworth, however, refused to drink the toast. Shelley and Coleridge would also have demurred if they had been there. The romantics were not opposed to the study of nature, just to the way that study was commonly carried out.

To Gillispie's verdict that romanticism is "the wrong view for science," they would respond that it depends on what you mean by "science." Coleridge was prepared to devote his life to explicating the proper meaning of "science." For him, science was "any chain of truths which are either absolutely certain, or necessarily true for the human mind, from the laws and constitution of the mind itself."[108] His notion of "science" had nothing to do with experiment and analysis. Science was lodged in the faculty of Reason, which he defined as "the power of universal and necessary convictions, the source and substance of truths above sense, and having their evidence in themselves."[109] The "faculty judging according to sense" was the Understanding, and it was this faculty, masquerading as science, that had unjustly usurped the proper province of Reason and true "science." He warned his countrymen that "in no age since the first dawnings of science and philosophy in this island have the truths, interests, and studies which especially belong to the reason, contemplative or practical, sunk into such utter neglect, not to say contempt, as during the last century."[110] Central to this moral collapse was the corruption of natural philosophy.

For the romantics, science was primarily a spiritual and moral quest—a search for those truths, ordained by God, that direct our lives and his Creation. Instruments were of value to the extent that they aided in this quest. Seconding Coleridge, Wordsworth made it clear that he was not opposed to science that "raised the mind to the contemplation of God in works."[111]

> . . . Science then
> Shall be a precious visitant; and then,
> And only then, be worthy of her name:
> For then her heart shall kindle; her dull eye,
> Dull and inanimate, no more shall hang
> Chained to its object in brute slavery;
> But taught with patient interest to watch
> The processes of things, and serve the cause
> Of order and distinctness.[112]

Just as the child on the beach in Wordsworth's allegory listened to the sound of the sea in a shell, so the poet listened to the Aeolian harp in order to hear the "authentic tidings of invisible things, of ebb and flow, and ever-during power."[113] The Aeolian harp was made by human hands, but played by nature's fingers. It was emblematic of that unity of man in nature which the romantics so earnestly sought. Because it expressed the sublimity of nature directly, it was the romantics' scientific instrument.

CHAPTER SIX

Science since Babel: Graphs, Automatic Recording Devices, and the Universal Language of Instruments

INSTRUMENTS have a rhetorical purpose. They teach, explain, persuade, and even command. Instruments have authority, they speak for nature, but *how* they speak and in what language is far from obvious.

Instruments are like languages because they mediate between the observer and what is being observed—between the subjective mind and the objective natural world. Both languages and instruments give us signs for things. In the case of language the signs are words; in the case of instruments the signs are images, sounds, numbers, graphical traces, or other representations.

In the seventeenth century the same debate arose concerning instruments that arose concerning language. Is the instrument telling us anything about the real essences of the objects being observed or is it merely a convention by which *we*, rather than nature, assign meaning to the representations that the instrument produces? The major criticism of the new instruments of experimental philosophy was that they presented distorted images and played tricks on the senses. The tricks might be "natural" as in natural magic, but they were products of human artifice and thus did not fairly image nature.

Words also could misrepresent nature, because they, like instruments, were constructed by humans. John Locke, because he did not believe that we could ever know essences, took an extreme conventionalist position and held that words were completely arbitrary. They could have no meaning beyond what we assign to them. Thus he opposed the use of analogy and other figurative language, which, to his mind, confused words with things and suggested real correspondences in nature that existed only in language.[1] Because of this parallel between words and instruments, it is, perhaps, not surprising that Locke, in spite of his empiricism, doubted the validity of the microscope, and that Thomas Hobbes, who also saw danger in the misuse of language, condemned the air pump, calling it a dangerous chimera.[2] Just as words could misrepresent nature, so could instruments. They could convince the observer that a natural phenomenon was real when, in fact, it was only an artifact of the instrument.

Others, however, like Athanasius Kircher and Gottfried Wilhelm Leibniz believed that real correspondences between words and things did exist, or could exist if words could be made to correspond to the essences of things. If they did exist, these correspondences would be found in the original language given to Adam by God and since corrupted at the Tower of Babel, but

Fig. 6.1. The Tower of Babel and the confusion of tongues. From Kircher, *Turris Babel*, p. 40. Courtesy of the University of Washington Libraries.

still shadowed in the languages that we speak today. If they could be made to exist, they would be in a universal language that could become the common language of the whole human race (see figs. 6.1 and 6.2). The instruments of natural magic supposedly shared this same secret correspondence with the essences of things.

There was a real difference, however, in the ways that natural magicians like Kircher and experimental philosophers like Robert Boyle saw the connection between instruments and language. For Kircher, instruments and language operated by *analogy* to reveal the secrets of nature. His instruments produced wonders analogical to the wonders of nature. For the experimental philosophers of the seventeenth and eighteenth centuries instruments operated *analytically*, taking nature apart to reveal the rules by which

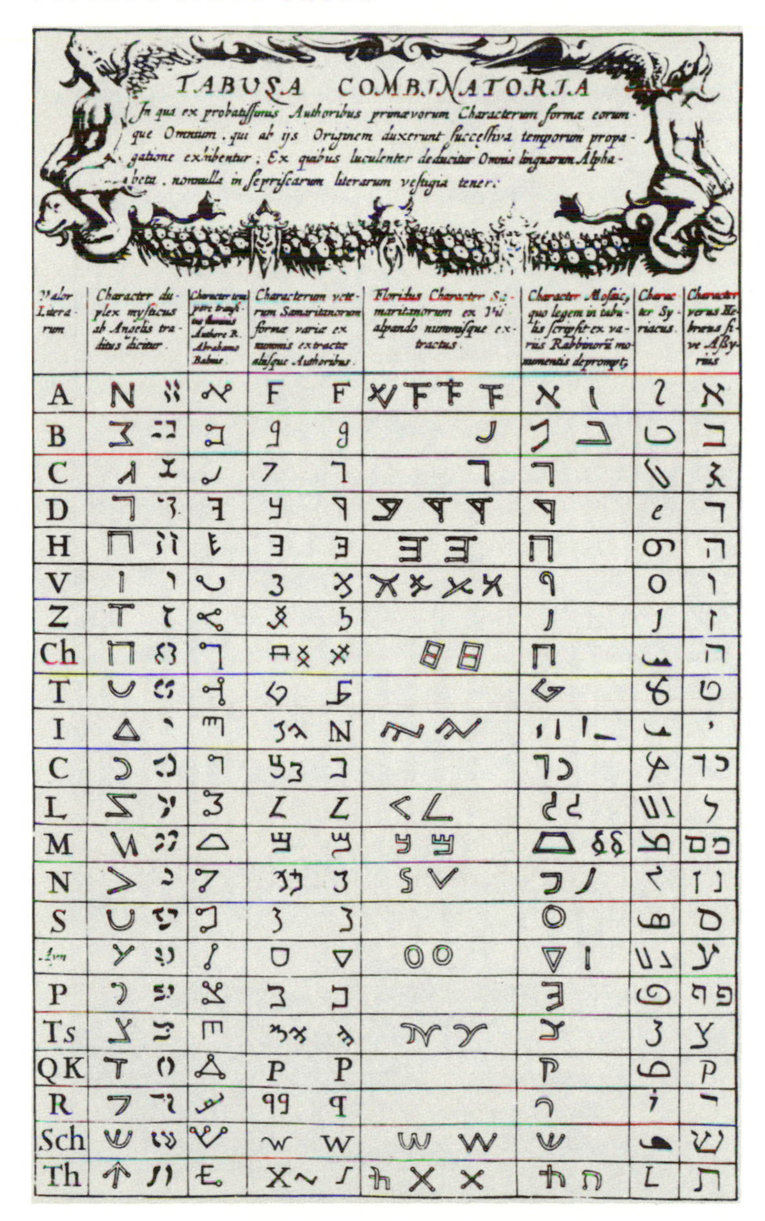

Fig. 6.2. Athanasius Kircher's table of characters from the original angelic language to Latin. From Kircher, *Turris Babel*, p. 157. Courtesy of the University of Washington Libraries.

it operated. This difference can also be exemplified by the kinds of pictorial representations that they used to describe correspondences in nature. The kind of pictorial representation that Kircher preferred was the emblem, a recognized trope of rhetoric that described through symbols and allegory the

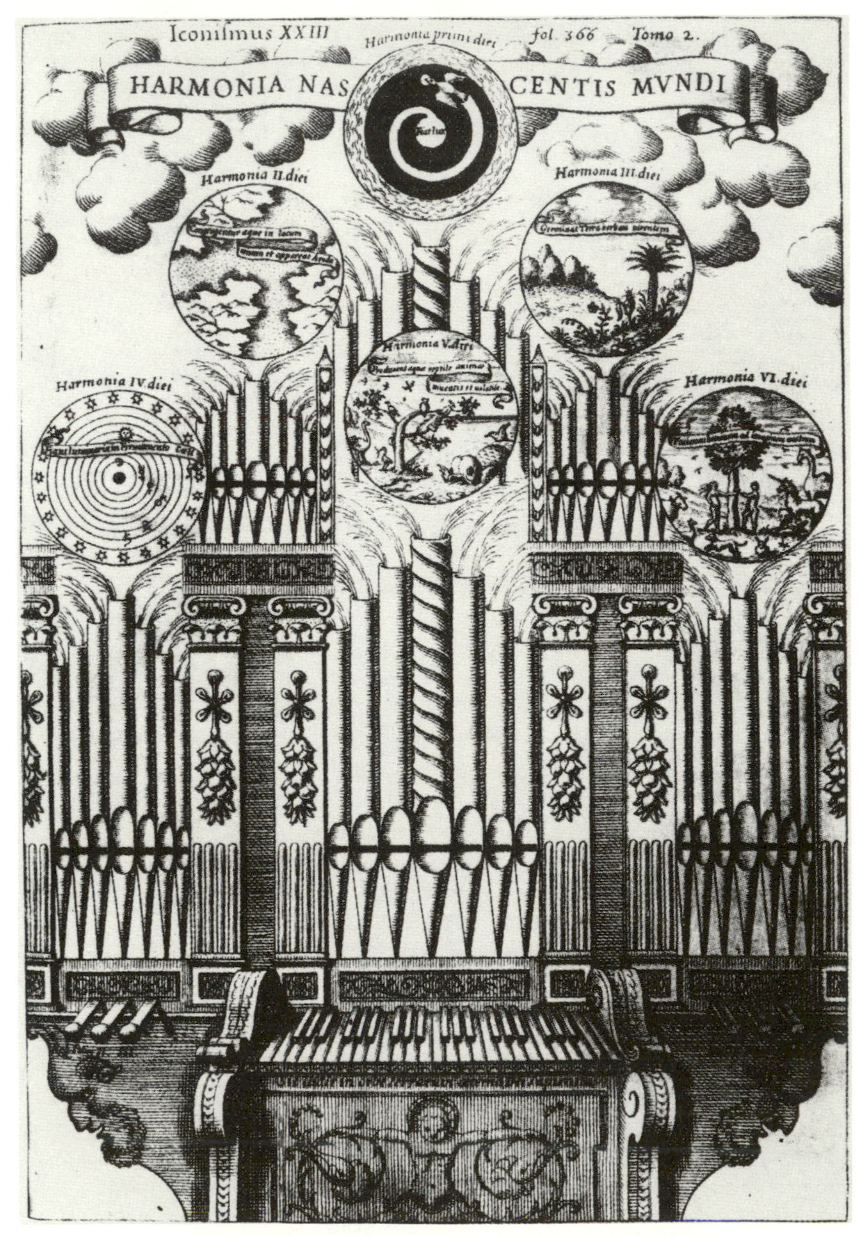

Fig. 6.3. Instrument as emblem: Kircher's great celestial organ of the Creation, each day represented by a separate register. From Kircher, *Musurgia*, 2:366. Courtesy of the University of Washington Libraries.

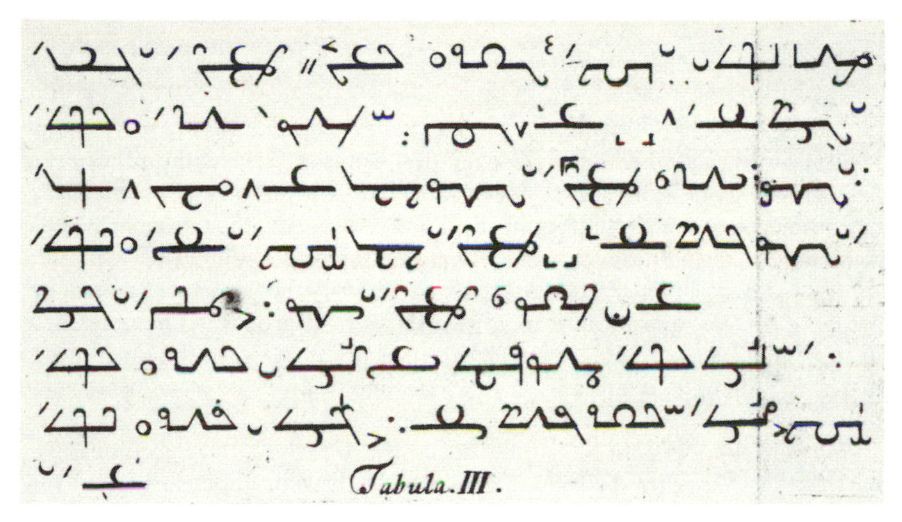

Fig. 6.4. An early example of a universal language was that of John Wilkins. This essay on "pocket watches" by Robert Hooke was written in Wilkins's universal characteristic. From Robert Hooke, "A Description of Helioscopes and some other Instruments," table 3, reprinted in Gunther, *Early Science in Oxford*, vol. 8, facing p. 152. Courtesy of the University of Washington Libraries.

meaning of the natural world. It was a kind of figurative language, which used both images and words.[3] Many of Kircher's emblems were drawings of real or imagined instruments represented allegorically (see fig. 6.3). His illustration of the sunflower clock is a good example. The sunflower is the emblem of constancy, because it follows the sun faithfully. The emblem acts like the instrument to expose a hidden meaning in nature.

In the wake of the Enlightenment's fascination with analysis there appeared a new analytical pictorial representation, and that was the graph. The graph did not develop from the emblem—its origins were quite separate—but it served a similar purpose. Graphs and emblems were both manifestations of the way that instruments and language mediated between the observer and nature. They were symbolic images revealing an order and meaning in nature that were concealed from the immediate senses.

Most of the earliest graphs were either drawn by instruments directly—these were the first recording instruments—or they represented numerical data from instruments.[4] They appeared in conjunction with the new emphasis on quantitative measure at the end of the eighteenth century and with the renewed search for an analytical language (see fig. 6.4).[5]

The graphical method was born of two strains that merged in the nineteenth century. One of these was analytical geometry, in which a functional relationship between two variables was described by a curve. The other was automatic recording instruments.[6] Both kinds of graphs were often referred to as scientific languages. It is the connection between graphs and language that we explore in this chapter.

Graphs and the Exact Sciences

Graphs are so ubiquitous that we take them for granted, but the historian will naturally ask when and why they first appeared. We can find a partial answer to the question *when* by searching back through the scientific literature.[7] There is some ambiguity in the answer, because it is not clear what is to count as a graph (do we include maps, mathematical diagrams, and illustrations of experiments?). Graphs, unambiguously recognizable as such, appeared in the last quarter of the eighteenth century, probably independently, in three places—in the indicator diagram of James Watt, in the lineal arithmetic of William Playfair, and in the scientific writings of Johann Heinrich Lambert.

Why graphs appeared when they did is not so certain, but the coincidence between their appearance and the new enthusiasms for quantitative measure and universal language schemes leads one to suspect a connection. In this regard Lambert is the most obvious source, because he thought deeply and wrote extensively about method in natural philosophy.

In that portion of his *Neues Organon* (1764) entitled *Semiotik* Lambert emphasizes the importance of symbols and signs in our formation of concepts. Like other language reformers he declares that common language is a "tyranny" that is forced on us by those uneducated persons who create it.[8] And yet "symbolic perception" (*symbolische Erkenntniss*) is absolutely essential for the clarification of concepts. According to Lambert, concepts can be retained and clarified only by a renewal of the sensations through which they were first formed. Because we seldom can repeat the events that led to the concepts in the first place, we substitute signs for them. We can do this because the moving hand, eye, or other organs of sense can distinguish the outlines or shadows of things that serve as signs for them. Such signs include gestures (the deaf can communicate entirely through gestures), articulated tones (such as music and the chimes of clocks), and visual images including numbers, words, emblems, heraldry, maps, and metaphors. These "symbolic perceptions" are absolutely indispensable for thought.[9] If we cannot attach signs to our concepts, we cannot use them.

If the impressions of objects on our senses were the same at all times and for all people, then we would have a single natural language. But this is not the case, and any attempt to discover a universal root language will be fraught with difficulty. We need to reform language, but in some cases it might be better to use symbols other than words. There is the danger that our language may become empty "word stuff" (*leeren Wortkram*), as happened in Scholasticism.[10] Therefore there is the need to control language or substitute for it more "scientific" signs. If we could obtain proper scientific signs, then we could free ourselves from the ambiguous words of common language and express all knowledge by a demonstrative figurative method.[11]

This was the dream of a universal characteristic that Lambert took up from Leibniz and Christian Wolff and hoped to carry to a new level of perfection.

According to Lambert, signs of concepts and things are "scientific" when "in the narrow sense, . . . they not only represent the concepts and things, but also reflect such relations that the theory of the matter and the theory of the signs can be interchanged."[12] Lambert's prime example of scientific signs is, of course, mathematics, but he begins with less abstract examples such as musical notes, the signs for choreography, and for the directions of the wind. These signs are "scientific" because they fully correspond to the concepts that they represent (e.g., ENE completely determines a certain wind direction).[13]

Lambert argues further that concepts have extension (*Ausdehnung*) and can therefore be represented by lines.[14] The *individua* from which a concept is composed are represented by points, while the entire species, composed of many *individua*, is a line. Moreover, the *individua* have degree.[15] Lambert uses his lines to create a symbolic logic somewhat akin to set theory in which the relative lengths and positions of the lines indicate how and to what degree one concept may be included in another.[16] In this effort Lambert continues Leibniz's search for a new symbolic logic that would bring together mathematical logic and syllogistic logic in a new, more extensive method of reasoning.

Lambert obviously believes that scientific knowledge can be represented with greatest precision by signs other than common language, and when he turns to the physical sciences experimental graphs do indeed appear. Lambert's scientific method is highly mathematical, because he believes that physical law can be expressed only quantitatively. He does not favor mathematical hypotheses, however. He argues instead for the collection of quantitative data with precise measuring instruments, followed by the identification of the regularities in those data that reveal natural law.

His own experimental researches were very wide-ranging, including photometry, hygrometry, and pyrometry, that is, the measurements of the intensities of light, humidity, and heat. Some of Lambert's data came from the work of other researchers, but he recorded many of the measurements himself using instruments that he made. In 1759 he met and lived in the house of the famous instrument maker Georg Friedrich Brander, who specialized in the manufacture of "mathematical" measuring instruments, so he was completely familiar with the problems of compiling and analyzing data from experiments. It was in the processing and presentation of these quantitative data that experimental graphs first appeared.[17]

Lambert's most elaborate graphs are in his *Pyrometrie oder vom Maasse des Feuers und der Wärme*, published posthumously in 1779. Because they are posthumous, it is difficult to establish when he first formulated the idea of expressing his data graphically, but one can, nevertheless, get some idea of the sequence of his thoughts.

In the third and fourth chapters of the *Pyrometrie* Lambert considers the amount of heat from the sun reaching any part of the earth during a day, month, or year. If one discounts the absorption of heat by the atmosphere, this is a mathematical problem, because it depends only on the length of the day and the elevation of the sun. The sixth chapter, however, is on the application of his mathematical theory to observations, and here one sees the discrepancies between theory and observation caused by the action of clouds and by the variable absorptivity of the earth's surface. Lambert tabulates data on temperatures at different latitudes collected by René Réaumur and others, but some of these tables have an unusual aspect. For instance, one table shows the distribution of temperatures at Algiers for each month from June 1735 through November 1736.[18] Lambert gives the number of days during each month that the thermometer reaches any given degree. The vertical columns of the table represent different degrees on the Réaumur scale. The result is a snakelike series of numbers that moves from higher to lower temperatures as the months move through the seasons. It is a table, but it looks very much like a graph. Most striking is a similar table for the temperatures at Pondicherry. Again the numbers snake back and forth with the changing seasons in a most graphlike manner (see fig. 6.5).[19] The block of numbers in the table takes on a geometrical form irrespective of the magnitudes of the numbers themselves. Finally Lambert includes actual graphs of the heat reaching the earth's surface throughout the year at different latitudes, and graphs showing the temperature fluctuations throughout the year at the earth's surface and at different depths below the surface (see figs. 6.6 and 6.7). The graphs show in striking fashion the decreased amplitude and the delay in temperature fluctuations as one moves below the surface.

Lambert uses graphs not only for presenting experimental data, but also for correcting observational errors. In his "Theorie der Zuverlässigkeit der Beobachtungen und Versuche" he describes how one should draw a curve through data points in such a way that the points fall equally on either side of the line with no point outside the range of experimental error (see fig. 6.8).[20] In graphs, Lambert finds a new method of symbolic representation for experimental results. Graphs not only display data "figuratively," but also, by showing smooth curves averaging the data, they reveal the mathematical regularities in a mass of data, in spite of the errors of observation.

Lambert's graphical method did not catch on immediately, which may be attributed in part to the obscurity of much of his writing and in part to the unfamiliarity of graphs themselves. We do not have any contemporary reactions to Lambert's graphs, but the graphs of William Playfair, which became much better known than Lambert's, brought forth the criticisms that they "lacked rigor," that they were mere "plays of the imagination" and "without importance" outside of pedagogy.[21] The concept of a graph is abstract, and its meaning will seem obvious only to those who are familiar with it. Those who were used to working with tables of numbers could persuade themselves that in drawing graphs one lost the precision of the numbers

	20	21	22	23	24	25	26	27	28
1733. Apr.				1	2	12	11	4	
May			4	7	9	10	1		
Jun.		6	10	10	4				
Jul.	1	4	18	8					
Aug.	9	15	5	1		1			
Sept		9	17	4					
Oct.	1	8	12	8					
Nov.	1	1	4	6	12	6			
Dec.			1	5	9	11	3		
1734 Jan.			2	6	5	9	5	3	1
Febr.				1	3	9	8	7	

1735	11	12	13	14	15	16	17	18	19	20	21	21	23	24	25	26	27
Jun.						1		4	8	9	4	4					
Jul.												21	7	3			
Aug.													9	10	5	1	
Sept.											1	15	8	5			
Oct.										1	6	16	7	1			
Nov.					4	9	3	1	6	7	3						
Dec.				5	14	12											
1736																	
Jan.		3	4	12	10	2											
Febr.	1	4	8	11	4	1											
Mart			1	5	17	5	3										
Apr.				1	5	7	10	5	2								
May						1	2	5		13	3	7					
Jun.										1	6	18	2	3			
Jul.												4	4	7	7	8	1
Aug.													1	7	14	8	1
Sept.											3	5	11	8	3		
Oct.								2	8	6	5	7	2	1			
Nov.					5	3	6	16									

Anwendung der Theorie auf Beobachtungen.

		21	22	23	24	25	26	27	28	29	30	31	32
1736.	Sept.					4	6	7	3				
	Oct.		1	1	3	8	6	10					
	Nov.	1	9	11	3	5							
	Dec.	5	15	6									
1737.	Febr.			4	13	11							
	Merz						16	14					
	Apr.						1	8	15				
	May							1	14	15	1		
	Jun.									10	9	8	2
	Jul.							2	7	18	3		
	Aug.						4	12	8	5	1		
	Sept.					1	5	10	5	6			
	Oct.	1	4	3	3	3	8	5					
	Nov.	3	6	11	6								
	Dec.		4	10	8								
1738.	Febr.				7	11	7						
	Merz					11	19						
	Apr.						5	15	5				
	May								9	6	4	7	3
	Jun.								3	19	5		
	Jul.								5	13	12		
	Aug.					4	2	6	8	8			
	Sept.						7	5	14	1			

Fig. 6.5. Johann Heinrich Lambert's tables of temperatures. The tables record the number of days in each month that a given temperature is reached. The smaller tables are for Algiers and Madagascar; the larger table is for Pondicherry. Note how the figures snake back and forth. From Lambert, *Pyrometrie*, pp. 350–353. Courtesy of the Science and Technology Research Section, Science, Industry and Business Library, The New York Public Library, Astor, Lenox and Tilden Foundations.

themselves. It is probably for these reasons that experimental and statistical graphs did not become popular until the 1830s.

Although Lambert searched for a universal characteristic for all of science, he did not draw any direct connection between his logical symbols and his experimental graphs. He referred to his graphs, both logical and experi-

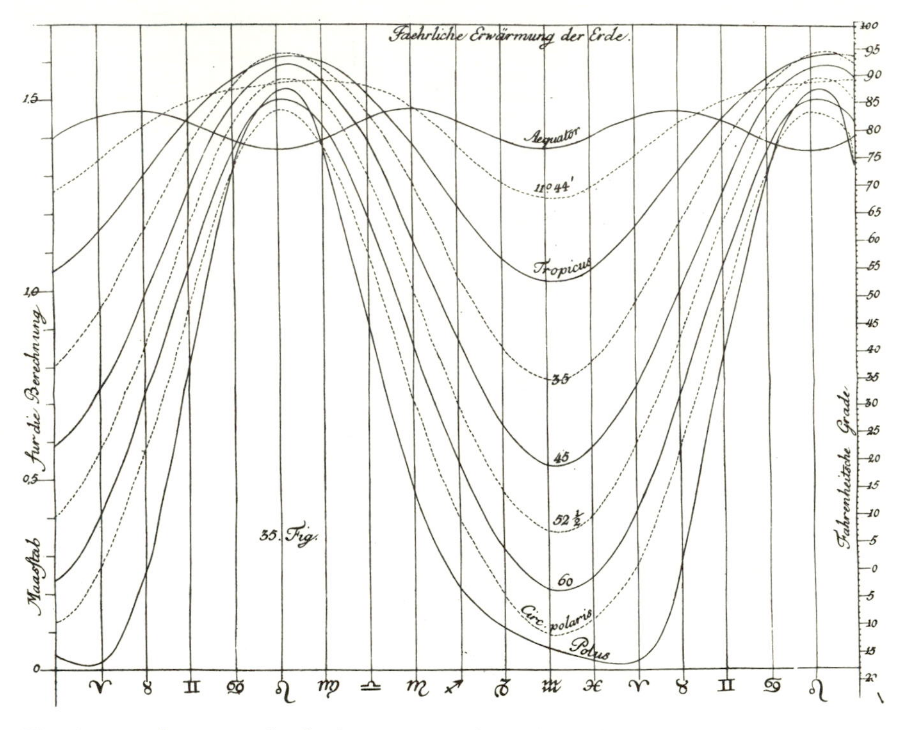

Fig. 6.6. Lambert's graph of solar warming throughout the year at different latitudes. From Lambert, *Pyrometrie*, fig. 35. Courtesy of the Science and Technology Research Section, Science, Industry and Business Library, The New York Public Library, Astor, Lenox and Tilden Foundations.

mental, as "figures" (*Figuren*). One would think that as an exponent of "figured" and "extended" concepts he would have associated his semiotics directly and unambiguously with his use of experimental graphs. We can only conjecture as to why he did not make the connection.

One possible explanation is that Lambert, like all of his predecessors, saw the search for a universal characteristic as a problem of logic. It was a problem of identifying the objects of knowledge and assigning appropriate signs to them so as to get them in the right categories with the proper headings and subheadings; it was a problem of creating the correct arrangement of pigeonholes and putting each object of knowledge in the correct one. In pursuing this taxonomical effort the projectors of universal languages created numerous tables. In the words of Mary Slaughter:

> Tables abound in the literature of the seventeenth century, from the tables of topoi and commonplaces to those of mnemonic systems, to those of the logic books, to those of anatomies, isogoges, natural histories, and so on. Many will recall here the innumerable tables of Ramus. The table is the graphic representation of taxonomic discourse, just as written words are the graphic representation of spoken discourse; and just as writing/printing are graphic representations of speech, so

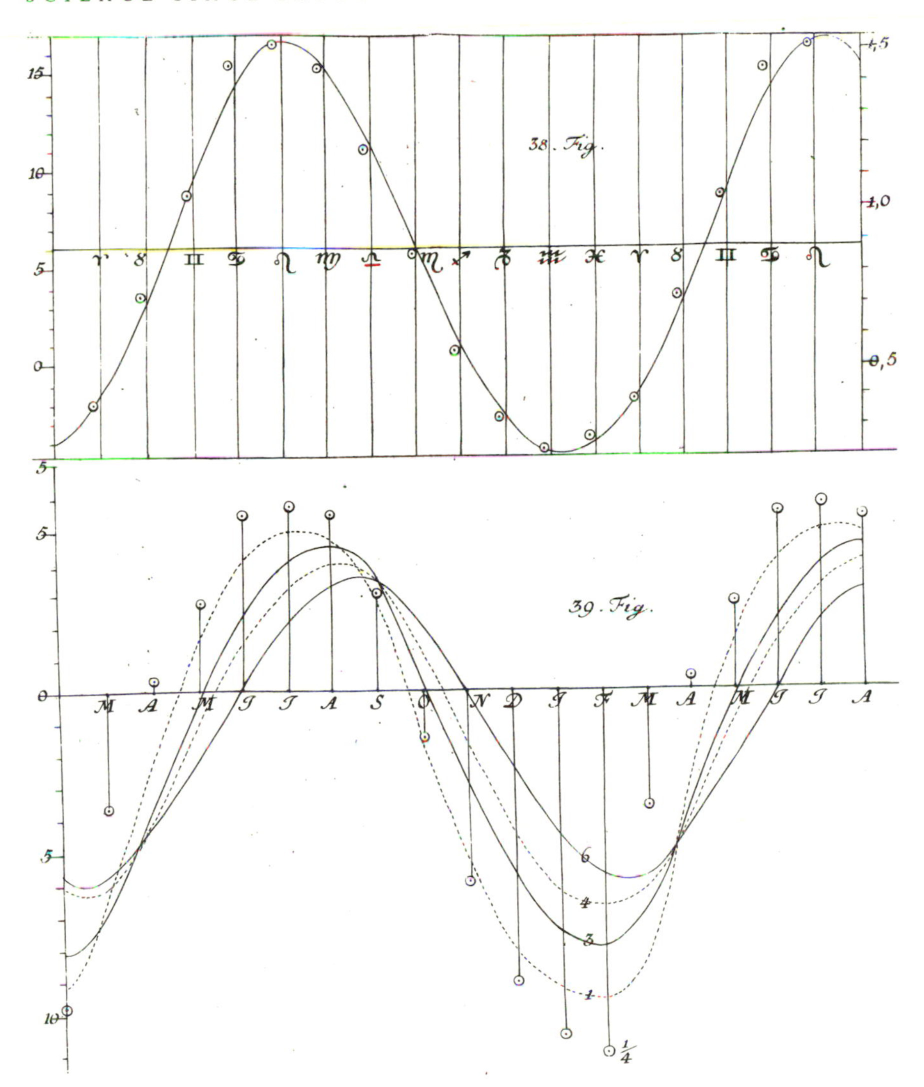

Fig. 6.7. Graphs of variation in soil temperature. From Lambert, *Pyrometrie*, figs. 38 and 39. Courtesy of the Science and Technology Research Section, Science, Industry and Business Library, The New York Public Library, Astor, Lenox and Tilden Foundations.

> the table is the graphic representation of classification and of the taxonomic structure.[22]

These traditional tables were all taxonomic systems of the pigeonholing variety.

However, tables are not all alike. Seventeenth-century books also contained tables of numbers, most notably astronomical tables. Tables of numbers were typically predictions based on theory—predictions of planetary

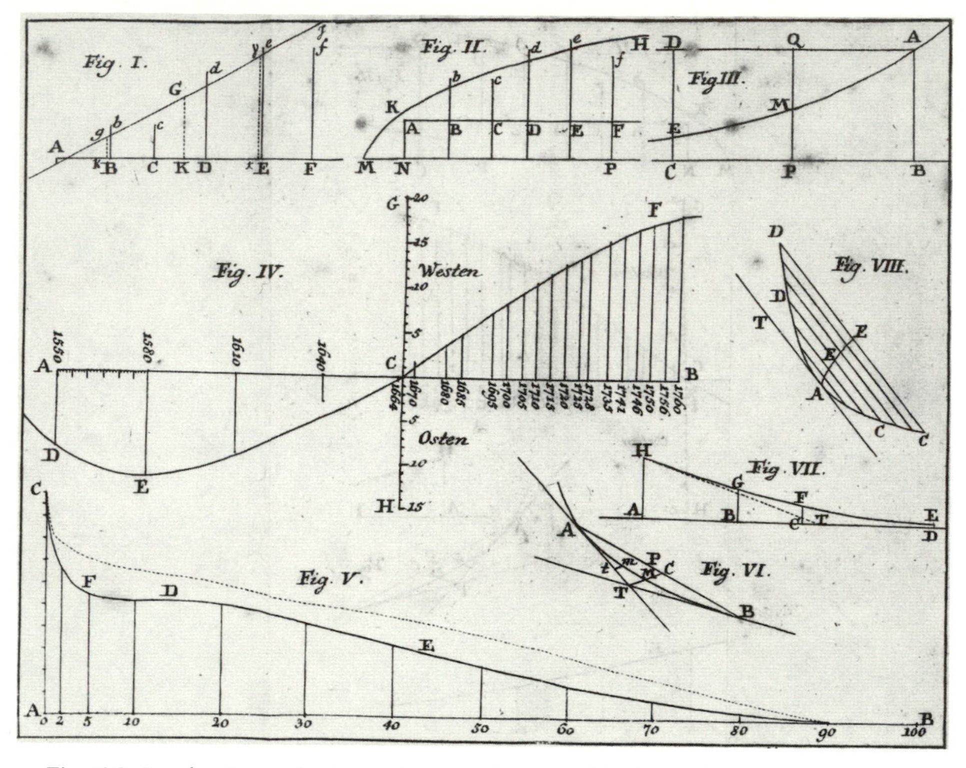

Fig. 6.8. Lambert's graphs from observed data. Fig. I is the variation in the seconds pendulum plotted against the square of the cosine of the latitude (theory indicates that the graph should be a straight line). Fig. IV is a plot of the variation of the compass declination with time, and Fig. V is a plot of the mortality rates in London between 1753 and 1758 as a function of age. The latter two graphs cannot be predicted from theory and must be drawn through the given data points. From Lambert, "Zuverlässigkeit der Beobachtungen und Versuche," in *Beyträge*, vol. 1, table 5. Courtesy of the Science and Technology Research Section, Science, Industry and Business Library, The New York Public Library, Astor, Lenox and Tilden Foundations.

positions, predictions of the tides, predictions of the length of the seconds pendulum at different latitudes, and the like—and they expressed some functional relationship between variables.

A third kind of table was quantitative data from experiments performed with measuring instruments. In this kind of table one looked not for the numbers themselves, but for some regularity or anomaly in the table that would reveal an unknown rule or functional relationship between the measured quantities. It was in this search for regularities or anomalies that experimental graphs were most useful.

Lambert's failure to describe a connection between his semiotics and his graphs indicates the conceptual difficulty of moving from a taxonomic approach to a functional approach, or, put in another way, from tables of the pigeonholing variety to tables that could be expressed as linear graphs. Always alert to the possibilities for extending mathematics into new areas, Lambert attempted to create a new mathematics of ordered systems. In his "Essai de taxéometrie" he proposed to measure by a fraction the extent to which any collection of items departed from an ordered system.[23] Thus Lambert carried his mathematical ideal even into taxonomy and systematics.

One can argue that a growing admiration of mathematics during the seventeenth and eighteenth centuries was the major spur to systematics and taxonomy during the same period. If this is true, the difference between the taxonomical and the functional approaches did not necessarily depend on whether one used mathematics. The difference came from contrasting ways of defining order.[24] Although Lambert was unable to resolve this contrast, his belief in the mathematical method and the universality of signs led him to express his results graphically, both in logic and in experiment.

William Playfair was more explicit in connecting graphs with language. His graphs first appeared in his *Commercial and Political Atlas; Representing, by Means of Stained Copper-plate Charts, the Exports and General Trade of England at a Single View* (1785). The book contained beautiful graphs of British trade over the previous twenty years (see figs. 6.9 and 6.10). The advantage of graphs as Playfair saw it was not that they gave a more accurate statement than tables, but that they gave "a more simple and permanent idea of the gradual progress and comparative amounts, at different periods, by presenting to the eye a figure, the proportions of which correspond with the amount of the sums intended to be expressed."[25] Playfair called his method "lineal arithmetic" and considered it an application of "the principles of geometry to matters of Finance." It was, therefore, a mathematical method, but he also recognized it as a universal language that made it possible to record and comprehend a great deal of information in a single diagram. He lamented that the ancients had not mastered the method, for "had records, written in this sort of shape, and *speaking a language* [emphasis added] that all the world understands, existed at this day, of the commerce and revenue of ancient nations, what a real acquisition would it not have been to our stock of knowledge?" Playfair claimed that his graphs were drawn for posterity, written in a language that any person might understand "even though a native of another country." They exhibited "the most extensive mercantile transactions that ever took place in the world, in a manner the most simple, easy, and comprehensive," and thus they preserved a financial record of the British Empire the like of which was, regrettably, unavailable for the only comparable empire, that of the Romans.[26]

In the above quotations Playfair referred to two important aspects of previous language schemes. One was the search for a figurative language, like Chinese or Egyptian hieroglyphics, that used figures instead of syllabic words. Pictures, or more abstract pictographs, could be "read" in any language and would thus be truly universal. The other aspect was the substitution of numbers for pictographs. Robert Boyle noted that numbers had the

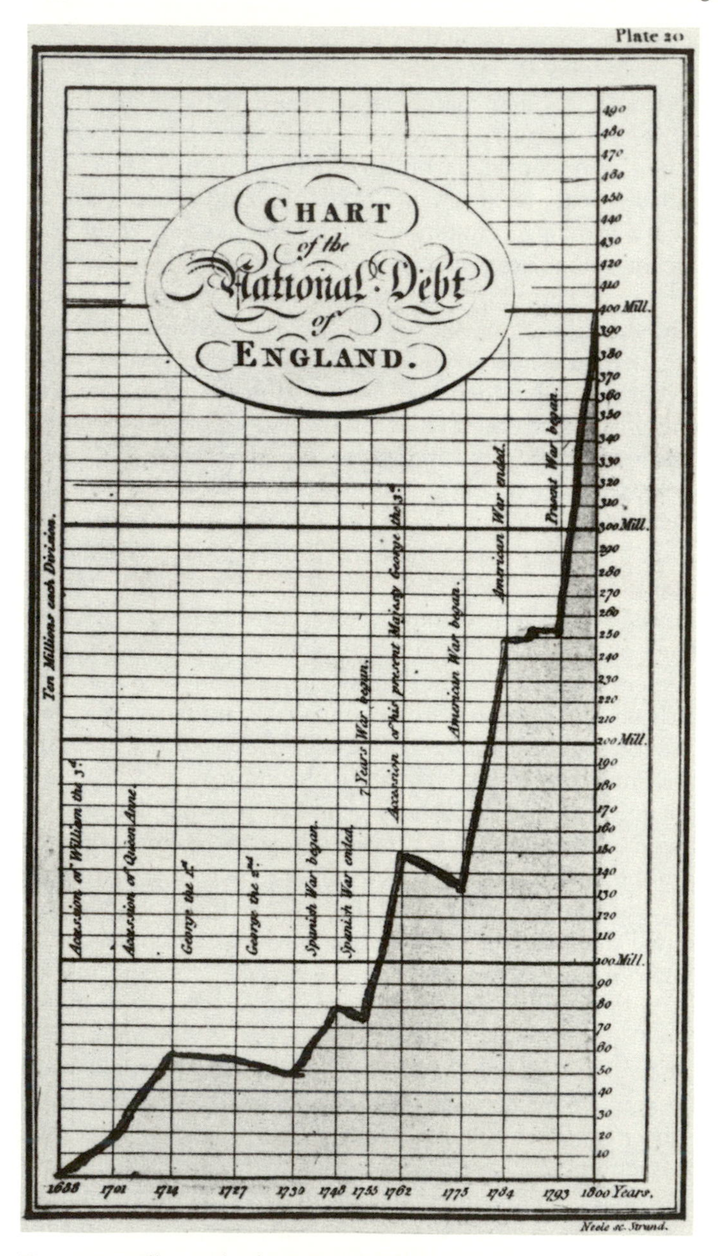

Fig. 6.9. William Playfair's graph of the British national debt from 1699 to 1800. From Playfair, *The Commercial and Political Atlas*, pl. 20, opposite p. 83. Courtesy of the University of Washington Libraries.

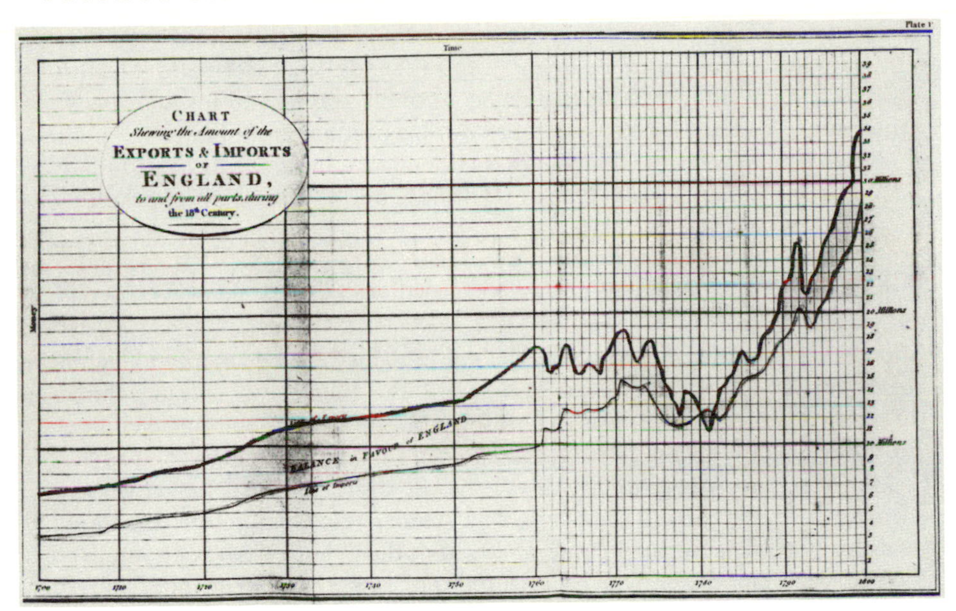

Fig. 6.10. William Playfair graph of England's balance of payments during the eighteenth century. From Playfair, *The Commercial and Political Atlas*, pl. 1, opposite p. 1. Courtesy of the University of Washington Libraries.

same meaning for all humans. Numbers could be used as signs, which could then be translated through a code book or dictionary into any language.[27] Robin Rider points out that the only universal language which was ever actually adopted was this kind of numerical language. It was used in semaphore systems that appeared in the last decade of the eighteenth century, first in France and then in England.[28] Playfair, who was in Paris during the early days of the Revolution, claimed to have brought the semaphore to England, although that honor is usually credited to Richard Lovell Edgeworth.[29]

Playfair also claimed to have searched for graph-drawing predecessors; he concluded that he was the first, at least for matters of finance. Others had constructed charts of weather data and chronologies, but nothing comparable to his lineal arithmetic, which would indicate that he had no direct knowledge of Lambert's earlier graphs. He may have had indirect knowledge of them, however, because he said on occasion that his older brother, John, had taught him that whatever could be expressed in numbers might be represented by lines, and had made him keep a temperature record that represented degrees by lines on a divided scale.[30] John Playfair was a mathematician, who is better known for his *Illustrations of the Huttonian Theory of the Earth* (1802). It is quite possible that he was familiar with Lambert's *Pyrometrie*, because Lambert's graphs were also temperature data. Moreover, William Playfair had had a career as a machinist and had worked as a draftsman for Bolton and Watt beginning in 1780. He may have taken the idea of graphs from Watt, although this seems unlikely, considering the early date (1785) of Playfair's *Commercial and Political Atlas*.[31] William Playfair was a shadowy (one might even say shady) character, and it is difficult to find any definite connections between his graphs and his many other enterprises. Perhaps it took someone with his peculiar combination of interests—that is, in universal languages, mathematics, mechanical recording instruments, and finance—to come up with the notion of statistical graphs.

Graphs and Automatic Recording Devices

Automatic recording instruments have a long history, but made no significant impact on natural philosophy until the nineteenth century. James Watt's indicator diagram was the most significant example of this technique in the eighteenth century, but Watt and Boulton kept it secret as long as they could and it was not widely known.[32] The indicator card moved in one dimension by linkage to the piston of a steam engine. Therefore its motion was proportional to the volume of steam in the cylinder. A pencil attached to a spring manometer, which measured the pressure in the cylinder, moved at right angles to the motion of the card (see fig. 6.11). The area of the figure drawn by the pencil on the card was proportional to the work done by the steam on each stroke of the piston. Watt's first "indicator" was a simple pressure gauge with which he noted the pressure in the cylinder when the piston was at different positions. He recorded this information in the form of tables. When he (or John Southern, to whom the invention is usually credited) devised the indicator card as an automatic recording instrument, the information was of necessity graphic, and the table gave way to a graph—a very important graph, since the pressure-volume diagram was at the heart of Émile Clapeyron's elaboration of the new thermodynamics.[33] In this case the use of a recording instrument required the transition from tables to graphs.

Around 1800, recording instruments became recognized as more than convenient ways of accumulating data. The automated weather stations of the seventeenth and eighteenth centuries relieved the tedium of taking observations, but they did not reveal any new information. The new recording devices of the nineteenth century revealed "secrets" that could not be obtained in any other way—certainly not by direct observation.[34] Automatic recording instruments appeared in two renovated sciences at the beginning of the nineteenth century and, to a large extent, created them. These were acoustics and experimental physiology. Instruments in both of these fields produced graphs, but the graphs produced by recording instruments differed from the graphs of Lambert and Playfair. Lambert and Playfair used their graphs to reveal relationships between two variable quantities, while recording instruments gave a *figure* that was not in the first place a mathematical relationship. As the nineteenth century progressed, these two graphical methods became less distinct and merged into a single method in

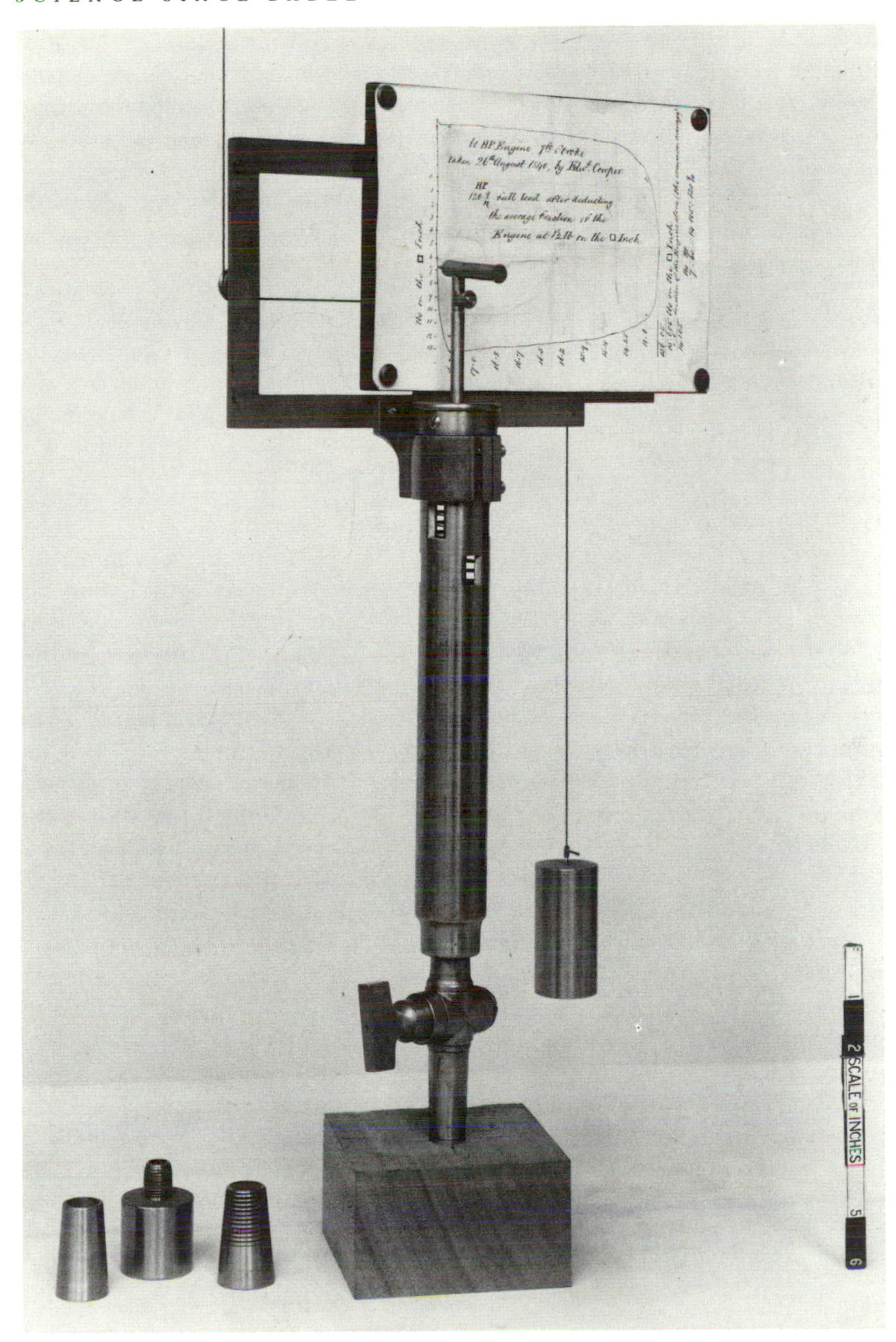

Fig. 6.11. James Watt's indicator diagram. Courtesy of the Science Museum/Science & Society Picture Library.

the hands of Étienne-Jules Marey, as we shall see. Additionally, as the value of these techniques for modern science came to be appreciated, numerous critical investigators and inventors of recording devices conceived this brand of experimental data as a *new scientific language*—a language inscribed by nature.

Galileo authored the most famous appraisal in the history of science of the sort of text that nature offered its students. "Philosophy," he wrote in *The Assayer*,

> is written in this grand book, the universe, which stands continually open to our gaze. But the book cannot be understood unless one first learns to comprehend the language and read the letters in which it is composed. It is written in the language of mathematics, and its characters are triangles, circles, and other geometric figures without which it is humanly impossible to understand a single word of it; without these, one wanders about in a dark labyrinth.[35]

Galileo has also been credited with the earliest attempt to create an automatic record of an acoustic phenomenon. In his *Discourses Concerning Two New Sciences*, Salviati (the character serving as Galileo's spokesman) described how "scraping a brass plate with a sharp iron chisel" occasionally produced a "whistling" sound and formed on the plate "a long row of fine streaks parallel and equidistant from one another."[36] However, Galileo did not regard these chisel marks as the true language of the phenomenon. Their importance in his discussion was the *countability* of the streaks and the relationship between their *number* or density and the particular pitches of whistles with which they were associated. Unlike Galileo's example, the novel characteristic of later researches in acoustics (and in physiology) was the intrinsic value attached to the *shapes* that inscriptional apparatus could produce. In addition to the measurements that recording devices could yield, their images and inscriptions offered a qualitatively different type of scientific information.

The work of Ernst Florens Friedrich Chladni is often considered the starting point of modern acoustics. His principal discovery—his sand figures (announced in 1785)—marked the start of a sustained tradition in which the visual representation of sound became a ubiquitous feature in acoustical research. Chladni's *Klangfiguren* were formed on variously shaped glass or metal plates that had sand spread over their surfaces. When the experimenter set the plates in vibration, by running a violin bow against their edges, the sand would be thrown off the quivering areas and deposited along stationary nodal lines. These sand patterns, which could become quite complex and visually stunning, became a popular demonstration throughout the following century (see figs. 6.12 and 6.13).[37]

Chladni conducted his experiments in Wittenberg, where he wrote several works on acoustics, including his compendious *Die Akustik* (1802). He hit upon his technique while trying to find an acoustic analogue to Georg

Fig. 6.12. Making a Chladni figure. From Deschanel, *Elementary Treatise on Natural Philosophy*, p. 787. Courtesy of the University of Washington Libraries.

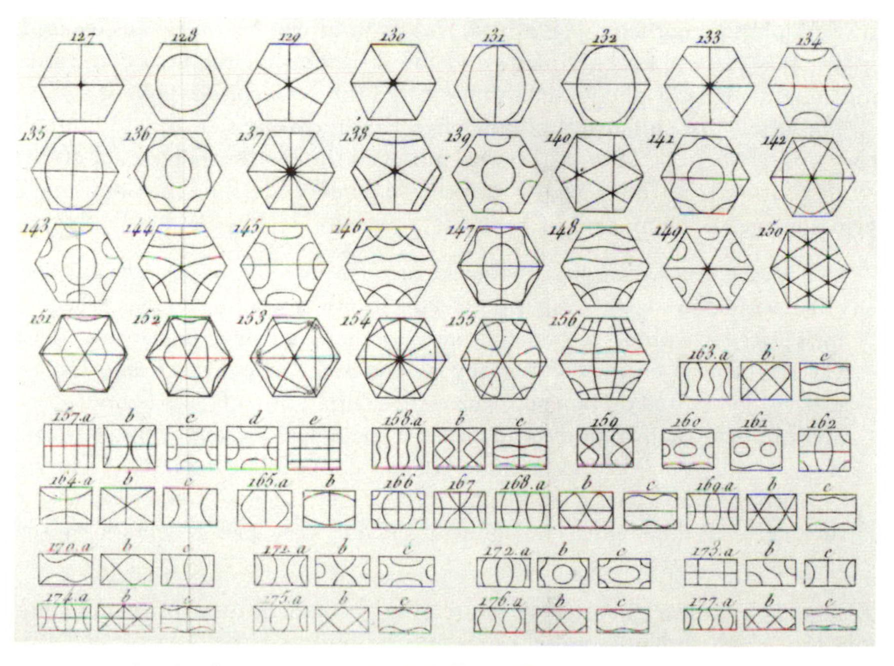

Fig. 6.13. Chladni figures. From Chladni, *Traité d'acoustique*, pl. 6. General Collections, Princeton University Libraries.

Lichtenberg's electrostatic figures—patterns of dust particles that collected on a charged cake of resin.[38] Such an aim coincided with a *naturphilosophische* search for symmetries and signs of hidden relationships among natural forces. This very problem was explored in 1806, when the prominent *Naturphilosoph* Hans Christian Oersted—who would become famous for his discovery of electromagnetism in 1820—used Chladni's technique in a further effort to disclose a connection between sound and electricity.[39] Both Lichtenberg's and Chladni's figures intrigued another romantic physicist, Johann Wilhelm Ritter. In the *Klangfiguren*, Ritter saw nature's own form of script—ur-images and hieroglyphs that constituted the true alphabet of the "Book of Nature."[40] Ritter wrote to Oersted: "Each sound has a letter associated with it, and we wonder if we do not hear writing—read when we hear . . . and is not every seeing with an inner sight hearing, and every hearing a seeing from within! . . . Let us ask ourselves, how do we transform our thoughts, our ideas, into words; and do we ever have a thought or an idea without its hieroglyph, its character, its script?"[41] He held the opinion that material images, like Chladni's figures, entailed the true language—a *pictorial* language—of science. Ritter reveled in the pure multiformity of the *Klangfiguren*, their symmetry, and their relationship to other forms in nature. While the mathematical approach to sound was by no means excised, it was this respect for the image and the attitude that pictures could give meaningful signs of phenomena that excited the *Naturphilosophen*.

The gifted British natural philosopher, Rosetta stone sleuth, and undulatory optical theorist Thomas Young embraced the pictorial approach to the study of sound. In 1800, Young introduced a new technique for obtaining a visual image of the motion of a vibrating string.

> Take one of the lowest strings of a square piano forte, round which a fine silvered wire is wound in a spiral form : contract the light of a window, so that, when the eye is placed in a proper position, the image of the light may appear small, bright, and well defined, on each of the convolutions of the wire. Let the cord be now made to vibrate, and the luminous point will delineate its path, like a burning coal whirled round, and will present to the eye a line of light, which by the assistance of a microscope, may be very accurately observed. According to the different ways by which the wire is put in motion, the form of this path is no less diversified and amusing, than the multifarious forms of the quiescent lines of vibrating plates discovered by Professor Chladni.[42]

Young also pioneered a means for creating permanent inscriptions of sonic vibrations, which he described in his natural philosophy textbook of 1807:

> The situation of a particle at any time may be represented by supposing it to mark its path, on a surface sliding uniformly along in a transverse direction. Thus if we fix a small pencil in a vibrating rod, and draw a sheet of paper along, against the point of the pencil, an undulated line will be marked on the paper, and will correctly represent the progress of the vibration.[43]

In the nineteenth century, Wilhelm Weber and Guillaume Wertheim, as well as many other investigators, devised related ways to preserve the traces of styluses attached to sounding bodies, such as rods and tuning forks.[44]

The British physicist Charles Wheatstone came from a family of musical instrument makers, and his early scientific writings dealt exclusively with acoustics. He created images of vibration with an instrument that, like Young's piano wire, required the persistence of vision of a lustrous point. Wheatstone's "kaleidophone"—its name an homage to David Brewster's kaleidoscope—consisted of silvered glass beads, and other reflective objects, fixed to the ends of prismatic or circular rods. When set in vibration, the tips of these rods produced "a great variety of pleasing and regular forms," thus providing a "combination of philosophy with amusement," a feature found in many of Wheatstone's creations.[45]

While Young and Wheatstone did not expressly identify their "graphs" as languages, their acoustic techniques coincided with and informed their studies in language, speech, and vision. Young began his career with an attempt to understand the physiology of vision and the nature of vowel sounds; it ultimately embraced his effort to decipher Egyptian hieroglyphics.[46] Wheatstone also studied cryptography and invented communication devices including the telegraph. Their creation of instruments to give visual representations of sound was part of a much broader investigation into the nature of language, speech, vision, and hearing. It is worth noting that if one wants to measure and record the quantitative features of sound, one must employ a visual image. The ear detects pitch, loudness, and timbre, but not the frequency, amplitude, and shape of sound waves. Recording instruments give us this information by representing the sound visually.

One of the most important instruments of nineteenth-century acoustics was the *phonautograph* of Édouard-Léon Scott de Martinville. This device was conceived in an attempt to fuse instruments, language, and mechanical inscriptions. Scott was a typographer who became interested in the preservation of speech *written* in its own natural language, rather than in the artificially constructed characters that appear on a printed page. A manifest expression of his passion may be seen in his 1849 book *Histoire de la sténographie*, which traced the development of conventions for shorthand and bespoke his initial efforts to create a precise and universal inscription of speech sounds.

In his introduction to the *Histoire*, Scott mourned the loss of the multitude of spoken words that never become fixed on the printed page—a loss that must have seemed especially sharp to a typesetter.

> What more beautiful satisfaction could indeed be offered to the savant, to the man of letters, than that he would receive a means of recalling instantaneously that which strikes him in a discourse, in an improvisation, in a scenic representation—a means that would permit the poet, the dramatist, the novelist, to fix at will his inspirations, brilliant but always so fugitive, which sometimes come to illuminate

> his mind, and which he is unable to relocate in his memory in their first color. To fix these thoughts as quickly as they are presented would be for him a means of making himself their master and of increasing the activity of his imagination. If one adds to this the fact that stenographic writing occupies little space and can be used as a secret writing, one will understand why the creation of such a precious instrument has been found worthy of study by our greatest savants, such as Leibniz, Porta, Condorcet, etc., sustained perhaps by the hope of resuscitating a skill formerly so flourishing.[47]

Scott supposed that a reformation of the stenographic art would allow for the instantaneous preservation of one's thoughts and observations.

As his subtitle indicates, his goal was to create a means of writing words as rapidly as they were spoken. The guiding principle of this program was that "the pen should not have to make more movements to trace the words than the vocal organ does to pronounce them."[48] Scott would eventually make this ambition a reality.

In a later book Scott's efforts are closer to linguistics and philology than to acoustics. *Les noms de baptême et les prénoms* (1857) reviewed the history of personal first names as an exercise in history and ethnography. Yet Scott also entertained the hoary philosophical notion of a relationship between the essence of one's individual character and one's name. Plato's thesis, presented in the *Cratylus*, that names have a prophetic power in dictating the course of an individual's life, would be, Scott explained, "without doubt little in harmony with the doctrines of our century, and we will be careful not to support it. Nevertheless, we believe that there is in the choice of the name one bears an imperceptible influence, which has its source, not in philosophical mysticism, but rather in the profound and secret order of things and in the very constitution of our moral self."[49] Scott did not purport to unravel the question of a fundamental connection between names and personalities or between words and things, and yet he did believe that words had a special significance that went beyond mere convention. As a stenographer and typesetter he saw as his task the rapid and precise recording of these precious spoken words in an unambiguous written form.

Since 1854, Scott had been planning the construction of a device to automatically transcribe vocal sounds. He had produced a functioning model by the beginning of 1857, and he submitted a discussion of his results in a sealed packet to the Académie des sciences on January 26 of that year. The packet contained his essay "Principes de phonautographie," which described his researches on *l'écriture acoustique*. His comments echoed the earlier poignant plea for the preservation of speech. "Is it possible," he asked,

> to achieve for sound a result analogous to that attained presently for light by photography? Can one hope that the day is near when the musical phrase escaping from the lips of the singer will come to write itself . . . on an obedient page and

> leave an imperishable trace of those fugitive melodies that the memory no longer recalls by the time that it searches for them? Between two men joined in a quiet room, could one place an automatic stenographer that preserves the conversation in its most minute details. . . . Could one conserve for future generations some traits of diction of our eminent actors, who now die without leaving after them the feeblest trace of their genius? The improvisation of the writer, when she rises in the middle of the night, could she recall the day after with all her freedom, that complete independence of the pen so slow to translate an ever-fading thought in her struggle with the written expression?[50]

That the air carried the vibrations of articulate sounds was no mystery to physicists. However, the problem at hand was, according to these principles, "to construct an apparatus that reproduces by a graphic trace the most delicate details of the movement of sonorous waves. . . . [and] with the help of mathematics, to decipher this natural stenography."[51]

Scott's invention achieved this inscription by mimicking the structure of the human ear—a model that occurred to him while he was proofreading a plate of drawings of auditory anatomy for a physics textbook.[52] The phonautograph consisted of a paraboloid collecting chamber, one end of which was open, while the other was covered with a thin elastic membrane—his surrogate tympanum. The acoustic stimulation of this diaphragm activated a system of ossicle-like levers and a stylus whose motion would be traced on a steadily moving paper, wood, or glass surface coated with lampblack (see figs. 6.14 and 6.15).[53]

Describing his invention to the Société d'encouragement in November 1857, Scott explained his ambition "to force nature to constitute herself a general written language of all sounds."[54] He knew that phonautograph traces did not translate easily to readable words, but he did believe that he had made an important first step. "I saw the book of nature open before the gaze of all men and, as small as I am, I believed that I would be able to read it."[55] He ultimately realized that these traces would not provide the "natural stenography" that he sought, yet he thought the instrument and its representations could be used in the study of sound and in the analysis of timbre.[56] The scientific utility of the phonautograph increased when the acoustic instrument maker Rudolph König contributed some improvements to the collecting horn, membrane, and recording drum. In the study of vowel sounds, the phonautograph came to be widely used by many scientists, including F. C. Donders, Heinrich Schneebeli, and René Marage.[57] The phonautograph also spawned a new generation of imaging tools for acoustical analysis, such as König's manometric flame, which guided his investigation of timbre.[58]

The only "natural stenography" that approached the ultimate goal of Scott's work was the phonograph recording. Edison contrived the phonograph in 1877, twenty years after the introduction of the phonautograph, and the American inventor was undoubtedly familiar with the older device.

Fig. 6.14. Scott's phonautograph. From Pisko, *Die neuren Apparate der Akustik*, p. 73. Courtesy of the University of Washington Libraries.

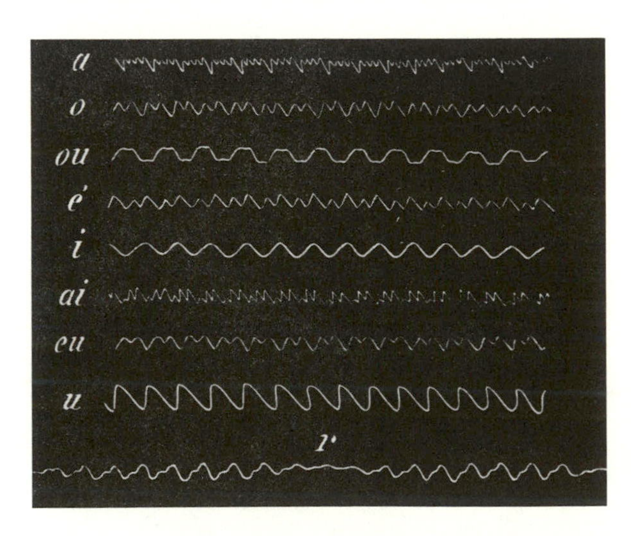

Fig. 6.15. Schneebeli's phonautograph vowel traces. From Schneebeli, "Expériences avec le phonautographe," p. 81. Courtesy of the University of Washington Libraries.

However, Edison conceived the phonograph in a slightly different technological context—his aim was the preservation of incoming telegraphic messages.[59] Not surprisingly, Scott viewed Edison's "invention" as technological plagiarism. Although his patents had expired, Scott defended his priority in a book he published in 1878, *Le problème de la parole s'écrivant elle-*

même—a collection of his publications on the phonautograph. In his embittered introduction, Scott insisted that he, as well as another French inventor, Charles Cros, had presented the essential pieces of the phonograph before Edison had.

Furthermore, Scott's criticism of Edison was marked by an enduring obsession with *writing*. According to Scott, Edison's work failed to attain its self-described goal. Although named a "phonograph," the device merely reproduced sound—it was not a *sound-writer*.[60] "The impression produced by the stylus of the phonograph," Scott argued, "is a singular hieroglyph that will wait a long time for its Champollion. I propose to call these microscopic traces *phonéglyphes*."[61] Scott's mind was still fixed on his belief that the printed transcription of speech—not the reproduction of sounds—would be the greater benefit to civilization.

Recording instruments became important in experimental physiology and medicine at approximately the same time that they became important in acoustics. Instruments such as the microscope, the thermometer, and the stethoscope had already replaced or augmented the judgment of the human sense organs in the practice of medicine. The reliance on artificial probes into the nature of living things increased during the course of the nineteenth century with the addition of devices to record what was being found. As in the case of acoustics, these instruments produced graphs.

The application of graphical techniques in physiology stems from Carl Ludwig's *kymograph* of 1846—an instrument that replaced the trained fingertips of physicians in the study of the pulse.[62] With this device, a mercury manometer attached directly to the artery of a dog would rise and fall according to the arterial pressure. A stylus attached to a float on top of the mercury column traced the undulations on a rotating smoked drum.[63] Subsequently, noninvasive techniques were developed for the study of the pulse, beginning with Karl Vierordt's *sphygmograph* in 1855.[64] In 1859, the French physiologist Étienne-Jules Marey greatly improved this device. His sphygmograph was small and was strapped to the patient's arm. Its central feature was a spring that rested on the radial artery and transmitted movements to a light recording arm, which inscribed the motion on a moving glass plate (see fig. 6.16).[65] Marey's sphygmograph was reliable and convenient enough to be used as a tool in clinical medicine. These devices constituted the first technological steps in a sustained tradition of recording graphically the motions and phenomena of circulation—efforts that included the development of the electrocardiograph.[66] Researches on other physiological questions also resorted to graphical recording devices. Within a decade of the appearance of the kymograph, Carlo Matteucci and Hermann von Helmholtz (with a device he dubbed the *myograph*) each began to use inscription techniques in their studies of muscular contractions.[67]

As the case had been in acoustics, the introduction of recording devices did not simply provide quantitative data that had once been inaccessible. The graphic trace presented this information in a new guise—one whose

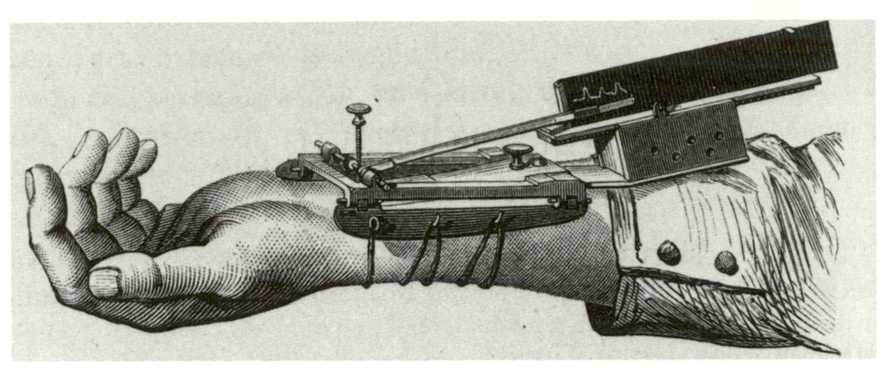

Fig. 6.16. Étienne-Jules Marey's sphygmograph recording the pulse. From Marey, *La méthode graphique dans les sciences expérimentales*, p. 560. Courtesy of the University of Washington Libraries.

meaning was embodied in its *shape*. Furthermore, recording devices in physiology, like those of acoustics, were often viewed as the mediators of a newly discovered scientific language. In 1867, John Burdon-Sanderson and Francis Edmund Anstie, British physicians who modified the sphygmograph and helped to promote its clinical utility, described in the following way problems involved in using the instrument: "The difficulty lies in the fact that the record is written in a language which we are only beginning to understand. Without a proper knowledge of the physiological facts, of which they are the transcript, the oscillations of the lever are quite as meaningless as the vibrations of the telegraphic needle to one who is not furnished with a proper alphabet."[68]

Étienne-Jules Marey gave the most notable description of the traces of recording instruments as a new language.[69] He was also the nineteenth century's most strident advocate of their use. Marey began his work with the study of circulation but later extended his field of inquiry to include all human and animal motion. The technologies he helped to develop took him from graphical inscription to the beginnings of cinematography. The fullest articulation of Marey's scientific goals was spelled out in the introduction to his voluminous 1878 survey of this burgeoning new technology, *La méthode graphique dans les sciences expérimentales et principalement en physiologie et en médecine*.

According to Marey, two factors impeded the development of science. The first of these was the fallibility of the human organs of sense and their lack of subtlety for detecting truths about nature. The second was the insufficiency of language for expressing and transmitting this knowledge. Through the approach he called the *graphical method*, Marey confronted both of these obstacles at the same time.[70]

Instrumentation, he explained, had disclosed and corrected the limits of the human senses, but precision instruments still could not follow rapid dynamic processes. "Movements, electric currents, variations of gravity or temperature," Marey wrote, "such is the field to explore."

> In this new enterprise, our senses, with perceptions that are too slow and confused, can no longer guide us, but the graphical method substitutes for their insufficiency; in this chaos, it reveals an unknown world. Inscription apparatus measure the infinitely small pieces of time; the most rapid and the most weak movements, the slightest variations of forces cannot escape them. They penetrate the intimate function of organs where life seems to express itself by an incessant mobility.[71]

With instruments that recorded their own inscriptions, it was possible to follow the faintest and fastest changes in the organs of living things. The graphical method also could solve the problem of communicating this new information unambiguously. Marey claimed that language has been a source of confusion in the practice of science. "Born before science and not having been made for it," he wrote, "language is often improper for the expression of exact measures, of well defined relationships."[72] An advanced level of civilization was marked by its use of "graphic expression." With this term, Marey was not referring simply to *writing*, which fixed the signs of language on stone or paper. He was also concerned with "natural writing [*le graphique naturel*]: that which, in all epochs and among all peoples, has represented objects in the same manner, which allows us to follow on the stelae of Egypt the scenes of a civilization that has disappeared. This graphic representation, if it were applied to the representation of ideas as to the figuration of objects, would constitute the true universal language."[73]

The graphical method provided precisely the sort of clear, accurate, and unambiguous communication that Marey's universal language demanded. With the graphical method, the heuristic value of graphs was combined with the painstaking exactness and sensitivity of automatic recording devices. Thus, Marey wrote, the graphical method "translates" natural phenomena into "a striking form that one could call the language of phenomena themselves, as it is superior to all other modes of expression."[74]

Marey recognized that many would criticize his attempt to substitute machines for human intelligence.[75] For Marey, however, ordinary language—no matter how elegantly employed—could not match the scientific utility of his new idiom: "Let us reserve the insinuations of eloquence and the flowers of language for other needs; let us trace the curves of phenomena that we want to know and compare them; let us proceed in the manner of the geometers whose demonstrations are not discussed."[76]

In the tracings of recording instruments, Marey discerned the fulfillment of Scott's dream of a "natural stenography." The rolls of paper and the pieces of smoked glass may not have contained a perfected form of human language, but they promised something greater. Etched on these surfaces were signals from nature more accurate than human senses could detect, and in a new universal, unambiguous, and precise language of science.

Of course Marey was overly optimistic. The images produced by instruments are thoroughly constructed and mediated through the skills and conventions incorporated in the design of the apparatus. Marey presumed that both pictorial and graphical representation entailed a perfect correspondence between the object or phenomenon and its resultant image. Like established instrumental techniques, styles of depiction, or the nonuniversal languages that humans speak and write, the language of the graphical method is embedded in a network of conventions. The signs and images produced by these means can be neither automatically understood nor eternally meaningful. Their intended value can be grasped only when one possesses, as Burdon-Sanderson and Anstie insisted for initiates to the sphygmograph, "a proper alphabet."

Graphs and Semiotics

At the same time that Marey was attempting to establish a nonverbal graphical method for the natural sciences, Charles Sanders Peirce in the United States was creating a new philosophy called "semiotics" that saw all of human experience as a system of signs. According to Peirce, "the woof and warp of all thought and all research is symbols, and the life of thought and science is the life inherent in symbols."[77] Like Lambert, Peirce emphasized that the signs used in communication were not limited to language. Gestures, cries of alarm, lightning before thunder, pictures—all are signs that have significance for the individual perceiving them. In Peirce's view, logic in its general sense, that is, the method by which we reason on our perceptions, "is . . . only another name for *semiotic* (σημειωτική), the quasi-necessary, or formal, doctrine of signs"; it is "quasi-necessary" because semiotics also covers reasoning that is not deductive.[78] During the twentieth century semiotics has grown into a major field of study that has focused primarily on linguistics, cultural anthropology, and literature. For its founder, however, its major value was in logic, mathematics, and the method of the natural sciences.

Central to Peirce's semiotics is his system of logical diagrams or "existential graphs." He calls it "my chef d'oeuvre."[79] In a brief historical introduction, he credits Lambert with having created the first system of logical graphs in the *Neues Organon*, which was followed eight years later by Leonhard Euler's similar and better-known system. Peirce owned a copy of Lambert's *Neues Organon*, which included the *Semiotik*, and it is probably from Lambert that he took the name and at least some of the inspiration for his new philosophy.[80] Peirce and Lambert had much in common although their lives were separated by a century. Both studied logic and language as a system of signs, and both were familiar with instruments and precise measurement. Beginning in 1872 Peirce directed gravimetric measurements for the U.S. Coast and Geodetic Survey, which meant making extremely precise pendulum measurements throughout the United States and Europe. He also undertook photometric measurements at the Harvard Observatory in order to determine more precisely the magnitude of important stars. In both cases he made measurements with delicate instruments and used his mathematical skill to improve the underlying theory.[81] It was precisely this kind of measurement, especially photometry and its mathematical reduction, that had occupied Lambert.

Peirce was one of the first to call his diagrams "graphs." The word "graph" was apparently coined by the mathematician J. J. Sylvester in 1878 in peculiar circumstances. Sylvester drew parallels between diagrams of chemical bonds in molecules and graphical representations of mathematical invariants and covariants of binary quantics (see fig. 6.17). The idea came originally from W. K. Clifford, but it was Sylvester who named these algebraic-chemical diagrams "graphs."[82] In 1906 Peirce stated, "By a *graph* (a word overworked of late years), I, for my part, following my friends Clifford and Sylvester, the introducers of the term, understand in general a diagram composed principally of spots and of lines connecting certain of the spots."[83] These "spots and lines" were obviously not Playfair's "lineal arithmetic" or Whewell's "method of curves." "Graphs," as that word was used in the late nineteenth and early twentieth centuries, referred to chemical diagrams, logical symbols, and the mathematical forms that became known as "graph theory."[84] Peirce's complaint that the word was "overworked of late" meant that by 1906 its use had begun to spread beyond its original meaning to include experimental and statistical graphs.

Peirce's own mental processes must have been extremely graphic, because he concludes that in all reasoning, the mind forms diagrams in the imagination that are then "experimented" on through the addition of new constructions until the diagram represents a conclusion to the proposed problem:

> We form in the imagination some sort of diagrammatic, that is, iconic, representation of the facts, as skeletonized as possible. The impression of the present writer is that with ordinary persons this is always a visual image, or mixed visual and muscular. . . . This diagram, which has been constructed to represent intuitively or semi-intuitively the same relations which are abstractly expressed in the premisses, is then observed, and a hypothesis suggests itself that there is a certain relation between some of its parts—or perhaps this hypothesis has already been suggested. In order to test this, various experiments are made upon the diagram, which is changed in various ways. . . . and it is seen that the conclusion is compelled to be true by the conditions of the construction of the diagram. This is called "diagrammatic, or schematic, reasoning."[85]

Peirce refers to Kant's schemata as examples of this kind of reasoning, but it is apparent that Peirce's own semiotics most closely describes it.[86] According to Peirce, experiments on diagrams "take the place of the experiments upon real things that one performs in chemical and physical research. Chemists have ere now, I need not say, described experimentation as the putting of questions to Nature. Just so, experiments upon diagrams are questions put to the Nature of the relations concerned."[87] Thus Peirce sees all reasoning as a kind of mental "experiment" with diagrams.

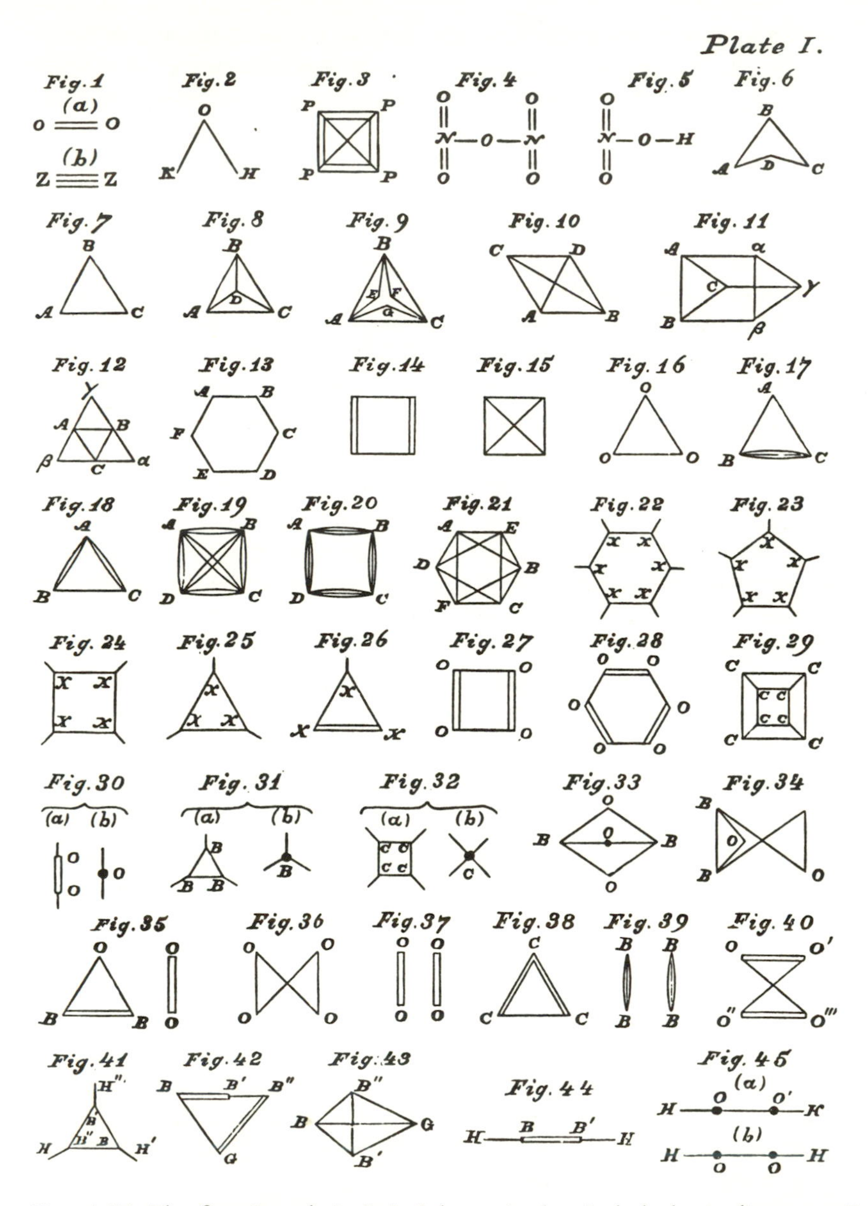

Fig. 6.17. The first "graphs": J. J. Sylvester's chemical-algebraic diagrams. From Sylvester, "On an Application of the New Atomic Theory to the Graphical Representation of the Invariants and Covariants of Binary Quantics," facing p. 82. Courtesy of the University of Washington Libraries.

Peirce believes that any diagram is primarily "iconic" and therefore different in kind from language, which is primarily "symbolic."[88] He defines a diagram as "a representamen which is predominantly an icon of relations and is aided to be so by conventions. Indices are also more or less used. It should be carried out upon a perfectly consistent system of representation, founded upon a simple and easily intelligible basic idea." What we have been calling an experimental graph is, for Peirce, a diagram, not a graph.

Peirce uses experimental graphs as would be expected of any working scientist in the late nineteenth century, but never to our knowledge does he analyze them in terms of his semiotics. This is the same unexpected disparity that we saw in the case of Lambert. Peirce, like Lambert, probably saw experimental graphs as being obvious compared to the logical problems raised by his existential graphs. Both men regarded semiotics as a new approach to logic and epistemology. Their major quarry was the operation of the mind, not the proper method of expressing experimental or statistical data.

Practicing scientists, however, know nothing of Peirce's existential graphs, while experimental and statistical graphs have become a major, if not *the* major, mode of scientific communication. In fact experimental and statistical graphs have become the hallmark of "scientific" explanation and are often used rhetorically to bring the authority of natural science to any argument, just as, in the seventeenth and early eighteenth centuries, arguments stated in the form of geometrical proofs benefited from the authority of geometry.[89] Experimental and statistical graphs appear regularly in newspapers and popular literature because they are easy to understand and because they purport to present "fact."

While the significance of a graph is easy to grasp (once the convention has been mastered), the semiotic structure of a statistical or experimental graph is complex. Its purpose is to expose a lawlike relationship, or an anomaly, in the midst of a cacophony of complex signs. It makes one sign stand out above all others. This is true for graphs made by recording instruments as well as for experimental and statistical graphs. The physician listening to the patient's heartbeat detects a telltale murmur—a sound that he knows from his experience is a "sign" of illness. He calls for an electrocardiograph. An instrument draws a trace on paper. The shape of the graph is an "indexical representamen" of the illness. The physician does not draw the graph. An instrument draws it. Nor does the physician know in detail how the instrument operates. And yet he knows what the sign drawn by the electrocardiograph is telling him. The signs that made his diagnosis possible include words and diagrams in the physician's textbooks in medical school, words communicated between the physician and the patient, electrical signals in the patient's body and in the electrocardiograph itself, and marks on the strip chart coming from the machine, but the physician need focus only on the graph and see how it differs from a given norm. Rather than penetrate into the labyrinth of signs that led to this graph, he notes the anomaly and completes his diagnosis.

In applying semiotics to language, literature, myth, and kinship one may conclude with many modern critics that these structures are human creations and that any attempt to find an objective reality at the heart of the labyrinth of signs is an exercise in futility. It is like unrolling a ball of string. The string is a sequence of signs, all of which signify other signs by conventions assigned to them. At the end of the string one hopes to find the object that is being signified, but when one comes to the end of the string there is nothing. It is string all the way down, and the ball is gone.

The argument that our world is of our own construction raises the question "Does a recording instrument write its own signs, or does it write only what the operator tells it to write?" When I write a letter I use an instrument—a pen. The pen lays the ink on the paper, but we say that I am doing the writing. A draftsman uses a pair of compasses to inscribe a circle of a particular radius. The instrument allows the draftsman to draw a more perfect circle than he could otherwise draw. The instrument adds something to the sign, but the draftsman knows that it is circles that he wants. He has circles "in mind," and therefore it is he who draws the circles. If one seeks the objects signified by these signs, they are to be found in the mind operating the instrument. These signs are reflexive in the sense that the instrument presents the operator with a sign that signifies an object in his mind. The object being signified and the idea created in the mind by the sign (the "interpretant" in Peirce's nomenclature) are in the same place. Logical diagrams, like Peirce's existential graphs, are also reflexive in that they indicate to the mind (and to other minds) how the mind works.

Experimental graphs and the instruments that produce them are quite different. The natural scientist can accept the notion that our experience is all signs. He can even accept the possibility that the relations between the object and the sign representing it are largely human conventions, but he cannot accept the argument that they are *all* convention. The electrocardiograph is a human construction; the theory that allows it to be built is also a human construction; the concepts that make its graph intelligible are human constructions; and yet the telltale anomaly in the trace drawn by the instrument points to an object that is not a human construction. In fact the purpose of the instrument is to signify that external object as clearly and unambiguously as possible and to "black-box" all the man-made structure in between.

Peirce liked to draw mazes that mirrored the complexity of semiotic systems. In his drawing of the labyrinth of signs the Minotaur stands at its heart (see fig. 6.18). The literary critic or cultural anthropologist may conclude that the Minotaur is another human creation—part of the structure of myth. But a scientist will argue that if the labyrinth represents the semiotic system behind an experimental graph, the object that the Minotaur signifies is beyond our ability to construct or deconstruct.[90]

Fig. 6.18. C. S. Peirce's sketch of the labyrinth of signs. By permission of Houghton Library, Harvard University, MS CSP 1537.

The Aesthetics of Graphs

The word "graph" became increasingly common as a suffix in the nineteenth century as it served to complete the names of more and more recording instruments, such as the kymograph, telegraph, seismograph, phonautograph, phonograph, photograph, and stereograph. After Marey, the identification of recording instruments and graphs with language became less obvious, but the association was not entirely lost. For example, German philosophers such as Friedrich Nietzsche, Walter Benjamin, and Theodor Adorno carried the pursuit of graphical "ur-languages" from Chladni and Ritter into the aesthetics of recorded music. In a 1934 essay entitled "The Form of the Phonograph Record" Adorno describes the phonograph record as "covered with curves, a delicately scribbled, utterly illegible writing."[91] The language that is automatically inscribed on the record is a secret one, decipherable only by another instrument. The lifeless art of the machine preserves the art that would otherwise die. The justification of the phonograph is that it "reestablishes by the very means of reification an age-old, submerged and yet warranted relationship: that between music and *writing*."[92] For Adorno, the mechanical reproduction of music reverses the process of turning signs (the musical score) into music and instead turns music into language:

> This occurs at the price of its immediacy, yet with the hope that, once fixed in this way, it will some day become readable as the "last remaining universal language since the construction of the tower," a language whose determined yet encrypted expressions are contained in each of its "phrases." If, however, notes were still the mere signs for music, then, through the curves of the needle on the phonograph record, music approaches decisively its true character as writing. Decisively, because this writing can be recognized as true language to the extent that it relinquishes its being as mere signs: inseparably committed to the sound that inhabits this and no other acoustic groove.[93]

The phonograph record has the advantage over the musical score in that it has written on it a language, not "mere signs." There is irony in the fact that the reification of music by machine in a most inhuman manner brings it "mysteriously closer to the character of writing and language."[94] The machine avoids the trap of semiosis—the "mechanical" assignment of mere signs to music—and preserves its aesthetic value in a new language. For Adorno, music and language retain a mysterious connection with the essences of things that is missing in "mere signs." Although he detests the "machine age," Adorno admits that an instrument like the phonograph can preserve this mysterious connection. There is no doubt about the source of Adorno's argument, because he immediately attributes it to Chladni and Ritter, who first saw the possibility of "inscribing music without it ever having sounded."

All of this takes us back to the dilemma of the language projectors of the seventeenth century. Can words, or signs, or graphs, or instruments, or even phonograph records get at the essences of things? For the natural philosopher the answer must be a qualified "yes." We can never know essences in the sense of Kant's *Ding an sich*, but we do believe that our experience is not all of our own making and that when we use our instruments to interrogate nature, nature talks back to us; we are not just talking to ourselves.

The graph produced by a recording instrument goes beyond the pure phenomena to reveal, by a sign, relationships within the phenomena. If it does not reveal such relationships, we say that the graph does not "signify"—it is not "significant"—which means (following Peirce's terminology) that it creates no interpretant in the mind beyond the pure phenomena that it describes. If the graph is significant, we can use the relationship that it reveals to construct theory and sometimes to assign causes.

In creating graphs, instruments almost seem to reason. This is especially true of computer-constructed graphs that reduce large amounts of information and create images far different from our direct experience.[95] In the nineteenth century Marey had already anthropomorphized even his simple instruments: "Patient and exact observers, endowed with more numerous and more perfect senses than our own, they work by themselves for the edification of science; they accumulate documents of an irrecusable fidelity, that the mind grasps easily, for which the comparison is easy and the memory durable."[96]

For both Marey and Peirce the graph is a superior scientific language, because it "speaks" diagrammatically in a way that most closely approximates the operations of human reasoning. Linguists may contest this claim. One can never know whether or not instrumental mediation can effect perfect correspondences between the signs of things and their essences. To achieve this would effectively undo the damage wrought at Babel, by recapturing the lost primordial language and healing the rift between man and nature. This is too much to ask from even the entire enterprise of natural science. Graphical expression does, like any language, depend upon a system of conventions in order to function. But this dependence does not reduce scientific results to mere illusions of meaning. Rather, the reliance effects the possibility of meaning that could not be gained in any other way. Thus, one has to concede that the graph has become an indispensable means of reasoning and communicating in modern science.

CHAPTER SEVEN

The Giant Eyes of Science: The Stereoscope and Photographic Depiction in the Nineteenth Century

In 1838, Sir Charles Wheatstone published his "Contributions to the Physiology of Vision."[1] This paper announced his explanation of the significance of the interocular discrepancy for binocular space perception. Prior to Wheatstone's researches, a number of individuals had observed an essential component of Wheatstone's innovation: in binocular vision, the two eyes receive slightly different images.[2] Kepler and Descartes had surmised that the muscular sensations arising from the convergence of the eyes in binocular vision might play a role in measuring the distances of objects.[3] But Wheatstone was the first to propose that the sensorium fathoms visual space by combining the information from a pair of two-dimensional, monocular pictures. "It being thus established," Wheatstone wrote, "that the mind perceives an object of three dimensions by means of the two dissimilar pictures projected by it on the two retinae, the following question occurs: What would be the visual effect of simultaneously presenting to each eye, instead of the object itself, its projection on a plane surface as it appears to that eye?"[4]

Wheatstone's paper introduced an instrument that facilitated this test. The apparatus employed two mirrors mounted in a right angle in order to present the reflection of one perspectival drawing to each eye, thus creating a single perception of marked relief (see figs. 7.1 and 7.2). Wheatstone called his device "a Stereoscope, to indicate its property of representing solid figures."[5]

Since its invention, the stereoscope has served as a tool for the study of vision. In this capacity, Wheatstone's investigation of the mental aspect of depth perception offered a fundamental contribution to experimental psychology, a field that became prominent in American and European universities in the late nineteenth century.[6] Yet, despite its crucial role in the laboratory, the stereoscope is perhaps more immediately recognized as the consummate Victorian amusement. The stereoscope belonged to the class of "philosophical toys" such as the kaleidoscope and the zoetrope that provided entertainment but also illustrated scientific principles.[7] The stereoscope occupied a curious cultural position during the second half of the nineteenth century. As Robert Hunt, a British photographic chemist, noted in 1856, "The stereoscope is now seen in every drawing room; philosophers

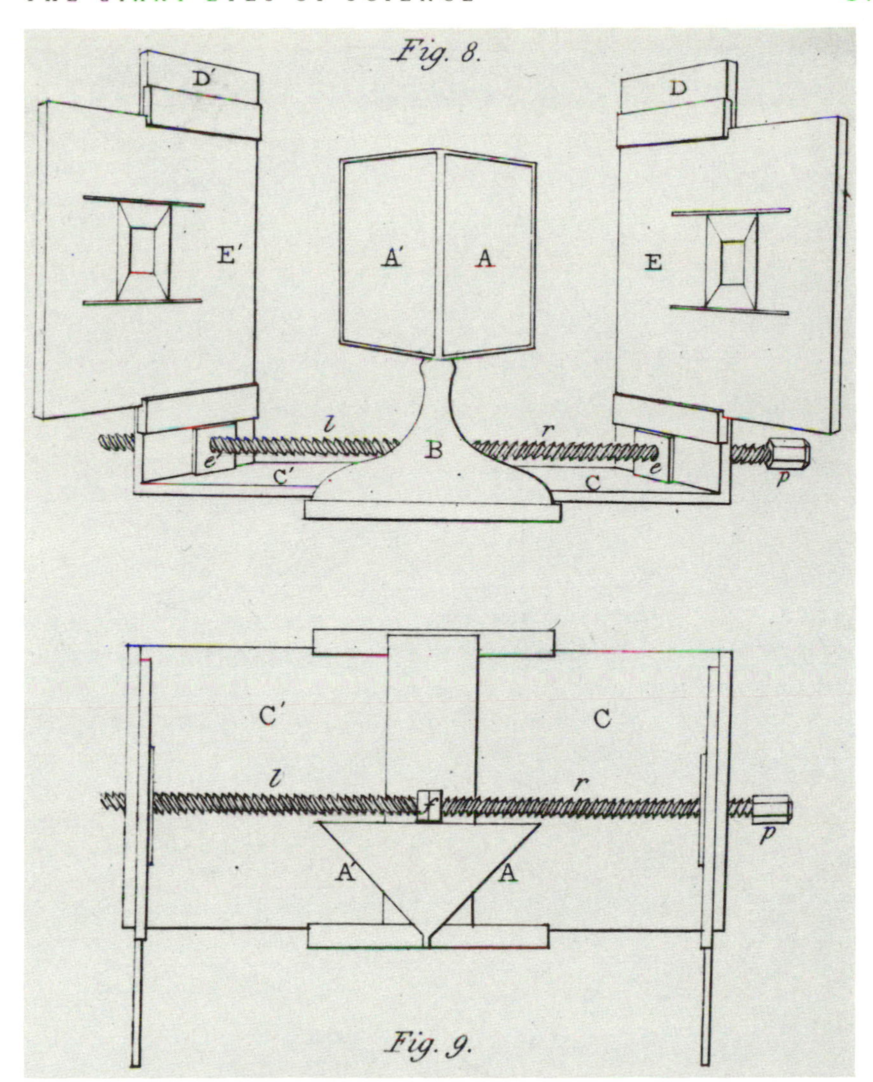

Fig. 7.1. Wheatstone's reflecting stereoscope. Front and top views. From Wheatstone, "Contributions to the Physiology of Vision—Part the First. On some Remarkable, and hitherto Unobserved, Phenomena of Binocular Vision," *Philosophical Transactions of the Royal Society* 128 (1838): 371–394. Courtesy of the University of Washington Libraries.

talk learnedly upon it, ladies are delighted with its magic representations, and children play with it."[8] The aim of the London Stereoscopic Company, founded in 1854—"A stereoscope for every home"—was nearly realized.[9] Several years later, an American source claimed that "a home without an instrument and a collection of views is almost an anomaly."[10] The instru-

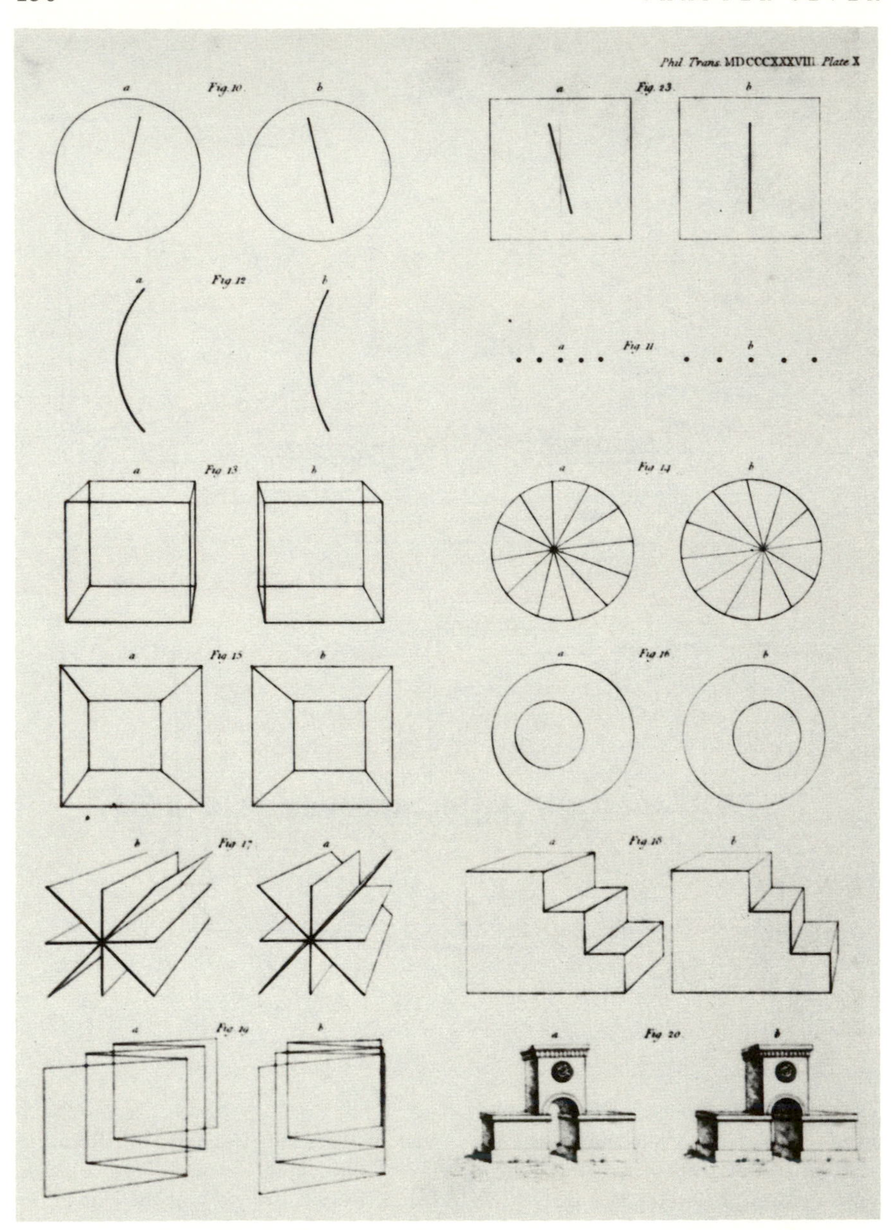

Fig. 7.2. Stereoscopic drawings. From Wheatstone, "Contributions to the Physiology of Vision—Part the First. On some Remarkable, and hitherto Unobserved, Phenomena of Binocular Vision," *Philosophical Transactions of the Royal Society* 128 (1838): 371–394. Courtesy of the University of Washington Libraries.

ment was discussed in newspapers and magazines, in art journals, and in scientific treatises. The whole range of society peered through its oculars.

The widespread prominence of the stereoscope creates advantages as well as difficulties for the historian. Although contemporary discussions of the instrument are plentiful, one may doubt the seriousness and forthrightness of some of these accounts. Many of these writings intended to sell stereoscopes and to excite curiosity about their dazzling spectacles. Exaggerations of the stereoscopic performance were quite common, and one may be tempted to dismiss such items as mere puffery. Nevertheless, historians are challenged to assess this verbiage and to try to use it to learn about the past. Evaluating the contemporary opinions about the stereoscope becomes a part of the larger problem of deciphering the hyperbolic rhetoric of nineteenth-century popular writing about science. Furthermore, in the case of the stereoscope and photography, the terminology and argumentation found in this literature remained surprisingly consistent. Both the "popular" and the "scientific" judgments of the stereoscope shared a coherent nexus of ideas about representation, visual physiology, and the philosophy of human perception.

In the nineteenth century, the tenets of "natural theology" defined the terms for arguments concerning the machinery of vision, photography, and stereophotography. This conception exalted the perfect design of the human sense organs as the basis for a truthful representation of nature. The best-known work in this tradition is William Paley's *Natural Theology* (first edition 1802), which regarded the *eye* as the ideal optical instrument, as well as the supreme piece of evidence that the universe and its inhabitants were deliberately designed by God.[11]

Many works of the period contained this theme. The argument was restated vigorously in the fifth Bridgewater Treatise, *Animal and Vegetable Physiology Considered with Reference to Natural Theology* (1834) by Peter Mark Roget, later of *Thesaurus* fame. Roget wrote:

> On none of the works of the Creator, which we are permitted to behold, have the characters of intention been more deeply and legibly engraved than in the organ of vision, where the relation of every part to the effect intended to be produced is too evident to be mistaken, and the mode in which they operate is at once placed within the range of our comprehension. Of all the animal structures, this is, perhaps, the one which most admits of being brought into close comparison [with] the works of human art; for the eye is, in truth, a refined optical instrument, the perfection of which can never be fully appreciated until we have instituted such a comparison; and the most profound scientific investigations of the anatomy and physiology of the eye concur in showing that the whole of its structure is most accurately and skilfully adapted to the physical laws of light, and that all its parts are finished with that mathematical exactness which the precision of the effect requires, and which no human effort can ever hope to approach,—far less to attain.[12]

Although not all writers on visual themes stated natural theological arguments as stridently as Paley or Roget, such a teleological conception of the human body and its function was held widely in the nineteenth century, and it provides a crucial piece of the intellectual context required for the historical estimation of the stereoscope.

In this chapter, stereoscopic photography will be considered as a means of depiction—a role that cuts across the boundaries of this instrument's uses as either a plaything or a tool in the psychological laboratory. As art historians such as Michael Baxandall have shown, past methods of making pictures raise an intricate problem in cultural history, one requiring the modern student to learn to "read" the various ways that pictures, like texts, can generate meaning.[13] For example, the conventions and devices of quattrocento painting—the gestures, facial expressions, and pigmentation—that were tacitly understood in their own cultural context cannot be understood today without historical effort. However, the present ubiquity of photographs may lead one to believe that the photographic process offers an unambiguous pictorial style—a technique unburdened by intellectual constraints that provides a perfect mirror of the present, as well as a clear window on the past. Therefore, the purpose of the present exploration of mid-nineteenth-century photographic—and especially stereographic—theory and practice is to recapture the richness of the cultural and scientific assumptions involved in this form of picture making.

The central feature of nineteenth-century stereography was its relationship to human binocular vision. Pictorial "realism," however, had been based on vision since Alberti's fifteenth-century articulation of linear perspective.[14] Svetlana Alpers's *The Art of Describing* argues that the style of picture making illustrated by Kepler's theory of the retinal image—in which the eye is treated as a camera obscura—provided the model of scientific depiction for Dutch artists in the seventeenth century.[15] Artists have also explored the vicissitudes of this tradition. Baxandall describes how Chardin's paintings may have incorporated the subjective visual phenomena investigated by eighteenth-century Lockean writings on perception.[16] Explicitly "impressionistic" works can also appeal to the experience of vision as a standard.[17] The nineteenth-century photographer P. H. Emerson, for example, employed soft focus to simulate actual vision.[18]

The network of natural theological presuppositions, which informed both popular and scientific accounts of the stereoscope, established the human eyes as the ideal instrumentation for visual representation. This led photographers, scientists, journalists, and art critics to evaluate the apparatus of stereoscopic depiction as a substitute for the innate fidelity of the eyes. Nineteenth-century writers also debated the merits of using this instrumentation to surpass the capacity of the human eyes. This discourse embodied contemporary attitudes toward the role of instruments in science and their value as a means of studying nature.

The Human Model of Representation

The tremendous popularity of the stereoscope would have been impossible without the aid of photography. The advent of stereoscopic double photographs, called "stereographs," dramatically extended the range of stereoscopic subjects. These had previously been limited to simple drawings, like those contained in Wheatstone's paper. Living in an age when photographic images are ever-present, the modern observer can scarcely appreciate the amazement produced by the first photographs. This new medium created permanent images by a purely mechanical process whose detail and accuracy surpassed any effort of art. Daguerre's camera successfully froze reality on its chemically sensitized plates in 1839—the year after Wheatstone introduced the stereoscope. Another photographic pioneer, William Henry Fox Talbot, invented the calotype technique in 1840 and demonstrated the process in his book *The Pencil of Nature* (1844–1846). This title echoed the contemporary attitude that photography was nature revealing herself at humanity's behest.[19] Stereoscopic photographs shared this fidelity, but they added the sensation of depth and solidity.

Soon after Daguerre's announcement, Wheatstone considered the possibility of photographic stereoscopic pictures. He called on Talbot and his associate, Henry Collen, to produce some stereoscopic calotypes for the stereoscope. The difficulty involved in aligning photographs in Wheatstone's cumbersome mirror arrangement, along with the cost of the instrument, diminished any chance of popular interest in the reflecting stereoscope.[20] But in 1849, the Scottish natural philosopher and steadfast opponent of the wave theory of light Sir David Brewster came up with a convenient and inexpensive lenticular stereoscope (see fig. 7.3).[21] George Lowden of Dundee constructed several models based on this design for Brewster. After a disagreement with Lowden—a common event in many of Brewster's professional relationships—he unsuccessfully searched for another British manufacturer. During the spring of 1850, Brewster took one of Lowden's models with him to Paris, where he showed the device to the opticians François Soleil and Jules Duboscq. Within a short time, they began producing Brewster's stereoscope and accompanying stereoscopic daguerreotypes.[22]

The Soleil-Duboscq version of Brewster's lenticular stereoscope created a sensation at that celebration of Victorian progress, London's Great Exhibition of 1851. Queen Victoria herself praised Brewster's work, and the craze ensued. An *Illustrated London News* correspondent at the Crystal Palace marveled at the utter precision of the stereoscopic productions and contemplated their value for the artistic field.

> We may have in future galleries of portraits no fictions of painters, but the people as they were—not flat and framed, and hung along the walls, nor in cold marble, but round and real as they looked in life: and so with buildings and scenery, we

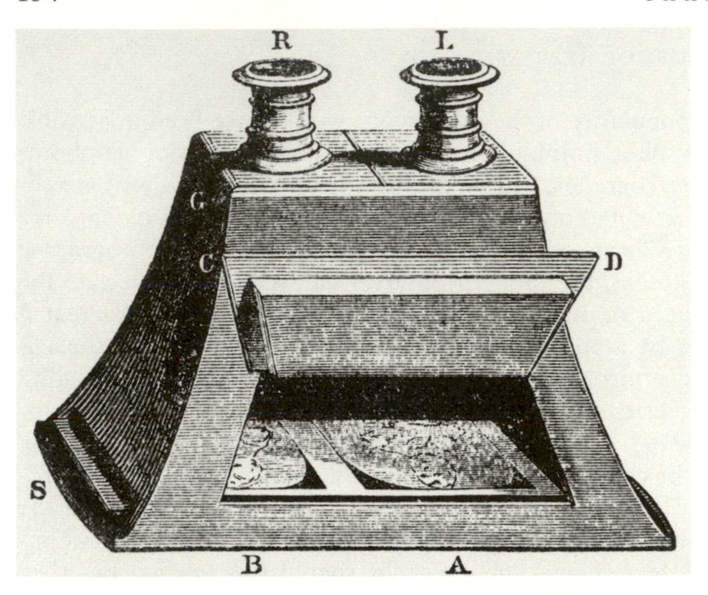

Fig. 7.3. Brewster's lenticular stereoscope. From Brewster, *The Stereoscope*, p. 67. Personal copy.

> may have, at a cheap rate, our hall of antiquities—Pompeii as it is, Ninevah as Layard sees it—scenery in foreign lands, in our own, in all the minuteness, grandeur, and beauty of nature. Neither Claude nor Turner could have given any more than half such physical or aerial perspective. The artist may carry in his stereoscope the immortal works of the genius-inspiring masters of every age and country, and wherever the highest living beauty is to be found he may have in an instant his models, subject to no errors of his pencil, but in the full rich roundness of reality.[23]

One day, the accurate sunbeam might replace clumsy human fingers and re-create its subjects in their natural solidity and depth.

The first glimpse through the stereoscope lenses startled many viewers. The twin pictures did not produce the sensation of staring into a box but rather the feeling of actually witnessing the captured spectacle. "The stereoscopic view of a city shows not a mere drawing; the *real city itself* seems presented to the sight. So, too, with the portrait: the flat outline disappears, and the *living subject* seems to stand before the eye."[24] Stereographs, drawn through the aid of the mathematically precise camera lens, by the "unerring hand of Nature," could neither add to nor detract from the visible scene.[25]

> But so long as mere drawings by hand were used, it might be held that the effect, however wonderful, was but some trick of art by which the senses were cheated. But the Daguerréotype admits of no trick: the silvered plate has neither line, nor light, nor shade, but such as the sun gives it: the two plates in the two cameras stand truly for the two eyes, and receive each just such picture, no more, no less,

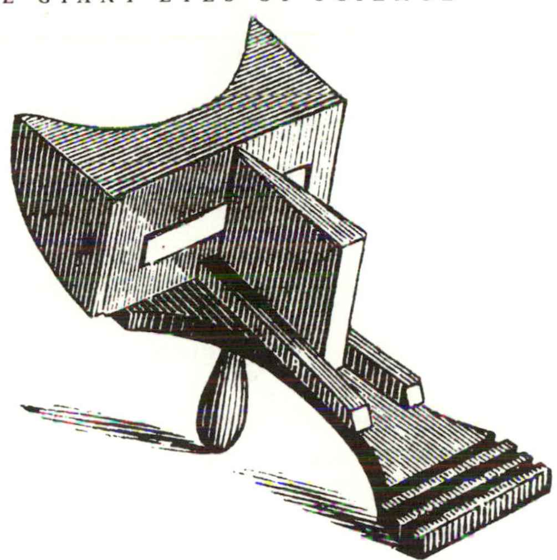

Fig. 7.4. Holmes's handheld stereoscope. From "The 'Holmes' Stereoscope," *Philadelphia Photographer* 6 (1869): 24. Courtesy of the University of Washington Libraries.

> as each eye receives. There is, therefore, no further room for doubt as to the need for two eyes: we have taken by the Daguerréotype the very picture from each, and have made them tell their secret. Our double vision is but perfect vision.[26]

Based on sturdy Victorian scientific principles, the stereoscope re-created three-dimensional perception with perfect fidelity.

In the United States, the author and physician Oliver Wendell Holmes became an outspoken champion of the stereoscope.[27] Holmes made two principal contributions to the cause. In 1861, he designed a handheld version of Sir David Brewster's lenticular stereoscope (see fig. 7.4). The "Holmes Stereoscope" became the overwhelmingly predominant type used in America—indeed, in England it was known as the "American Stereoscope."

Holmes's second contribution consisted of three enthusiastic, albeit anonymously published, essays in the *Atlantic Monthly.* "The Stereoscope and the Stereograph" (1859), "Sun-Painting and Sun-Sculpture" (1861), and "Doings of the Sunbeam" (1863) trumpeted the new photographic technology and its most promising application—the stereograph. Holmes described Daguerre's invention in Promethean terms, and he contemplated a brilliant future for the new art.

> We are looking into stereoscopes as pretty toys, and wondering over the photograph as a charming novelty; but before another generation has passed away, it will be recognized that a new epoch in the history of human progress dates from the time when He who
>
> —never but in uncreated light
> Dwelt from eternity—

took a pencil of fire from the hand of the "angel standing in the sun," and placed it in the hands of a mortal.[28]

"If a strange planet should happen to come within hail," Holmes supposed, "and one of its philosophers were to ask us, as it passed, to hand him the most remarkable material product of human skill, we should offer him, without a moment's hesitation, a stereoscope containing an *instantaneous* double-view of some great thoroughfare."[29] Holmes expressed limitless zeal for the popular scientific marvel of the day. His assessment of the stereoscope's possibilities surpassed mere praise for the stunning representation of the visible world. Through the means of the photograph and the stereograph, Holmes explained, *form* became a distinct intellectual entity—independent of physical objects—in the same way that the printing press had liberated *thought*. Thus, the stereoscope could become the "card of introduction to make all mankind acquaintances."[30] "*Form is henceforth divorced from matter*," Holmes observed. "In fact, matter as a visible object is of no great use any longer, except as the mould on which form is shaped. Give us a few negatives of a thing worth seeing, taken from different points of view, and that is all we want of it. Pull it down or burn it up if you please."[31] In the age of the stereoscope, the pyramids, the Pantheon, and all other human and natural creations would be expendable. "We have got the fruit of creation now," Holmes explained, "and need not trouble ourselves with the core. Every conceivable object of Nature and Art will soon scale off its surface for us. Men will hunt all curious, beautiful, grand objects, as they hunt the cattle in South America, for their *skins*, and leave the carcasses as of little worth."[32] As the printing press facilitated the transmission of verbal ideas across space and time, the stereoscope rendered feasible the dissemination of binocular information. The stereoscope had potential beyond the realm of amusement. It could prove itself a priceless tool for communication, education, art, and philosophy. "This is no *toy* . . . ," he declared, "it is a divine gift, placed in our hands nominally by science, really by that inspiration which is revealing the Almighty through the lips of the humble students of Nature."[33] Perhaps Holmes would have moderated his excessive comments if the thin veil of anonymity had been lifted. Yet even his extreme estimate of the instrument's potential hinged on a principle expressed by many others in his day—the unique stereoscopic medium captured the visual essence of nature.

The rationale for Holmes's estimate was articulated by the *Illustrated London News* writer who claimed that the stereoscope duplicated the operation of the human visual organs: the twin cameras "stand truly for the two eyes."[34] Thus, nineteenth-century photochemistry produced an ironic corollary to Kepler's 1604 discovery that the eye behaves like a lifeless mechanical instrument—a camera obscura.[35] By replacing the retina with a sensitive plate, the camera had become an eye. This did not remain a casual metaphor. For nineteenth-century students of stereophotography, the camera as

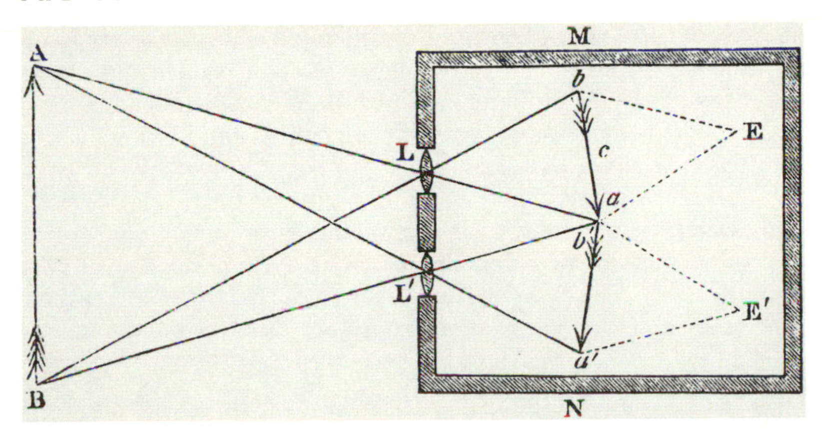

Fig. 7.5. Brewster's binocular camera. From Brewster, *The Stereoscope*, p. 146. Personal copy.

eye—or more exactly, the binocular camera as *pair* of eyes—became a potent leading principle.

In 1849, Sir David Brewster designed a binocular camera that imitated nature by maintaining a lensatic separation of 2½ inches—the average human interocular distance. His camera photographed a pair of images that retained the same parallactic discrepancy present in human binocular vision. In his 1856 treatise, *The Stereoscope*, Brewster showed that this binocular camera produced two pictures identical to those seen by the two eyes (see fig. 7.5).[36]

Three years after describing his binocular camera, Brewster announced a conclusion that advanced his attempt to reproduce mechanically the physical circumstances of human sight. "The object of photographic art," he contended, "is to obtain an accurate representation of nature, as it appears when seen either with one or with two eyes."[37] Following this ethic, Brewster argued:

> As the pupil of the human eye is little more than two-tenths of an inch in diameter, we may regard the picture on the retina as a correct representation of external objects, in so far, at least, as its correctness depends upon the size of the lens which forms the picture. In like manner we may consider the image of objects formed by a lens the size of the pupil of the eye as a correct representation of the object.[38]

At the time Brewster wrote this, his apparently innocent recommendation would have entailed a major change in photographic practice. In the early years of photography, even the most rapid photochemical processes required long exposures. The portrait subject in a daguerrean studio could expect to sit uncomfortably motionless for several minutes. Many photographers employed clamps and braces to help their patrons maintain the desired pose without fatigue. It is not surprising that a trip to the daguerrean parlor often produced less-than-flattering results.

Despite these interminable exposures, photographers required very large apertures, usually ranging from 1½ to 6 inches in diameter. This produced a lens area many times greater than that which Brewster suggested in 1852, and he bluntly stated what disregard for his principles would bring: "Every addition to the area of the lens introduces parts of the object which have nothing to do with the picture, and when we use lenses of two, four, or six inches in diameter, we obtain, though a common eye may not discover it, monstrous representations of humanity, which no eye and no pair of eyes ever saw or can see."[39] The oversize lens surveys more perspective area than the human eye could allow, "and a monstrous portrait of the human bust is thus obtained by the photographer, the monstrosity increasing with the size of the lens."[40] The camera, improperly used, would desecrate, rather than emulate, human sight and human form.

Although photographic studios during the 1850s scarcely enabled subjects to retain a relaxed appearance, Brewster insisted that the size of the lens, rather than the unsteadiness of the sitter, was the primary agent in the "hideousness" of photographic portraits.

> The photographer, therefore, who has a genuine interest in the perfection of his art, will receive these truths with gratitude; and by accelerating the photographic processes, with the aid of more sensitive material, he will be able to make use of lenses of very small aperture, and thus place his art in a higher position than that which it has yet attained. The photographer, on the contrary, whose sordid interests bribe him to forswear even the truths of science, will continue to deform the youth and beauty that may in ignorance repair to his studio, adding scowls and wrinkles to the noble forms of manhood, and giving to a fresh and vigorous age the aspects of departing or departed life.[41]

By following the prescriptions of nature and scientific truth, decent, upright photographers may create honest representations of their patrons.

In 1855, John F. Mascher—a Philadelphia photographer who, in 1853, had obtained the first American patent for a stereoscopic viewer—independently arrived at exactly the same conclusions as Brewster had.[42] Mascher wondered why a pair of his photographs, taken from two points separated horizontally by 2½ inches, failed to reproduce stereoscopic relief. He attributed this shortcoming to the fact that his camera lenses, which were much larger than human eyes, admitted extraneous views that ruined the perspective discrepancy between the two pictures.[43] Mascher described his investigations of images formed by various apertures on a sensitive plate; he learned that

> a picture, taken with a camera, with lenses larger than the human eye, will show more of the object than what the human eye, placed in the same position, will be able to see. . . . Such pictures are anti-stereoscopic; distortions; disfigurations intolerable in proportion to what the lense, with which it is taken, exceeds in diameter the size of the human eye. Such pictures will do for owls to look at! . . . We

> might with the same propriety call the hide of an ox, when spread upon a flat surface, a portrait of that animal, as to call a picture, taken in a camera with such large lenses, a portrait of the "human face divine."[44]

In the stereoscope, Mascher found that pictures produced by cameras with the same dimensions as human eyes—a ⅛-inch diameter aperture and lenses separated by 2½ inches—truthfully reproduced the binocular sensation of relief. Mascher did not regard his discovery as an application of principles of geometrical optics. Rather, he cited divine intelligence as the origin for the arrangement of the human visual organs, thus making proper stereophotography a moral imperative.

> In the human eye we find, as in all other parts of the body, the most extraordinary wisdom displayed, and it is only the hand of Omnipotence that could have designed and constructed such a wonderful organ. Not only do we find a single eye perfect in all its parts, but we also find the two eyes arranged in such a manner as to give the greatest possible amount of effect to binocular vision. Who can devise anything better? To imitate and equal, it ought to challenge our undivided attention.[45]

Mascher thus revealed the natural theological dimension of his method for producing stereoscopic photographs. Although he experimented with apertures at least as small as 1/66-inch diameter and discussed a "theoretical eye" that "occupies no more room than a *mathematical point*," Mascher selected the human visual apparatus as the ideal.[46]

For many nineteenth-century scientists and photographers, the eye had become the archetype for the camera. This analogy was particularly appealing to those who investigated the stereoscope and binocular vision, in which the relationship between the picture-making instrumentation and the human model was especially close. Brewster employed a geometrical analysis of the images created by various apertures and Mascher conducted an empirical study, but both agreed that the dimensions of the human eyes lead to photographic truth.

Nearly every nineteenth-century discussion of photography and vision claimed that two-dimensional representations—paintings or photographs—appear most lifelike when viewed with only one eye. Occasionally, a writer summoned Francis Bacon's name as a historical authority on this score, but the thrust of these statements hinged on the notion that monocular vision is suitable to discern all of the detail in a flat surface.[47] But three-dimensional views—these sources explained—require two open eyes for the appreciation of solidity and space. By dint of the precision of the photographic image, the stereograph acquired an uncanny facility for preserving its subjects "in all the roundness and solidity of nature and truth."[48]

Oliver Wendell Holmes provided an amusing gloss on the character of the stereoscopic image. His son and namesake—who achieved fame on the United States Supreme Court—would make a more determined effort than had his father in legal training. Yet the elder Holmes found some application for a rule that may have been a vestige from his lone year of law school: the "law of evidence"—a person shall not be convicted on the testimony of fewer than *two* witnesses (or, in this case, two *photographs*). Holmes pointed out that a single picture may benefit from some flattering touches. "A lady's portrait," he said, "has been known to come out of the finishing-artist's room ten years younger than when it left the camera. But try to mend a stereograph and you will soon find the difference. Your marks and patches float above the picture and never identify themselves with it."[49] Applying his judicial wit, Holmes reasoned: "No woman may be declared young on the strength of a single photograph; but if the stereoscopic twins say she is young, let her be so acknowledged in the high court of the chancery of the God of Love."[50] Holmes often rewarded his readers with such clever observations. But beyond the humor of this comment, one can read it as a sign that Holmes appreciated the fundamentally unique status of the stereoscopic medium. One could tamper with an ordinary photograph, but it would be impossible for human hands to falsify what nature's pencil had written on the stereograph.

Not all descriptions of the stereograph's truthfulness employed legalistic or literary arguments. Most discussions had a basis in the physiology of vision. The binocular camera captures as much information on its sensitive plates as would fall on two human retinae, had a person witnessed the same view. Joseph LeConte, an American authority on vision, returned to his comparison between the eye and the camera throughout his major work, *Sight* (first edition 1881). LeConte completed his comparison between the animate and the inanimate optical instruments with his discussion of binocular vision. He explained that

> there are certain effects which can not be produced by one camera or by one eye. As two cameras from two positions take two slightly different pictures of the same object or the same scene, which when combined in the stereoscope produce the clear perception of depth of space—but only *phantom* space—even so the two eyes act as a double camera in taking and a stereoscope in combining two slightly different images of every object or scene, so as to give a clear perception of a *real* space.[51]

In several descriptions of stereographic apparatus, the double camera nearly became living tissue. An essay in *Harper's Magazine*, "The Eye and the Camera," described the anatomy and function of these analogous devices. The writer claimed that an "ordinary camera resembles a single eye," and a stereoscopic camera

> is like a forehead with two eyes in it. The two round tubes in front contain the lenses, and the brass caps which fit over them when the exposure is complete are the eyelids. The diaphragm, which is inserted in each of these tubes to regulate the size of the aperture, is like the pupil of the eye that contracts and expands ac-

> cording to the degree of light. And this double instrument makes two pictures at the same instant, which differ from each other just as the images received by one eye differ from those received by the other in an observer standing at the same place.[52]

The appearance of a photographer at work encouraged the connection between man and camera and the anthropomorphic description of the photographic instrument. With the artist's head beneath the camera's hood, the human and the machine seem fused. Optically, both survey the same scene, sharing the camera's lens. The logo adopted by New York's great photographic supply house, E. and H. T. Anthony, accentuated this symbiosis: the actively engaged photographer's skinny, bent legs mimicked those of his tripod (see fig. 7.6).[53] A *Punch* cartoonist mistook a photographer with a camera for a new species of urban wildlife (see fig. 7.7).[54] Another contemporary illustration creates a visual conundrum (see fig. 7.8).[55] A pair of cameras stereograph a woman's countenance—but it is a statue, not a living face. Two representations of human eyes gaze at each other. One has the correct external features; on the inside, however, there is lifeless stone and not flesh. The opposing figure possesses none of the softness of the human form; yet its dual cameras nearly copy the optical relations of human sight. Whether for humor, for irony, or by accident, the two pairs of "eyes" together contain the physical equipment for sight, but none of the intellectual works. Both stare blindly into empty eyes.

One of the more animated descriptions of an anthropomorphized binocular camera is Holmes's account of his visit to a stereoscopic studio:

> A skeleton shape, of about a man's height, its head covered with a black veil, glided across the floor, faced us, lifted its veil, and took a preliminary look. When we had grown sufficiently rigid in our attitude of studied ease, and got our umbrella into a position of thoughtful carelessness, and put our features with much effort into an unconstrained aspect of cheerfulness tempered with dignity, of manly firmness blended with womanly sensibility, of courtesy, as much as to imply,—"You honor me, Sir," toned or sized, as one may say, with something of the self-assertion of a human soul which reflects proudly, "I am superior to all this,"—when, I say, we were all right, the spectral Mokanna dropped his long veil, and his waiting-slave put a sensitive tablet under its folds. The veil was then again lifted, and the two great glassy eyes stared at us once more for some thirty seconds. The veil then dropped again; but in the mean time, the shrouded sorcerer had stolen our double image; we were immortal. Posterity might thenceforth inspect us, (if not otherwise engaged,) not as a surface only, but in all our dimensions as an undisputed *solid* man of Boston.[56]

In this portrayal of the roles of man and machine, the camera was alive, while the human photographer had been reduced to subservience.

The analogy between the eye and the camera owed much of its power to the notion that the divinely constructed human form offered the model for

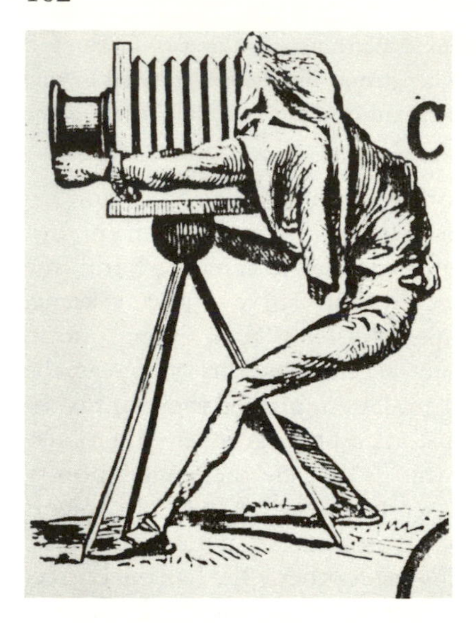

Fig. 7.6. The Anthony logo. From the cover of *Anthony's Photographic Bulletin*, 1870. Courtesy of the University of Washington Libraries.

Front and Back view of a very Curious Animal that was seen going about loose the other day. It has been named by Dr. Gunther " Elephans Photographicus."

Fig. 7.7. "Elephans Photographicus." From *Punch* 44 (1863): 249. Courtesy of the University of Washington Libraries.

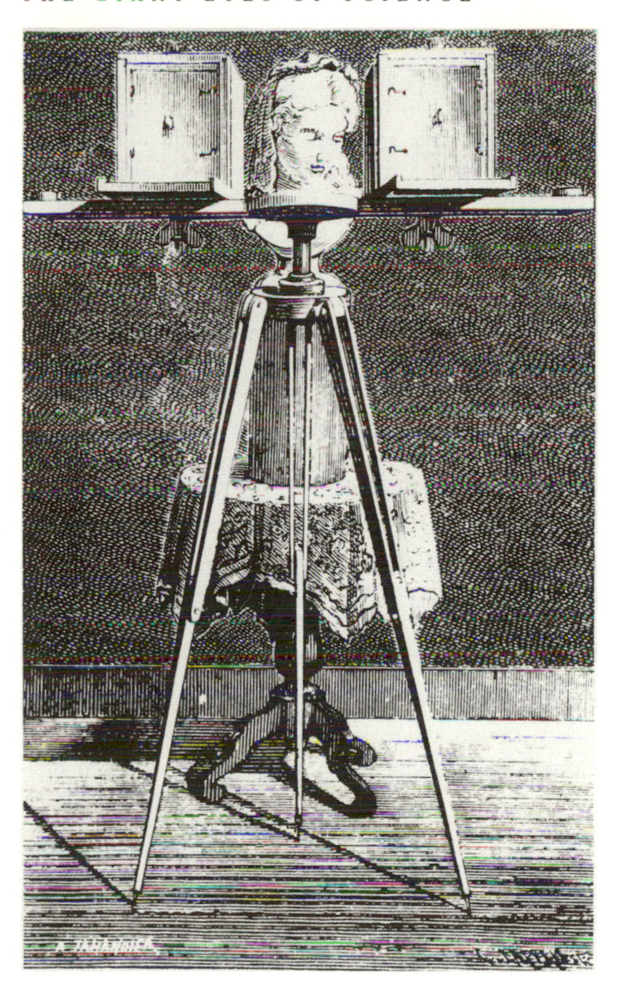

Fig. 7.8. Two pairs of eyes. From Tissandier, *A History and Handbook of Photography*, following p. 312. Courtesy of the University of Washington Libraries.

the most efficient application of physical principles. For natural theologians, the eye epitomized the perfection of God's design. The operation of binocular vision, and its expression in the stereoscope, shared this wise application of natural laws. "To produce the effect of nature," one source explained, "we must do as nature does: two pictures must be painted, one for each eye, and combined, to produce the sensation of one. *This is effected by the Stereoscope.*"[57] The specifications of nature—delineated in the human frame—dictated the standard for truthful representation. Because it duplicated the optical circumstances of human binocular vision, the stereoscopic camera functioned like a pair of surrogate eyes and could create truthful pictures of the world.

Instrumental Distortion and Enhancement

The twin lenses of the stereoscopic camera were often praised as infallible scribes that could etch reality on their photographic tablets. A critic for the London *Art Journal* in 1858 noted that in the stereoscope, "the actual is absolutely before us and we know it. There has been here no possibility of either adding or subtracting. The sun is a rare truth teller which cannot lie to produce effect, nor err to lead astray."[58] Yet despite these confident claims regarding the accuracy and honesty of photography and the stereoscope, there was a contemporaneous challenge to the veracity of the photographic medium. The complaint was put forward by many art critics and theorists who did not consider photography worthy to rank among the "fine arts." For both the supporters and the detractors of the new medium, the relationship between photography and human vision stood at the core of the debate. However, the critics claimed that photography was inherently flawed: it misrepresented human vision and, therefore, produced distorted images.

Lady Elizabeth Eastlake's well-known contribution on this matter, which appeared anonymously in a *Quarterly Review* for 1857, portrayed photography as a necessarily unfaithful means of depiction.

> Far from holding up the mirror to nature, which is an assertion usually as triumphant as it is erroneous, it holds up that which, however beautiful, ingenious, and valuable in powers of reflection, is yet subject to certain distortions and deficiencies for which there is no remedy. The science therefore which has developed the resources of photography, has but more glaringly betrayed its defects. For the more perfect you render an imperfect machine the more must its imperfections come to light: it is superfluous therefore to ask whether Art has been benefited, where Nature, its only source and model, has been but more accurately falsified.[59]

Photography's principal inadequacy was observed in the alleged incompatibility between the mechanical camera and human sensibilities. As explained in "The Photographic Portrait" in the American art journal the *Crayon*, "the camera, although obedient to the laws of physical nature, is quite indifferent to the laws of our intellectual nature; it is, in fact, a falsifying agent of that which we know to be true in nature."[60]

According to this line of reasoning, the camera failed because it could not exercise the tasteful judgment of the human artist. "Paradoxical as it may appear," the *Crayon* reported, "it would seem as if light, like man, lost its moral power, and wrought out deeds of evil, when it condescends to work in the dark of the camera."[61] The camera did not select the distinctive qualities of its subject; for "Mr. Photographer Light," *worth* was equated with *luminosity*. As a "true democratic leveller," Mr. Photographer Light attached "no more importance to expression than a politician does to truth when it interferes with the high lights of his argument."[62] The strongly illuminated features, rather than the significant ones, became prominent in the photograph.

Portraiture provided the most common, as well as the most troublesome, situation for the nineteenth-century photographer. In the words of Lady Elizabeth Eastlake, "of all the surfaces a few inches square the sun looks upon, none offers more difficulty, artistically speaking, to the photographer, than a smooth, blooming, clean washed, and carefully combed human head."[63] The *Crayon* contended that the female visage fared especially poorly under these circumstances, while men were better equipped to survive the rough operation of the instrument:

> Ladies are generally indifferent subjects for the photograph, owing to the delicate texture of their skin and its coloring, and to the fact that their expression of countenance is out of mechanical conformity with the mechanical workings of the instrument; whereas the brawny and materialized face of man, from the nature and calling of his life, is in direct harmony therewith, showing the inability of the instrument to conceive and render expression, and its inevitable reduction to a mere anatomical diagram. A lady whose complexion exhibits any marked tint of yellow, also one whose features approximate to an even plane, are specially unfitted for photographic portraiture. Bilious temperaments and brunettes we are confident will never contribute to erect a monument in honor of Daguerre.[64]

Many factors contributed to the odds against the camera's producing a complimentary representation of the human face. Brewster's complaint of the "monstrosities" that he attributed to oversize apertures has been noted above. The oddities of the photosensitive chemicals also did not conform to the expectations of human vision, because colors were not translated to the monochromatic representation with uniform intensity. Hence, blue and purple left the strongest imprint on the photographic plate, while orange and red became the most faint. The complications that the peculiarities of the actinic spectrum engendered for portraiture became evident in a demonstration by the photographic scientist Antoine Claudet. Claudet produced a daguerreotype of a painting of an absurdly colored female head. Its eyes were red, its lips blue; its face consisted of various shades of indigo, violet, and yellow. Yet its tones appeared perfectly correct in the daguerreotype. Conversely, another picture, which had been painted with apparently natural human coloration (but whose pigments had been intentionally, if unnoticeably, mixed with yellow, green, blue, indigo, and violet) became, in its daguerreotype, "as ridiculous in appearance as the party-coloured female head which gave a correct picture."[65] Eastlake knew of Claudet's experiment and she warned the reader of how this lesson applied to the representation of a living female face. "If the cheek be very brilliant in colour," she wrote, "it is as often as not represented by a dark stain. If the eye be blue, it turns out as colourless as water; if the hair be golden or red, it looks as if it had been dyed, if very glossy it is cut up into lines of light as big as ropes. This is what a fair young girl has to expect from the tender mercies of photography."[66]

Opinions varied as to how well the camera's mechanical eye treated its human counterpart. The American photographic chemist John William Draper was pleased by the outcome. He wrote: "The eye appears beautifully; the iris with sharpness, and the white dot of light upon it, with such strength and so much of reality and life, as to surprise those who have never before seen it. Many are persuaded that the pencil of the painter has been secretly employed to give this finishing touch."[67] Not everyone shared his admiration. Eastlake felt that "the spectrum or intense point of light on the eye is magnified to a thing like a cataract."[68] The appraisal of the *Crayon* was perhaps even more harsh:

> Instead of the diamond-like point of light which gleams with so much brilliancy, and is yet subdued like the serene reflection of a star in transparent water—intensified but not lost in the aqueous cavern of the eye-ball—we have a positive white spot surrounded by flat inky blackness, being a wholly external reflection without depth, and no more characteristic of the eye than if revealed to us from the slimy surface of an oyster.[69]

Photographic portraiture was often derided for its inability to preserve the fleeting marks of human emotion. This was especially the case in the first decade of the art, when the sitter could maintain only a silent and solemn physiognomy for the duration of the several-minute exposure. Because the human eye's "persistence of vision" remains for only a fraction of a second, it can "catch" smiles, laughter, and other sorts of motion that would have been a blur to an early photographic camera. Brewster offered a backhanded defense of this characteristic of photography. He described the photograph as something of an inverted picture of Dorian Gray, since, as the subject matures, "the grave and sombre, and perhaps ungainly, picture grows even into a flattering likeness."[70] Frederick Scott Archer's invention of the collodion process in 1851 reduced the required length of exposure from several minutes to several seconds. As Eastlake recognized, movements and emotions came within the camera's range.

> Under the magician who first attempted to enlist the powers of light in his service, the sun seems at best to have been a sluggard; under the sorcery of Niepce he became a drudge in a twelve-hours' factory. On the prepared plate of Daguerre and on the sensitive paper of Fox Talbot the great luminary concentrates his gaze for a few earnest minutes; with the albumen-sheathed glass he takes his time more leisurely still; but at the delicate film of collodion—which hangs before him finer than any fairy's robe, and potent only with invisible spells—he literally does no more than wink his eye, tracing in that moment, with a detail and precision beyond all human power, the glory of the heavens, the wonders of the deep, the fall, not of the avalanche, but of the apple, the most fleeting smile of the babe, and the most vehement action of the man.[71]

The advent of collodion entailed a marked reduction of exposure time, although it was not quite as dramatic as Eastlake suggested. However, the acceleration of photographic processes did eventually outstrip the limits of human vision. By the 1880s, Étienne-Jules Marey and Eadweard Muybridge managed to "stop" movements too rapid for the eye to see.[72]

In a harsh evaluation of photography, the French author and critic Charles Baudelaire leveled a vitriolic attack against the would-be art. In his "Salon de 1859," he disparaged the philistine crowd who would elevate photography—which possessed a coarse exactitude and lacked the invigorating touch of human imagination—above *true* art. As soon as the inversion was made, "our squalid society rushed, Narcissus to a man, to gaze at its trivial image on a scrap of metal."[73] These fanatics became "new sun worshippers," and "[a] little later a thousand hungry eyes were bending over the peepholes of the stereoscope, as though they were the skylights [*lucarnes*] of the infinite."[74]

Such arguments were part of a romantic tradition in literature and art that emphasized the mental and emotional aspects of vision. Oliver Wendell Holmes and the advocates of stereophotography considered an imitation of the mechanics of sight adequate to liberate the visual essence from its material bondage. By contrast, William Blake, William Wordsworth, and Ralph Waldo Emerson, among others, insisted that complete vision required more than mere optics—it also demanded an active intellect.[75] According to romantic writers, the bodily organs of sight *by themselves* were insufficient for perception, as the great nineteenth-century aesthetic theorist John Ruskin pronounced: "You do not see *with* the lens of the eye. You see *through* that, and by means of that, but you see with the soul of the eye."[76] To such an attitude, instrumental means of picture making seemed inherently offensive. Thus, photography could offer nothing but a perversion of human sight.

There is an element of irony in this strain of criticism. Despite their idealistic rhetoric, Eastlake and Baudelaire did not deny the value of photographic resemblances to fulfill the needs of "historic interest," which demanded "mere manual correctness."[77] Furthermore, the flaws disclosed in their essays related only vaguely and abstractly to the absence of a conscious component in photography. Instead, the authors pointed to photography's failure to replicate the experience of human perception. For example, the monochromatic medium could never provide an accurate representation of color, nor could the long photographic exposures preserve the fleeting expressions of human emotion.[78] Therefore, Eastlake and other art critics agreed essentially with the natural theological judgment: human vision provided the model for proper depiction. Errors and distortions were produced by departures from this standard.

However, nineteenth-century commentators did not unanimously agree that only falsity could arise from the transgression of ocular orthodoxy. Many writers praised the potential of photographs and stereographs to surpass the limitations of unaided sight. Like the telescope or the microscope, the stereoscope could be regarded as an instrument that created a more valuable representation of the world than the human eyes produced.

In an essay on his binocular camera, Sir David Brewster suggested a way to expand the capability of human sight. Brewster recognized that large objects—he mentioned buildings and "colossal statues"—must be seen from an extended distance to be viewed in their entirety. Because human eyes are separated by only 2½ inches—the same spacing that he had previously demanded for the separation in binocular cameras—viewing a large statue from afar allows for very little binocular parallax and therefore results in a diminished sense of stereoscopic depth. "As we cannot increase the distance between our eyes, and thus obtain a higher degree of relief for bodies of large dimensions," Brewster asked, "how are we to proceed in order to obtain drawings of such bodies of the requisite relief?"[79]

Brewster resolved this quandary by violating his own rule concerning the consistent use of a 2½-inch separation between the lenses of a binocular camera. In the case of large objects, he said, the interval should be expanded. In the stereoscope, two images procured in this fashion will create the impression that one is viewing a reduced copy of the oversize structure. Brewster claimed that this method provided "a better and more relieved representation of the work of art than if we had viewed the colossal original with our own eyes, either under a greater, equal, or a less angle of apparent magnitude."[80]

Furthermore, Brewster considered that the ideal method for viewing such bulky forms might be achieved if the cameras were separated by a distance equal to the breadth of the object under consideration. All objects would be "reduced with mathematical precision to a breadth of 2½ inches, the width of the eyes, which gives the vision of a hemisphere 2½ inches in diameter, with the most perfect relief."[81] In this scheme, all visual representations must conform to the measurements that prove best adapted to the physical configuration of the human eyes.

Brewster predicted that his technique would afford an impressive gain for humanity, especially in the arts.

> The art which we have now described cannot fail to be regarded as of inestimable value to the sculptor, the painter, and the mechanist, whatever be the nature of his production in three dimensions. Lay figures will no longer mock the eye of the painter. He may delineate at leisure on his canvas, the forms of life and beauty, stereotyped by the solar ray and reconverted into the substantial objects from which they were obtained, brilliant with the same lights, and chastened with the same shadows as the originals. The sculptor will work with similar advantages. Superficial forms will stand before him in three dimensions, and while he summons into view the living realities from which they were obtained, he may avail himself of the labours of all his predecessors, of Pericles as well as of Canova; and he may virtually carry in his portfolio the mighty lions and bulls of Nineveh,—the gigantic sphinxes of Egypt,—the Apollos and Venuses of Grecian art,—and all the statuary and sculpture which adorn the galleries and museums of civilized nations.[82]

Brewster's essay was one of the earliest discussions of a deliberate attempt to manipulate normal visual perception with the stereoscope. However, the arrangement of stereoscopic cameras became a hotly debated subject among photographers during the early 1850s. While these arguments encompassed a variety of opinions, the discourse can be characterized as a confrontation between two pictorial styles. One cast of mind maintained that the separation of the human eyes should be duplicated unwaveringly by the photographer. Appeals to *nature*, as one might gather, provided the most potent rhetorical device for those who maintained this stance.[83] Many advocates of this view also warned, in terminology that was quite uniform, that widely separated cameras would make objects appear like "models," which would seem "distorted" or even "monstrous" if the exaggeration became extreme.

On the other hand, many photographers—including some who agreed that art should emulate nature—relished this enhanced perspective and model-like appearance. This technique of separating the lenses of the stereoscopic camera by more than the human interocular distance—a practice that more recent stereographers call "hyperspace"—became standard among landscape photographers. (The camera separation for landscapes was usually on the order of a few feet.) The stereograph purchaser could observe deeper valleys and more dramatic cascades than existed in reality. Antoine Claudet's stereodaguerreotypes, which accompanied Brewster's stereoscope at the Great Exhibition, were produced in this manner. Several of his examples depicting the interior of the Crystal Palace, as an *Illustrated London News* writer remarked, revealed even remote items with "as full roundness and relief as those at hand."[84] These representations showed

> a view as if the pictures were taken from a small model of the building brought sufficiently near for the whole to be within the distance influenced by the angle of the eyes. In fact, instead of seeing the object itself, you see a miniature model of it brought close to the eyes; so that, in this instance, the stereoscopic Daguerréotypes actually surpass the reality. No one has ever seen the interior of the Exhibition from end to end with such clearness as it is seen in M. Claudet's pictures.[85]

The stereoscope brought before the eyes images that unaided sight could never have achieved.

This practice disgusted an 1854 correspondent to the *Photographic Journal* who vigorously endorsed the photographic imitation of nature. He detested the proliferation of "painfully exaggerated specimens, so repulsive to truth or good taste, ordinarily shown in the beautiful invention of Wheatstone."[86] This correspondent might have been even more disgusted (and a little embarrassed) to learn that Wheatstone himself could be seduced by the charms of stereoscopic enhancement. The second installment of Wheatstone's discussion of binocular vision described how stereoscopic pictures could be produced to show "their true relief" (2½-inch camera separation), but he also noted that "the mind is not unpleasingly affected by a considerable incongruity in this respect; on the contrary, the effect in many cases

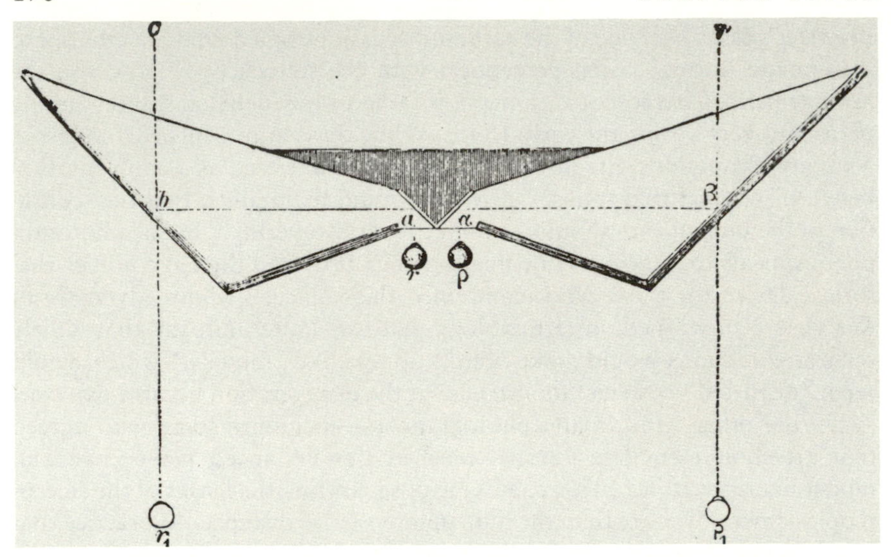

Fig. 7.9. Helmholtz's telestereoscope. Mirrors (a, α and b, β) allow the user's eyes (r, ρ) to see what they would see if they were as widely separated as r_1 and ρ_1. From Helmholtz, *Physiological Optics*, 3:311. Courtesy of the University of Washington Libraries.

seems heightened by viewing the solid appearance, intended for a determinate degree of inclination of the [optic] axes, under an angle several degrees less; the reality is as it were exaggerated."[87]

In 1857, Hermann von Helmholtz devised an instrument that produced this effect for observing natural, rather than photographed, subjects (see fig. 7.9).[88] Helmholtz noticed that stereoscopic pictures with artificially enhanced depth gave "a much clearer representation of the form of a landscape than the view of the landscape itself."[89] This increased appreciation for the layout of the terrain explained "why models of mountains with exaggerated heights please us better than such as represent the elevations on a correct scale."[90] Helmholtz's "telestereoscope" employed a system of mirrors to present images to the eyes from two widely separated points. While using the telestereoscope to study a landscape, Helmholtz observed that the resulting impressions "assume the same bodily appearance as in the stereoscope, and retain at the same time the whole richness of the natural colors, so that images of surprising beauty and elegance are obtained."[91]

The stereography of the moon offered the ultimate example of augmented relief. Two telescopic photographs, taken several months apart, could exploit the moon's libration (its "wobble" that exposes additional portions of the lunar surface to sunlight) in order to provide the parallactic discrepancy necessary for a solid-looking combination (see fig. 7.10). Stereographs of the moon, and of the sun as well, allowed astronomers to discern previously unrecognized details of these bodies' surfaces. The first successful stereograph of the moon was taken in 1858 by Warren De la Rue, at the Cranford Observatory.[92] De la Rue anticipated that some would disapprove of

Fig. 7.10. A lunar stereograph. Made in 1858–1859 by Warren De la Rue, printed by Charles Panknin, London. Courtesy of the Oliver Wendell Holmes Stereoscopic Research Library, Helen D. Moseley Collection, National Stereoscopic Association.

his "unnatural" productions, because they transcended the capabilities of human vision. Defending his work, however, De la Rue insisted that,

> to use Sir John Herschel's words, the view is such as would be seen by a giant with eyes thousands of miles apart: after all, the stereoscope affords such a view as we should get if we possessed a perfect model of the moon and placed it at a suitable distance from the eyes, and we may be well satisfied to possess such a means of extending our knowledge respecting the moon, by thus availing ourselves of the giant eyes of science.[93]

Oliver Wendell Holmes certainly enjoyed gazing through these magnificent oculi. After studying a few representations of the "spotty globe," with all its mountains and craters prominently displayed, Holmes delightedly related the charming effect of the sphere to round "itself out so perfectly to the eye that it seems as if we could grasp it like an orange."[94]

As De la Rue feared, however, not everyone admired such gigantic means for examining the moon. The author of the entry "Stereoscope," in *Chambers's Encyclopaedia* (1883), repulsed by the exaggerated rotundity, mentioned that the lunar surface showed "conspicuous relief" in the stereograph, which "no human eyes" could perceive.[95] The writer displayed little patience for attempts to enhance reality and wished that stereographers "would be content to adopt that exact relation of the two retinal pictures which subsists in ordinary binocular vision."[96]

Some of those who voiced an opinion on "proper" stereoscopic technique adhered rigidly to the standard defined by the human eyes. Yet most accounts contained startling inconsistencies between the recommendations of nature and the advantages gained when these limits were surpassed. Augmented stereographic landscapes delighted an *Edinburgh Review* essayist, for example, even though the productions were "untrue," since no pair of eyes could receive the widely separated aspects unless one's head had been expanded to "Brobdingnagian dimensions."[97] The urges to copy and to enhance the experience of human vision often went hand in hand.

This sort of self-contradiction became most blatant in Sir David Brewster's writings. A single chapter of his book *The Stereoscope* shows the full spectrum of his views. There, he complained that exceeding a 2½-inch lens separation gave "unreal and untruthful pictures, for the purpose of producing a startling relief."[98] He then challenged Antoine Claudet, whose recommendations for stereoscopic photography were flexible and who believed that "there cannot be any rule for fixing the binocular angle of camera obscuras. *It is a matter of taste and artistic illusion.*"[99] Brewster blasted this position. He declared: "No question of science can be a matter of taste, and no illusion can be artistic which is a misrepresentation of nature."[100] Later in the chapter, however, Brewster gave his advice for accommodating colossal statues and other large objects. Within a single internally inconsistent paragraph, he conceded that there may be a "special purpose" which would demand a distant placement of the cameras; yet, he insisted, the addition of "artificial relief is but a trick which may startle the vulgar, but cannot gratify the lover of what is true in nature and art."[101]

By expanding the interval between the dual cameras, the stereoscope joined a number of other popular devices that manipulated and enriched ordinary perception. The camera obscura and the magic lantern had existed for centuries. Publicly staged spectacles, such as the immense panorama and the diorama, were tremendously popular before 1850. Brewster had invented the kaleidoscope, named for its capacity to exhibit beautiful forms, which registered sales in the millions after its introduction in 1817.[102] Other "philosophical toys" of this era created illusions with motion, like the zoetrope or thaumatrope.[103]

There were those who did not appreciate the pleasures of increased perspective that could be had in the stereoscope. However, as indicated by recurrent references to the joy of examining reduced "models" of reality, methods of representing the natural world as a miniature appealed to many in the nineteenth century. Joseph LeConte mentioned that, by controlling the convergence of his eyes, he was able to combine stereographic images without an instrument. Thus, he reproduced the scene "in exquisite miniature, but with perfect perspective. The effect is really marvelously beautiful."[104]

The "Claude Lorraine Mirror" was another contemporary contrivance that supposedly enhanced the visible world.[105] Following the entry for "Wheatstone's Stereoscope" (which actually looks more like Brewster's design), Benjamin Pike Jr.'s *Illustrated Catalogue of Scientific and Medical Instruments* (1856) described its Claude Lorraine mirror:

> I don't know whether it was the invention of the famous Italian artist, who was in landscape paintings what Landseer is in the representation of animals; or whether the mirror was so called because, like Claude Lorraine, it is said to improve upon

> nature; but, at all events, it is a great curiosity. Its construction is the same with the ordinary looking-glass, except that jet is used in place of quicksilver, and it is intended to reflect only the inanimate world. The Claude Lorraine mirror derives its value from the principle that all objects are more beautiful in miniature, which renders their defects less apparent; for the unsightly strikes the eye with immediate pain, while that which is perfect grows upon us more gradually. With this mirror, you frame for yourself, as it were, little landscapes at every turn, in which the sky is softer, the grass richer, and the foliage more graceful, than anything you can see without it.[106]

Like the stereoscope, the Claude Lorraine mirror reworked and embellished the visible world.

Some of these quasi-magical properties were also available for three-dimensional viewing. In addition to designing a popular version of the lenticular stereoscope, Oliver Wendell Holmes invented what he called a "Claude Lorraine Stereoscope." This style featured a "gilded, slanting diaphragm with two oval openings, so that the effect was that of seeing the stereograph through a round window with a golden light on it reflected from the slanting surface of the diaphragm."[107] Holmes admitted that "a Claude Lorraine light on the stereograph, is, in many cases, very striking, but, for common use, the simpler form is preferable."[108]

Additional means of manipulating reality with the stereoscope were also popular in the nineteenth century. Stereographs were occasionally tinted or painted to bring color to the representations. Sir David Brewster has been credited with the development of "spirit" photography, which employs a partial double exposure, allowing ethereal figures to haunt the stereoscopic scene.[109]

A contrivance that created extreme alterations of the visible world was Wheatstone's "pseudoscope," which inverted binocular relief by means of a pair of prisms.[110] Wheatstone devised the pseudoscope in order to study the relationship between binocular and monocular cues, as well as the role of experience and tactile information, in space perception. Although he created the instrument to explore the psychology of vision, Wheatstone was also delighted by the appearance of "another visual world" through the pseudoscope, "in which external objects and internal perceptions have no longer their habitual relation with each other."[111]

Wheatstone appreciated the paradoxes of seeing the inside of a teacup rendered solid and convex, or a terrestrial globe transformed into a concave hemisphere, with the map on the inside. He claimed to know "nothing more wonderful, among the phenomena of perception, than the spontaneous successive occurrence of these two very different ideas in the mind, while all external circumstances remain precisely the same."[112] If more people had shared Wheatstone's excitement over these perplexing sights, the pseudoscope might have fulfilled the expectations of another writer who assumed that all opticians would soon sell the device, "as, from the infinity of its illusions, it is sure, even as a toy, to become popular."[113] There is no indication, however, that the pseudoscope ever enjoyed widespread appeal. Perhaps many found pseudoscopic confusion more disturbing than pleasing. The pseudoscope may not so much have enriched the unaided eye's view as offered a frustrating glimpse of a world turned inside out. Pseudoscopic distortions seem to have marked the limit of instrumental manipulation to produce amusing visual effects.

All of these "philosophical toys" harness elements from the study of physical or physiological optics to create a more satisfying visual impression. The "real world" is transformed into symmetrically arranged bits of color, in the kaleidoscope, and compressed into nicely framed scenes, with the Claude Lorraine mirror. Sir David Brewster showed how the stereoscopic camera might surpass the abilities of normal vision and create enhanced views of large and distant objects. The augmented views produced in the stereoscope, and in these other instruments, offered an improved picture of nature.

Stereoscopic Preservation, Travel, and Education

The stereoscope also offered a measure of control over nature. It could shrink the moon to the size of a piece of fruit. By fixing the desired image, photography and stereography could also, in some sense, stop the course of time. Oliver Wendell Holmes submitted to the camera's rigid requirements—at least for a thirty-second exposure—in order to preserve his three-dimensional presence for curious members of future generations. One of the many other writers who considered the stereoscopic wishes of posterity noticed that when

> Daguerreotype portraits are first seen with the Stereoscope a feeling of regret is common to all, that this discovery does not date from a more distant time. What would not be the value of a stereoscope portrait gallery of our greatest historical characters, including Shakespeare, presenting all the life-like character and resembling in every respect the reflection of the human face in a mirror. Unfortunately the examples of past wonders, a sight of which we must now more than ever lament the loss of, are far too numerous; but now we do possess the astonishing power, it behooves us to think of the future, and not allow coming generations to accuse us of a selfish negligence in not leaving to them a legacy which science has placed at our disposal.[114]

Holmes shared the hope that the vast efforts of stereographers would be organized and preserved. He proclaimed: "We do now distinctly propose the creation of a comprehensive and systematic stereographic library, where all men can find the special forms they particularly desire to see."[115] Through a comprehensive system of stereographic exchanges, "there may grow up something like a universal currency of these bank-notes, or promises to pay in solid substance, which the sun has engraved for the great Bank of Nature."[116] Holmes also suggested the standardization of the double pictures' size and the specifications of stereographic apparatus. Through this uniformity of scale and magnification, we should compare the productions "without the possibility of being misled by those partialities which might tend to make us overrate the indigenous vegetable and the dome of our native Michel Angelo."[117]

Stereoscopic travel represented perhaps the greatest consequence of the photographically disciplined physical world. Sir David Brewster recognized this stereoscopic capacity for visiting foreign lands without the bother of transportation. In *The Stereoscope*, he mentioned that the London Stereoscopic Company offered "no fewer than *sixty* taken in Rome, and representing, better than a traveler could see them there, the ancient and modern buildings of that renowned city."[118] One writer considered the simulation of the "enlarging and ennobling" experience of travel—especially for young persons—as the "highest mission of the stereoscope."[119] Stereoscopic views can disclose "correct ideas respecting that which has hitherto been vague and indeterminate," for those who cannot embark on lengthy voyages.[120] In the study of geography, for example, students will no longer identify a particular country as a "mere diagram upon the map, picked out with blue or yellow, with thin hairy lines marking out the rivers, something like a section of a caterpillar for a chain of mountains, a rough imitation of a wart for a volcano, and a quantity of names in microscopic letters to signify cities, towns, and villages."[121] Instead of such cartographical abstractions, children would come to recognize the landscapes, architecture, and lifestyles of foreign lands.

Antoine Claudet emphasized the hardship that he and his ilk endured in order to bring home the world's most precious sights. "By our fireside," he explained, we might examine these scenes, "without being exposed to the fatigue, privation, and risks of the daring and enterprising artists who, for our gratification and instruction, have traversed land and seas, crossed rivers and valleys, ascended rocks and mountains with their heavy and cumbrous photographic baggage."[122]

Stereographs could also aid the potential tourist. Charles F. Himes, who investigated binocular vision at Troy University (in New York State), illustrated the powerful sense of reality in binocular expeditions. Professor Himes had a friend who examined stereographs of Paris to prepare himself for a visit to that city. After arriving in Paris, Himes's friend reported that "many prominent places in the city had a familiar appearance, that he felt quite at home, and was spared much annoyance and great waste of time, as any one who has been suddenly dumped down in a foreign city can appreciate."[123]

The stereoscope domesticated all Earth. By stereographically capturing the visual form of even the most exotic locations, photographers neatly analyzed and preserved the entire expanse of the planet for the benefit of civilization. The binocular instrumentation provided a metaphorical means for controlling nature, and, as several historians have shown, the choices of stereoscopic subjects revealed a similar urge. The expeditions across the American west, the Arctic, and other uncharted wilderness; the construction of the transcontinental railroad; and the architecturally symmetrical world's fair cities—these were among the favorite themes for stereographs. Likewise, the stereoscope serenely presented the aftermaths of human and natural catastrophes—such as military battles, fires, floods, and earthquakes—within the safe and cozy surroundings of one's own parlor.[124]

The unique capacity of the stereoscope to convey information about distant places led to its service as a pedagogical technology. This application became especially popular in the early twentieth century, when stereoscope manufacturers and leaders of the progressive "visual education" movement jointly promoted the instrument's educational value.[125] In addition to 3-D cinema, the lineage of the stereoscope may be recognized in modern technologies such as holography and virtual reality—applications whose alleged importance for entertainment, education, and industry is reminiscent of the proclamations once made on behalf of the stereoscope.

Conclusion: Instruments and Conventions

In the history of depiction, nineteenth-century discoveries in the fields of photochemistry and binocular vision combined to produce a fantastically popular visual medium—one that sustained an unprecedented correspondence to the physiology of sight. Modern historians, philosophers, and scientists have debated whether such tools as perspective painting, the photographic camera, and the stereoscope do, in fact, duplicate the human visual field; and they ask whether these techniques constitute a means for producing indisputably correct portraits of the world or merely a set of representational *conventions*.[126] Discussing the camera obscura, Svetlana Alpers reprised the haunting question: "But why did such a model of the 'natural' picture prevail in the first place? And what is its nature?"[127] In the nineteenth century, this matter was solved by the dictates of natural theology, since this framework accepted the human visual organs as perfectly designed instruments that provided the template for an ideal representation of nature. Animated by the capacity to preserve images, the photographic plates became surrogate retinae and the stereoscopic camera was transformed into a pair of external eyes.

Although this notion formed the basis for the nineteenth-century discourse on photography and the stereoscope, there existed a difference of opinion regarding the proper implementation of this technology. There was a marked tension (even among the statements of an individual) between the position that altering the conditions of sight entailed a transgression of divine authority and the belief that such an alteration provided a valuable extension of one's ability to understand nature. Since stereoscopic enhancement and distortion could both arise from a single process—varying the placement of the cameras—one may ask whether *distortion* was a meaningful category, or merely a term applied to representational styles that were *unfamiliar* and *unconventional.*

This has been a continuing problem in the history of scientific instruments. Galileo, for example, identified the eye as an optical instrument, although not an ideal one. He recognized that the eye is *not* an immediate source of information about nature, and that one's conception of the physical world is dependent upon the means used to study it.[128] When Galileo suggested that the visual capacity of the naked eye could be improved with a telescope, he had to show that the new information available with his device was not a distortion. Similar conflicts surrounded other novel scientific instruments that challenged the prevailing conventions for the representation of nature. Seventeenth-century advocates of the air pump and the dispersion prism—like proponents of more recent mega-instruments, such as the radio telescope or particle accelerator—had to convince other researchers that the phenomena manifested by their apparatus were not artificial aberrations.[129] There is no way to separate the new phenomena from the tools used to study them.

Techniques for the representation and study of nature are always embedded in a social, aesthetic, and scientific matrix. Such tools never provide a neutral mediation between observers and the world. Rather, an instrument embodies an approach to nature, as well as a means for constructing knowledge. By appreciating the complex function of these systems of mediation, historians may use them to learn about the past. This method is particularly helpful in the case of the stereoscope, because the debates concerning the design and role of binocular devices reflected the spectrum of attitudes regarding the status of the human frame as the supreme model for learning about nature. Whether it copied the 2½-inch separation rule for truthful depiction or created visual effects beyond the capacity of any human, the stereoscope became a mechanical analogue for the nineteenth-century mind. It delineated both the human standard of accurate representation and the potential of instruments to improve or to distort the perception of nature.

CHAPTER EIGHT

Vox Mechanica: The History of Speaking Machines

THE DESIRE to imitate the human voice is as ancient as history and as pervasive as human culture. Because the goal has appeared in a variety of investigative contexts, we do not expect to find a single line of development stretching from the speaking heads of antiquity to modern computer synthesizers. Instead we find different groups concerned with different aspects of the problem: natural magicians using speaking tubes or ventriloquism to produce the *appearance* of artificial speech; students of physiology trying to understand the *mechanism* of speech; acousticians trying to *analyze and reproduce* vowel sounds; inventors creating apparatus to *record and transmit* speech at a distance; musicians attempting to *duplicate the timbre* of the voice in their instruments. For this reason, the manifest ability of the apparatus in question—that is, their success in imitating the voice—did not improve dramatically during the bulk of this history. However, the instruments' designs did undergo substantial changes. These changes reflected not only the particular contexts in which the devices were created, but also alterations in the theory of the origin of vocal sounds and in the criteria for what constituted such a theory.

As in the case of the stereoscope, there was a question as to how one should duplicate human function. Should a speaking machine copy the anatomy of the organs of speech, or should it merely re-create the sounds of speech? Should it speak with a normal human voice or should it magnify the voice? If the purpose of the machine was to teach the deaf to speak by allowing them to feel the positions of the lips and tongue, then the machine had to copy human anatomy. If its purpose was to show how speech sounds were created from a combination of overtones, then it should use tuning forks and resonators or, perhaps, a wave siren.

Much of the impetus for creating speaking machines came from the practitioners of phonetics, elocution, and stenography, who wished to make their sciences less subjective. A speaking machine, if successful, could serve as a standard for pronunciation; it could help to analyze speech sounds into their phonetic components; and it could aid in forming a truly phonetic shorthand, because a phonetic sound could correspond to a unique arrangement of the elements of the speaking machine. The most practical goal for a speaking machine during the preelectronic age was to amplify speech so that a single machine could address a large crowd. Such a machine could also deliver a sermon or a speech any number of times in any number of places. And finally, speaking machines had value as entertainment. In this arena, fraud was often more successful than science. It is from this confusion of approaches and motives that modern instruments for recording, communicating, and synthesizing the voice emerged.

Speaking Heads and Automata through the Eighteenth Century

Although the ancient speaking statues and the Greek head of Orpheus at Lesbos were "fakes"—their effect was produced by concealed priests whose words reached the statue's lips through a tube, or by ventriloquism—these examples deserve attention, by merit of their outward appearance and effect. The nineteenth-century Scottish physicist David Brewster portrayed these statues as a facade that empowered the elites to enslave the ignorant masses.[1]

The long history of mythical speaking mechanisms often involved the supernatural.[2] These tales usually connect mechanical dexterity with sorcery. Gerbert (Pope Sylvester III from 999 to 1003) constructed a speaking head of brass (whose vocabulary was limited to "yes" and "no"). The builder was subsequently accused of practicing magic. Albertus Magnus allegedly constructed a head of earthenware that could move and speak. Legend has it that Thomas Aquinas was so terrified when he saw the head that he smashed it, causing the maker to exclaim, "There goes the labor of thirty years." Another version says that Albertus was using strange tools and devices to construct a statue of a beautiful girl. When Thomas discovered her, she said, "Salve, salve," and he was convinced that the devil was involved in its fabrication. Robert Grosseteste supposedly constructed a speaking head of brass that could foretell the future. Roger Bacon and his cohort, Friar Bungay, crafted a brazen head that exactly copied the internal works of human anatomy. Yet to obtain for it the power of speech, they needed to seek advice from Satan.

The imitation of the voice also appeared in the works of Francis Bacon. In the *New Atlantis* (published posthumously in 1627), Bacon postulated his ideal plan for a facility that would produce practical knowledge for the improvement of civilization. The purpose of this fictional enclave, called "Salomon's House or the College of the Six Days' Works," was to discover "the knowledge of Causes and secret motions of things; and the enlarging of the bounds of Human Empire, to the effecting of all things possible."[3] A tour of Salomon's House shows how far it is from our own time. Its main goal was to imitate the phenomena of nature by artificial devices. "We have," Bacon wrote,

> large and deep caves . . . for the imitation of natural mines . . . artificial wells and fountains, made in imitation of the natural sources and baths . . . great and spacious houses, where we imitate and demonstrate meteors; as snow, hail, rain . . . thunders, lightnings. . . . We have also furnaces of great diversities . . . in imitation

> of the sun's and heavenly bodies' heats . . . artificial rainbows, haloes, and circles about light. . . . We imitate smells. . . . We make divers imitations of taste likewise, so that they will deceive any man's taste.[4]

As far as the present chapter is concerned, the most intriguing feature of Salomon's House is found in the "sound houses,"

> where we practise and demonstrate all sounds, and their generation. . . . We represent and imitate all articulate sounds and letters, and the voices and notes of beasts and birds. We have certain helps which set to the ear do further the hearing greatly. We have also divers strange and artificial echoes. . . . We have also means to convey sounds in trunks and pipes, in strange lines and distances.[5]

We can interpret Salomon's House as an embellished account of Bacon's hopes for the future of metallurgy, agriculture, brewing, textiles, and so on. His goal was essentially practical and he sought to control the natural world by imitating it artificially. These imitations could be either exact copies (as in the duplication of animal sounds) or improvements on nature (as in the transmission of the voice over great distances). In either case the artificial duplication of nature produced a sense of wonder in the beholder.

The seventeenth-century polymath Athanasius Kircher was notable for his exhibitions of startling effects. In his *Musurgia universalis* (1650), he claimed that it would be possible to create a dramatic speaking statue. Kircher said that its observers would

> certainly hear and see the wonder of the talking figure, but would not be able to penetrate the origin of the secret process. They would observe the motion of the eyes, marvel at the mobility of the lips and the tongue, and look at the structure of the whole living and breathing body with astonishment, but what art moved the figure and what hidden motive force it possessed, nobody would be able to discover since it would hover freely in the air, supported by nothing, not connected to a tube and not driven by any wheel, but brought into being quite naturally by the "ars combinatoria."[6]

The operation of this speaking head was pure artifice. Kircher intended to make it to entertain the queen of Sweden, but it was never completed. He passed the secret of the floating head to his student Gaspar Schott, who said that the illusion was simple but expensive to execute. The feat was probably achieved with mirrors cleverly arranged to conceal the speaking head's owner. In the nineteenth century, Gaston Tissandier described this prank in his *Popular Scientific Recreations* (1883).[7]

John Wilkins was one of the individuals who promoted Francis Bacon's ideas on the progress of science. Wilkins was at the center of the pre-Restoration natural philosophical activities at Oxford that led to the foundation of the Royal Society. Problems of language, communication, and cryptography fascinated Wilkins, and he explored these topics in his *Mercury, or The Secret and Swift Messenger* (1641). The dissemination of knowledge, he lamented, was fettered by the proliferation of vernacular languages—a situation he hoped to amend by the formation of a simple and universal "philosophical language," which he developed later in his most important book, *An Essay Towards a Real Character, and a Philosophical Language* (1668). A large part of the *Real Character* was devoted to phonetics—the exploration of the production of speech sounds. Wilkins may have learned a good deal of this material from his experience in helping John Wallis, who taught a deaf boy to talk. Gadgetry (especially the possibility of constructing a flying machine) was another of Wilkins's interests, and he presented the basic principles of mechanics in his *Mathematicall Magick* (1648).[8]

Given his leanings, it is not surprising that he touched on the issue of speaking machines. In *Mathematicall Magick*, Wilkins discussed some mythical attempts to imitate the voice, such as Roger Bacon's brazen head and Albertus Magnus's statue.[9] Wilkins also related a report of a "cold Countrey, where the peoples discourse doth freeze in the air all winter, and may be heard the next Summer, or at a great thaw."[10] "But this conjecture," he added, "will need no refutation."[11] "The more substantiall way for such a discovery," Wilkins argued,

> is by marking how nature her self doth imploy the severall instruments of speech, the tongue, lips, throat, teeth, &c. To this purpose the Hebrews have assigned each letter unto its proper instrument. And besides, we should observe what inarticulate sounds doe resemble any of the particular letters. Thus we may note the trembling of water to be like the letter *L*, the quenching of hot things to the letters *Z*, the sound of strings, unto the letter *Ng*, the jirking of a switch the letter *Q*, &c. By an exact observation of these particulars, it is (perhaps) possible to make a statue speak some words.[12]

John Evelyn's *Diary*, however, tells us that Wilkins learned a lesson from history on how to make a simple, but effective, speaking statue. Among his peculiar possessions, the "universaly Curious" Dr. Wilkins had "an hollow Statue which gave a Voice, & utterd words, by a long & conceald pipe which went to its mouth, whilst one spake thro it, at a good distance, & which at first was very Surprizing."[13] Wilkins kept such a statue in his garden, expressly for playing tricks on friends.[14] Although there is no definite evidence that Wilkins ever completed the more sophisticated device mentioned in the *Mathematicall Magick*, we find in Christopher Wren's list of the "new theories, inventions, experiments, and mechanic improvements" that he had shown to Wilkins's group in the 1650s, "A Speaking Organ, articulating Sounds."[15]

Robert Hooke, the curator of experiments for the Royal Society, conducted several investigations on sound and music.[16] One of his experiments involved toothed brass wheels that, upon rotating, imparted periodic blows to a card or other object and created sounds of various pitches. Félix Savart built an identical device in 1830 and believed that his was original. Hooke began these investigations in 1676, although he did not present his work to the society until 1681.[17] On July 27 of that year, his biographer Richard Waller recounted, Hooke "shew'd a way of making *Musical and other Sounds*, by the striking of the Teeth of several Brass Wheels, proportionally cut as to their numbers, and turned very fast round, in which it was observable, that the equal or proportional stroaks of the Teeth, that is, 2 to 1, 4 to 3, &c. made the Musical Notes, but the unequal stroaks of the Teeth more answer'd the sound of the Voice in speaking."[18] It does not seem that Hooke had any intention of imitating the voice. He simply hit upon this observation by accident.

In addition to a speaking statue, John Wilkins's garden had impressive waterworks, including one that produced a mist and created a dramatic rainbow.[19] Such fountains, especially in connection with statuary and pneumatic musical instruments (such as imitation singing birds), were common adornments to palaces and pleasure gardens in the seventeenth century. Salomon de Caus's *Les raisons des forces mouvantes avec diverses machines tant utiles que plaisantes* (1615) included many examples of this variety of hydraulic amusement. The basic elements of these devices had been employed for entertainment since ancient times, and there is a continuous history of amusing automata through the nineteenth century.[20]

Jacques de Vaucanson ranks among history's most celebrated automaton builders. Vaucanson earned his reputation in the 1730s, when he exhibited three mechanical wonders: a flute player, a pipe and drum player, and a duck that could eat, drink, and excrete (see fig. 8.1). Legal squabbles over the flute player's profits indicate that Vaucanson intended to make money from the entertainment value of his productions. However, André Doyon and Lucien Liaigre, Vaucanson's biographers, portray the mechanic as an unrecognized cybernetic theorist, and they view his automata (especially the duck) as demonstrations of the mechanical approach to physiology.[21] They also depict him as a frustrated scientist who reluctantly entered the marketplace to gain financial support for his projects.[22]

Doyon and Liaigre have traced a tradition of the deployment of functional anatomical models—"anatomies mouvantes"—as tools for discovery in the history of physiology. Bacon had suggested such a method with vague language in his *New Atlantis*, since the inhabitants of Salomon's House "imitate also motions of living creatures, by images of men, beasts, birds, and serpents."[23] Descartes employed automata—like those designed by de Caus—as analogies for the human mechanism, in his *Traité de l'homme*.[24] According to Marin Mersenne, working models could be used to demonstrate William Harvey's theory of the heart.[25] The heart's operation must have seemed a reasonable object of study by this method. In 1677 the *Journal des Savants* noted a statue planned by a Württemberg physician that demonstrated the circulation of blood.[26]

Vaucanson's interest in physiological simulacra was evident in his 1731 proposal to make "*a machine of physics containing several automata in which the natural functions of several animals are imitated by the movement*

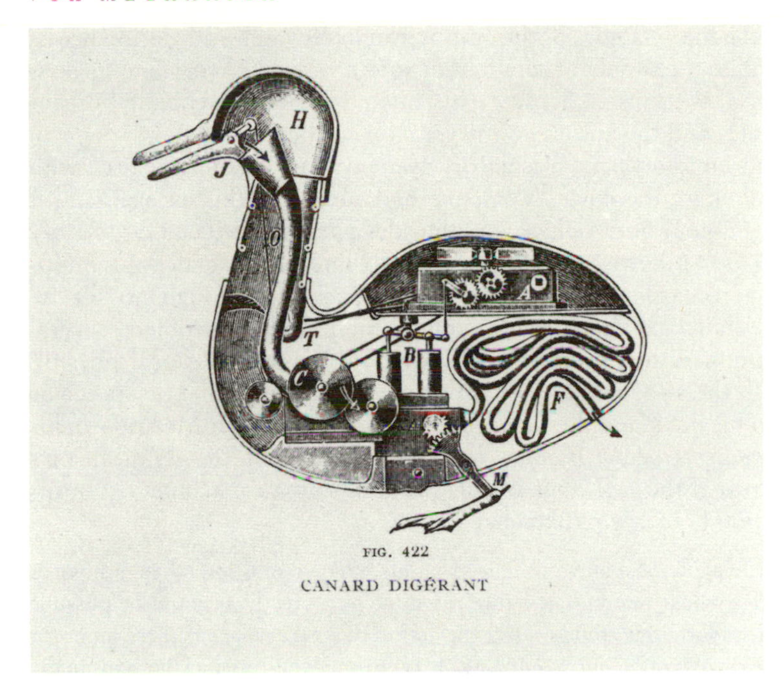

Fig. 8.1. Vaucanson's duck. From Chapuis and Gélis, *Le monde des automates.* Courtesy of the Science and Technology Research Section, Science, Industry and Business Library, The New York Public Library, Astor, Lenox and Tilden Foundations.

of fire of air and of water."[27] Ten years later, Vaucanson proposed a similar automaton—one that could be used for physiological experimentation.[28] His efforts may have been promoted by one of his acquaintances—the surgeon and anatomist Claude-Nicolas Le Cat, who treated living creatures as mechanical systems. In the late 1720s, Le Cat became involved in a debate concerning the role of therapeutic bleeding. Both Le Cat and his principal adversary—the physiocrat François Quesnay—proposed the use of working models of the circulatory system to determine the hydrostatic effects of bleeding. In 1744, Le Cat offered to the Academy of Rouen his "*Description of an automaton in which one can see the execution of the principal functions of the animal economy, circulation, respiration, secretions & by means of which one can determine the mechanical effects of bleeding, & submit to the judge of experience several interesting phenomena which do not seem perceptible.*"[29] No apparatus and no text survive from his proposal. About the same time as the suggestions of Vaucanson, Quesnay, and Le Cat, a machinist named Launois presented an artificial circulatory system at the St.-Laurent Fair.[30]

Doyon and Liaigre have documented Vaucanson's own lengthy, but ultimately fruitless, plans to build an automaton that would mimic the circulation of blood. The project intrigued Louis XV, who approved (and even demanded) the manufacture of the automaton in Guyana—a location necessary to maintain the supply of caoutchouc (india rubber). According to Condorcet's *éloge*, Vaucanson became frustrated with the attendant bureaucratic obstacles and the automaton never came to be.[31]

Projects to build artificial circulatory systems were often associated with attempts to imitate the voice. Mersenne had suggested this as well, and it was an alleged goal of the physician who designed an artificial circulatory system in 1677.[32] Likewise, Le Cat purportedly intended to endow his multifunctioned automaton with the power of speech.[33] Although no surviving evidence indicates Vaucanson's explicit interest in the problem, several writers summoned the mechanician to solve the mystery of vocal physiology. In 1738, the abbé Desfontaine invited Vaucanson to imitate speech as an encore to his duck and flute player.[34] Julien Offray de La Mettrie—in his 1748 materialist tract, *L'homme machine*—explained that Vaucanson's skill might reveal the workings of this piece of human machinery.[35] In his *L'art du chant* (1755), Jean Blanchet wrote:

> One could imagine & make a tongue, a palate, some teeth, some lips, a nose & some springs whose material & figure resemble as perfectly as could be possible those of the mouth: one could imitate the action that takes place in these items for the generation of words: one would be able to arrange these artificial organs in the automaton of which I have spoken. From then on, it will be capable of singing, not only the most brilliant airs, but also the most beautiful verse. Here is a phenomenon that would demand all the invention & industry of an Archimedes, or else a Vaucanson, & that would astonish all of learned Europe.[36]

Part of the interest in this problem arose from Antoine Ferrein's challenge to the prevailing assessment of vocal physiology. In 1700, Denis Dodart had offered a major revision of Galen's theory of the voice. While Galen had compared the vocal organ to a flute and had claimed that the length of the trachea determined the vocal pitch, Dodart dismissed the analogy and attributed the production of all sounds to the glottis.[37] In 1741, however, Ferrein introduced a new musical instrument analogy for the voice. (As the subsequent discussion will show, the comparison of the voice to various musical instruments continued long after Ferrein's time.) Ferrein believed that the folds of the glottis formed two true "vocal cords"—he coined the term—and that air rushing through the glottis produced sounds in the manner of a bow drawn across the strings of a *violin*. One of Ferrein's supporters, the physician Henri-Joseph-Bernard Montagnat, challenged Vaucanson and Castel to resolve the dispute with a functional mechanical model: "There is only the author of the color harpsichord, or that of the automaton flute player, who can succeed in giving us a pneumatic harpsichord whose sounds could imitate the voices of different animals, or only those of man, so varied in each individual."[38] Although no evidence exists showing that Vaucanson took up the matter, Doyon and Liaigre believe that

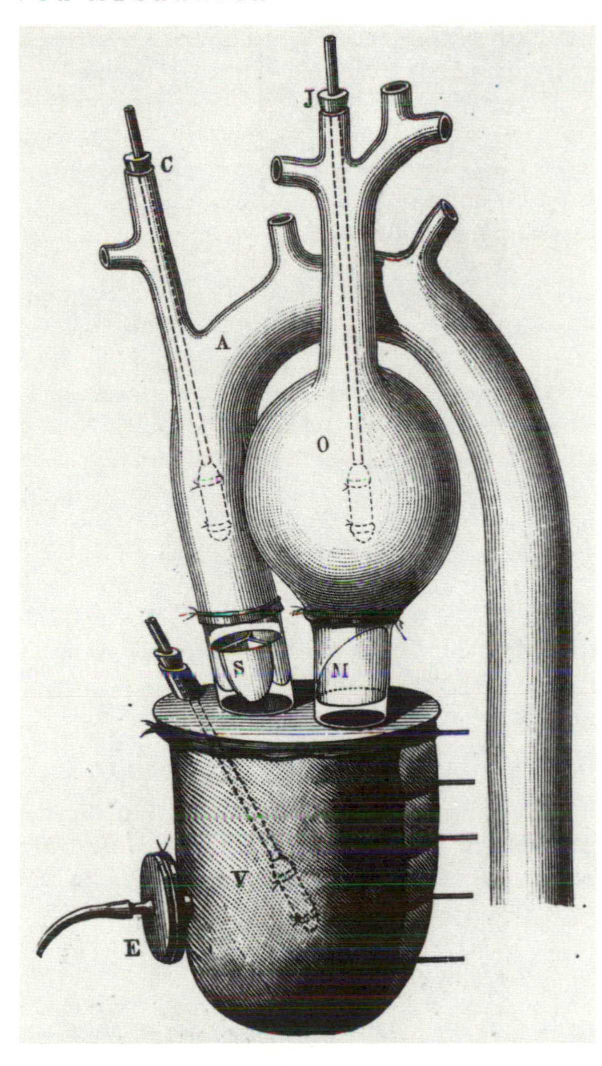

Fig. 8.2. Marey's artificial heart. From Marey, *La circulation du sang*, p. 712. Courtesy of the University of Washington Libraries.

"le climat intellectuel et philosophique" would have dictated Vaucanson's involvement.[39] In any case, eighteenth-century physiologists and philosophers regarded the voice as a reasonable subject for investigation by means of functional anatomical models.

Like Bacon's artificial world—which may seem pointless to a modern observer—imitating nature offered the possibility both to acquire and to demonstrate an understanding of phenomena.[40] For example, Étienne-Jules Marey constructed several physiological simulacra. Among his labors, Marey took up Vaucanson's unfinished problem—he made an artificial heart to aid his research on the circulation of blood (see fig. 8.2).[41] In this light, the efforts of Vaucanson, Le Cat, and Quesnay appear as attempts to demonstrate the truth of the Cartesian animal-machine and to show, as La Mettrie hoped to do, that the human body is nothing more than an elaborate mechanism.

Speaking Machines in the Late Eighteenth Century: Mical, Kratzenstein, Kempelen, and Darwin

With the speaking machines of the late eighteenth century, one is able to leave behind rumors and guesses and to discuss devices whose existence can be shown with certainty. Between 1770 and 1790, four persons—the abbé Mical, Christian Gottlieb Kratzenstein, Wolfgang von Kempelen, and Erasmus Darwin—produced functional speaking machines. Strangely, all of them worked in diverse parts of Europe with no apparent knowledge of each other.

The abbé Mical, who lived in Paris, had attained some personal wealth and spent his leisure time in the construction of mechanical amusements. He built two mechanical flute players and, eventually, a small ensemble of automata, which he destroyed after the figures' nudity was criticized. In 1778, he made a ceramic head, which could utter a few phrases. He destroyed this mechanism also, because he felt that it was unworthy of the praise it received in the *Journal de Paris*. His most elaborate production was completed in 1783—a pair of heads that exchanged sentences praising the king (see fig. 8.3).[42]

These heads were exhibited in Paris, and in 1783 Mical asked the Académie des sciences to examine his work. The anatomist Félix Vicq d'Azyr (who also studied the physiology of the voice) wrote a favorable, though not enthusiastic, review of the heads. The vocal sounds were produced by a bellows attached to several artificial glottises placed over stretched membranes, and gave a "*very imperfect* imitation of the human voice."[43]

Mical found an ardent supporter in the second-magnitude *philosophe* and Royalist propagandist Antoine Rivarol. Rivarol's enthusiasm came from the application he envisioned for the speaking heads—preserving examples of proper French pronunciation. Rivarol discussed this possibility in the notes to his "De l'universalité de la langue française," his offering on a subject proposed by the Berlin Academy in 1783. The speaking heads also appear in Rivarol's "Lettre à M. le Président de ***, Sur le Globe aérostatique, sur les Têtes-parlantes, et sur l'état présent de l'opinion publique à Paris" (dated 1783). These accounts of Mical's productions were vastly more laudatory than that of Vicq d'Azyr. Rivarol claimed that the heads pronounced their sentences "nettement" and in a voice that was "surhumaine."[44] Rivarol's postscript even suggested that such devices may provide a means for the deaf to communicate.[45]

Mical hoped to sell his speaking heads to the academy. However, upon the recommendation of Lieutenant of Police Lenoir—who deemed Mical's

Fig. 8.3. Mical's speaking heads. From Chapuis and Gélis, *Le monde des automates*, 2:205. Courtesy of the Science and Technology Research Section, Science, Industry and Business Library, The New York Public Library, Astor, Lenox and Tilden Foundations.

contrivance unworthy—the purchase was rejected. According to Rivarol, Mical (true to form) destroyed his masterpieces in a fit of despair and died riddled with debts in 1789. Another version held that the heads were sold. In any case, their fate is unknown today.[46]

Mical is clearly a part of the tradition of automaton makers. Rivarol bluntly portrayed the mechanician as the successor of Vaucanson.[47] The speaking heads were presented as entertainment, and even though Rivarol held lofty hopes for their service, one cannot say with any certainty that Mical sought anything but financial gain in presenting his work to the academy.

While the edifice of "official" science (the Paris Academy) was a source of frustration for Mical, the situation was precisely the opposite for Christian Gottlieb Kratzenstein. Born and educated in Halle, Kratzenstein became a professor of physics and medicine at Copenhagen University. His best-known research involved the therapeutic applications of electricity. The occasion of his publication on the sounds of the voice was a 1779 contest sponsored by the Imperial Academy of St. Petersburg, of which Kratzenstein was a member. The academy established two tasks: (1) determine the nature and character of the vowels *A*, *E*, *I*, *O*, and *U*; and (2) construct an instrument, like the *vox humana* pipes of the organ, that could accurately express the sounds of the vowels.[48] Neither Mical, Kempelen, nor Darwin mentioned this contest or seem to have known about it. Kratzenstein, however, earned the prize.

The great mathematician Leonhard Euler almost certainly instigated the competition. He was the driving force at the St. Petersburg Academy, and he had shown an interest in the problem in his *Lettres à une Princesse d'Allemagne*. One of these letters, entitled "The Wonders of the Human Voice" (dated June 16, 1761), sketched the very problem that the academy later proposed. "In many organs," Euler wrote, "there is a stop which bears the name of the human voice; it usually, however, contains only the notes which express the vocal sounds *ai* or *ae*. I have no doubt that with some change it might be possible to produce likewise the other vocal sounds *a*, *e*, *i*, *o*, *u*, *ou*."[49] But Euler continued his discussion and contemplated how one could duplicate mechanically all the sounds of speech and not simply the vowels. He concluded:

> The construction of a machine capable of expressing sounds, with all the articulation, would no doubt be a very important discovery. Were it possible to execute such a piece of mechanism, and bring it to such perfection that it could pronounce all words, by means of certain stops, like those of an organ or harpsichord, every one would be surprised, and justly, to hear a machine pronounce whole discourses or sermons together, with the most graceful accompaniments. Preachers and other orators, whose voice is either too weak or disagreeable, might play their sermons or orations on such a machine, as organists do pieces of music. The thing does not seem to me impossible.[50]

Since Kratzenstein had known Euler and had participated in the academy before 1779, it is possible that Euler had encouraged Kratzenstein's studies or that Euler's interest in Kratzenstein's work precipitated the contest.

In addition to his institutional setting, Kratzenstein's research tradition distinguished him from Mical. Neither automata nor accounts of natural or unnatural magic appeared in Kratzenstein's prize-winning essay. Rather, Kratzenstein aligned his efforts with those of the anatomists who had studied the voice: Jean Coenrad Amman (who employed his physiological efforts in the vocal instruction of deaf-mutes), Denis Dodart, Antoine Ferrein, and Albrecht von Haller.

Kratzenstein began his essay with a description of the anatomy of the throat and mouth, and he then turned to their function. He compared a child learning to use these structures to "an Organist who searches on his keyboard for the tone that he wants to play."[51] Kratzenstein recognized this same groping in a deaf boy who had begun to hear at the age of sixteen and learned to speak, as well as in the way humans learn to control their muscles, but then lose their awareness of conscious control.

> When we are adults, we act on our bodies as one who is blind since birth does on an organ whose structure he does not know at all: by force of practice, he succeeds in performing the music, without knowing that by his voluntary action, he makes different levers move, that he opens valves by means of which he supplies wind to the pipes; it is only by touch that he can learn it.[52]

Jamie Kassler has pointed to statements like this one to show that before the era of computer models, the playing of *musical instruments* was a commonly used analogy of cerebration.[53] But for the present chapter, it is instructive to consider Kratzenstein's remark as a metaphor for the *body* as well as the mind. He thought of the organs of the body—and those of the *voice*, in particular—as musical instruments.

Among his predecessors, Kratzenstein showed the most respect for Amman and Haller, and he agreed with their explanation of vowel sounds: "*The vowels are sounds modified by diverse openings of the mouth & the elevation of the tongue.*"[54] His turn toward the subject of speaking machines produced a striking allusion: "I have never, like Cl. Ammann, been occupied with teaching the mute to talk, nor like Pygmalion with animating an ivory statue; but for several years I have been occupied, in my moments of leisure, with a machine which can counterfeit the human voice, & which, like a musical instrument, can, by the help of the fingers, articulate some words."[55] From his observations of pronunciation, Kratzenstein constructed a table of the positions of the larynx, tongue, teeth, palate, and lips for each vowel. (Kratzenstein believed that the larynx played distinct roles in sounding the vowels *i* and *u*.) He surveyed the theories of the voice, noting Galen's flute analogy. Kratzenstein also explained that although Dodart rejected comparisons with common musical instruments, the French anatomist did suggest "a species of analogy with a paper flute, or what is called *chassis bruyant* (in

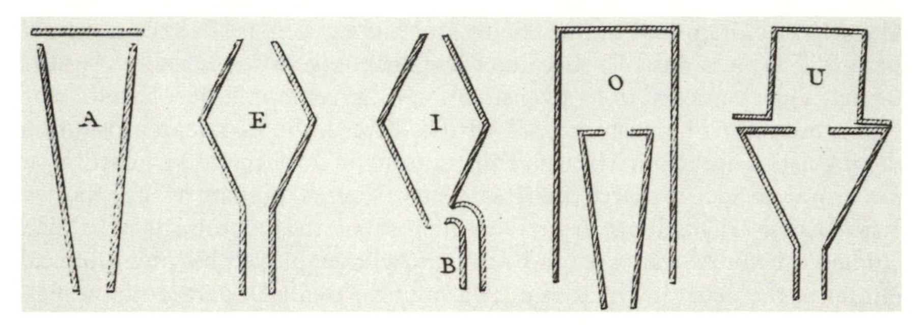

Fig. 8.4. Kratzenstein's vowel pipes. From Young, *A Course of Lectures*, vol. 1, pl. 26. Courtesy of the University of Washington Libraries.

German *papier schnarrwerk*)."[56] Kratzenstein described this as the common experience of blowing air between two strips of paper. He objected to Ferrein's violin-string metaphor and suggested that if one must make such a comparison, one should think of the glottis as a drum with a bisected membrane.[57] "In truth, such a membrane, more or less stretched, moved by the wind or by a stick, would not render a tone distinguished as a musical tone; but one could say that such a membrane covers the opening of the trachea, & allows the air to escape in sonorous undulations, in the same manner as that executed in the artificial glottis of the *voix humaine* of organs."[58] Using the natural organ of speech as his model, Kratzenstein described his artificial glottis—a contrivance that became known as a "free reed" (an innovation usually attributed to the organ builder Grenié).[59] By combining this reed with pipes shaped according to the situation of the tongue, lips, and so forth, Kratzenstein produced a set of peculiar-looking organ pipes that imitated the vowels (see fig. 8.4). As organ builders could easily reproduce these vowel pipes, Kratzenstein expected performances of their extraordinary effects to become widespread. He warned, however, that careless combinations of the sounds could produce disquieting diphthongs.[60]

While Mical and Kratzenstein were situated on opposite sides of the border between elite scientific societies and the public realm of theater and entertainment, one cannot as easily discern the scientific context of Wolfgang von Kempelen. Kempelen was a part of the court culture of eighteenth-century aristocracy. He served as privy councillor (*Hofrat*) for the Viennese court and he directed the Hungarian salt industry. Aside from these duties, Kempelen was renowned for his mechanical talents. The design of the hydraulic system for the pleasure garden and fountains at Schönbrunn would rank among his later achievements.

However, Kempelen's greatest source of fame was his automaton chess player, which he built in 1769. In that year, a Frenchman named Pelletier performed some tricks of magnetism (what we would call "animal magnetism" or spiritualism) for Empress Maria Theresa of Austria and the Viennese court. Kempelen told the empress that he could devise a far more intriguing demonstration than those of the visitor, and the mechanician was instructed to undertake his project. Six months later, he brought forth the chess player. This machine consisted of a seated figure dressed like a Turk, who would move his own chessmen on the board before him (see fig. 8.5). This mechanical marvel stunned and delighted the empress, her family, and a throng of visitors. Around 1773, deciding that more pressing matters than the Turk demanded his attention, Kempelen dismantled his creation.[61]

Kempelen was compelled to revive the chess player in 1781, when Maria Theresa's successor, Joseph II, requested a performance by the Turk in honor of the visit of the Russian royal couple, Grand Duke Paul and Grand Duchess Maria Feodorovna. The chess player earned the praise of its audience, and the duke and duchess encouraged Joseph to send Kempelen and his automaton on a European tour. This journey, begun in 1783, included the major cultural and political centers, such as Paris, Versailles, London, and Berlin.

Kempelen did not provide a full description of the machine's function. Cryptically, he confessed that "as a piece of machinery, it is not without merit, but its effects appear so marvelous only from the boldness of the conception and the fortunate choice of the methods employed to promote the illusion."[62] Fortunate choice, indeed! Numerous pamphlets and essays, attempting to expose the true secret of the Turk's operation, revealed how a person could fit inside the compartments of the case that was supposedly full of machinery (see fig. 8.6). (Such publications appeared with greater frequency in the nineteenth century, when Maelzel widely exhibited the chess player.)

Kempelen's behavior seems utterly audacious. How could he parade this deceit in front of the European nobility? Perhaps the Turk's performances were viewed somewhat like a modern magic show—in which the audience is aware that the performance is a trick, but the quality of the illusion makes it satisfying. In fact, many observers were certain that some sort of duplicity was involved.[63] Kempelen even allowed a large lodestone to sit on the device to thwart the suggestion that he controlled the Turk's movements with magnetic influence. However, numerous reports presented Kempelen as a new Prometheus, so at least some of the spectators may have believed that the chess player was entirely mechanical—a possibility that would have been encouraged by successful calculating machines, such as those of Leibniz and Pascal. Furthermore, there would have been no point in debunking the chess player if it were an obvious fraud.

The career of the chess player gives some insight into the context of Kempelen's activity, but the present study is concerned primarily with Kempelen's *second* most famous invention—his speaking machine. Kempelen describes his efforts toward this end in a volume that he published in French and German in 1791: *Le mécanisme de la parole suivi de la description d'une machine parlante* and *Mechanismus der menschlichen Sprache nebst der Beschreibung seiner sprechenden Maschine*.[64] In this book, Kempelen

Fig. 8.5. Kempelen's chess-playing automaton. From Friedrich, *Über den Schachspieler des Herrn von Kempelen*. Department of Rare Books and Special Collections, Princeton University Libraries.

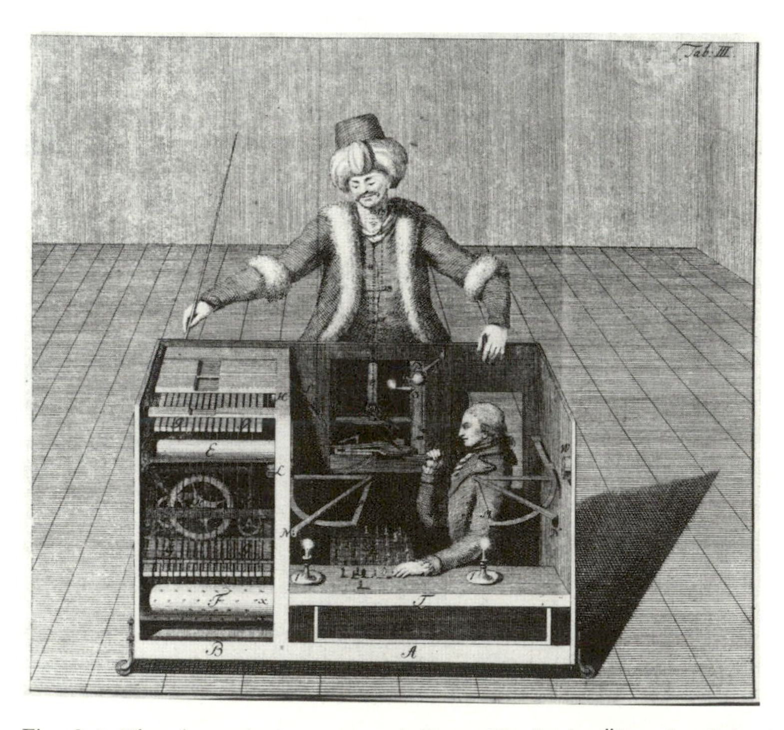

Fig. 8.6. The chess player—exposed. From Friedrich, *Über den Schachspieler des Herrn von Kempelen*. Department of Rare Books and Special Collections, Princeton University Libraries.

surveyed theories of the origin of language—a critical topic during the eighteenth century that had been examined by such authors as Jean-Jacques Rousseau and Johann Gottfried Herder.[65] In many ways, Herder's 1772 essay marked the endpoint of discussion on this subject, and Kempelen followed Herder in rejecting the possibilities of either a divine linguistic source or a natural or primordial human language.[66] Kempelen had also studied the works of Dodart, Ferrein, Haller, and other authorities on the physiology of the voice. The bulk of the book, however, was dedicated to phonetics and provided a detailed account of the formation of articulate sounds. Kempelen suggested that his labors might help in the instruction of deaf-mutes, as well as persons with speech impediments.

Kempelen's final chapter discusses his speaking machine. Although the topic constitutes a small portion of the book, it seems that Kempelen was led to the other linguistic topics through his interest in speaking machines. In a passage that is disheartening to a historian, Kempelen admitted, "I do not absolutely recall what was the first cause that furnished me with the idea of imitating human speech; I remember only that during the time that I was working on my chess player, in the year 1769, I began to examine various musical instruments with the intention of finding that one which most approached human speech."[67] He tried every instrument he could find: trumpet, "cors de chasse," oboe, clarinet, bassoon, and even *vox humana* organ pipes, which he judged as poor imitators of the voice.[68]

On an excursion, Kempelen heard a strange distant sound that resembled a singing child. Upon coming closer, he found that it was a peasant playing a bagpipe. Kempelen purchased a spare pipe from the musician and began some trials with his new instrument. He attached it to a bellows from his kitchen and connected it to a flute.[69] This did not produce any humanoid sounds. Kempelen then joined the bell of a clarinet to the arrangement, and he found that by cupping this funnel with his hand in various ways, he could create some vowel sounds.[70]

Even this "pitoyable" apparatus worked well enough to fool his wife and children who, upon hearing some of his experimentation, wondered what houseguest was praying excitedly in a strange language.[71] Kempelen believed that he had laid the foundation upon which he could build a complete system of human speech.[72] In order to make further progress, Kempelen explained that he needed to study formally the mechanics of articulation and "to always consult nature while following my experiments."[73] Thus, he added, his speaking machine and his theory of speech progressed equally and "the one served as a guide for the other."[74]

One may observe the marks of this mutual development in Kempelen's comparisons of human organs to mechanical structures, a notion found in the works of Le Cat and Vaucanson. For example, he considered the lungs as a pair of bellows (see fig. 8.7), and he represented the formation of consonants by the motions of boxes with hinged shutters portraying the lips and tongue (see fig. 8.8).[75] Likewise, Kempelen conceived the speaking machine

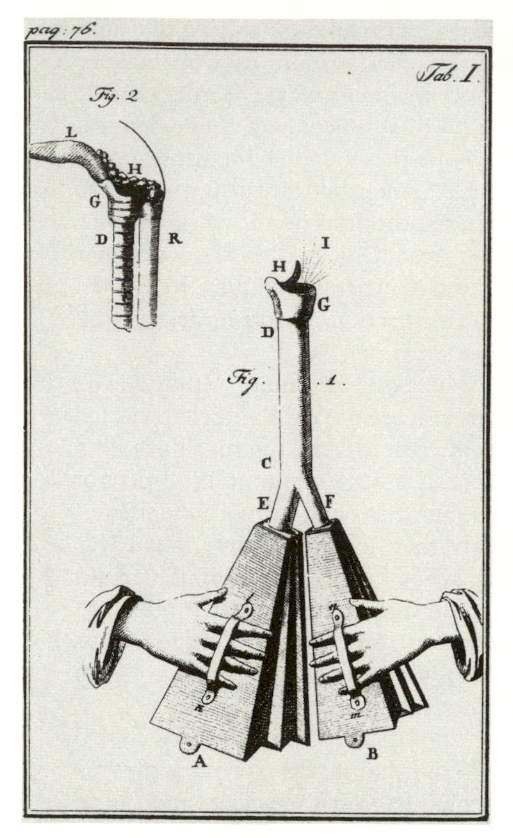

Fig. 8.7. Lungs and bellows.

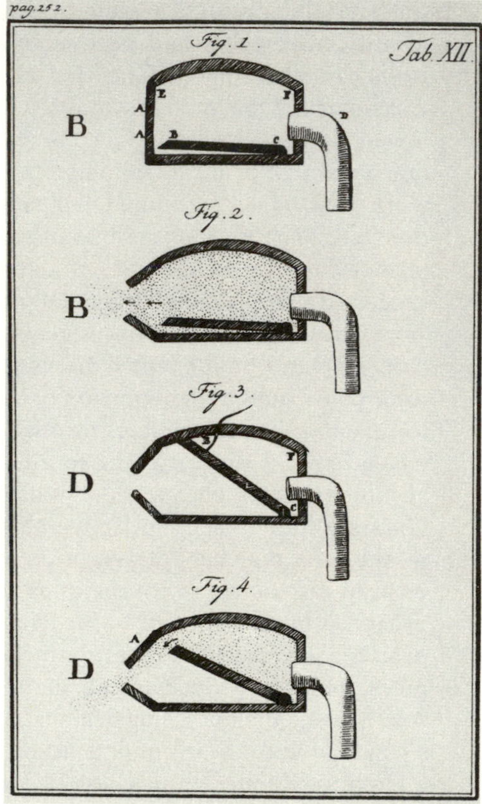

Fig. 8.8. Consonant boxes.

From Kempelen, *Mechanismus der menschlichen Sprache*, pp. 76 (fig. 8.7) and 252 (fig. 8.8). Courtesy of the University of Washington Libraries.

as the mechanical embodiment of all the organs of speech with an ivory reed for the larynx and special tubes for the nostrils.[76] Consequently, he concluded that the *configurations* of the throat and mouth defined the sounds of speech.[77]

At first, Kempelen considered the duplication of consonants—let alone speech itself—too difficult even to attempt. He hoped merely to make some vowel sounds.[78] But he must have been encouraged by his early success, because he eventually sought to reproduce words and sentences with his device. Kempelen initially tried a series of artificial orifices (like his hinged boxes), each one capable of uttering a single vowel or consonant. However, he could not conceive a means for joining these distinct sounds together into syllables and words. At that point, Kempelen realized that he must follow nature "absolument" and employ only *one* glottis and *one* mouth in the production of all sounds.[79] The result of his labor was the device shown in figure 8.9 (see also fig. 8.10).

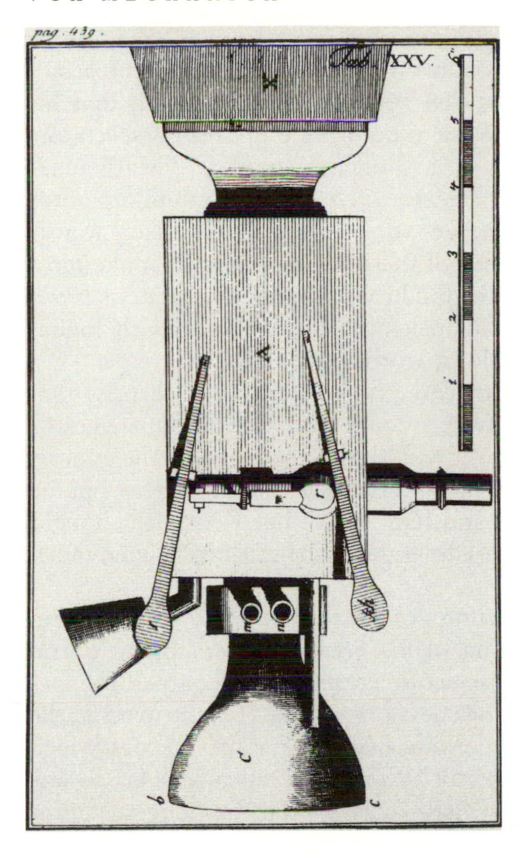

Fig. 8.9. Kempelen's speaking machine. From Kempelen, *Mechanismus der menschlichen Sprache*, p. 439. Courtesy of the University of Washington Libraries.

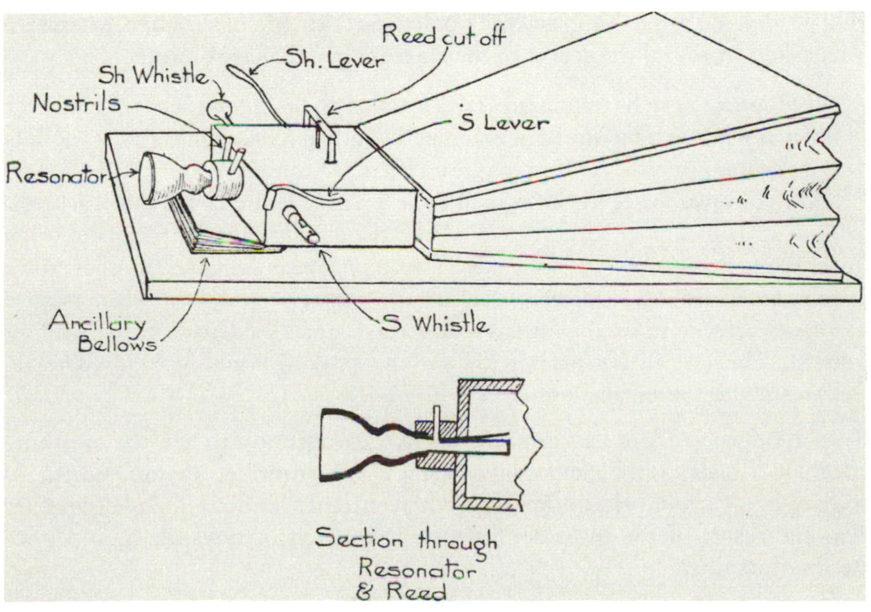

Fig. 8.10. Wheatstone's speaking machine. This was based on Kempelen's design. From Paget, *Human Speech*, p. 18. Courtesy of Routledge.

The earliest accounts of his speaking machine date from 1783, when he brought an improved version with him on the chess-playing automaton's European tour. At that time, Kempelen continually pointed out that his device was still imperfect; nevertheless, the machine impressed spectators with the pronunciation of words such as "papa," "mama," "Marianna," and "astronomie," as well as short phrases such as "Romanum Imperator semper Augustus" and "Maman aimez-moi." Although the machine was created by a Hungarian in the service of the Viennese court, it spoke most comfortably in French and Latin. Goethe heard the machine and, in 1797, wrote, "The speaking machine of Kempelen . . . is in truth not very loquacious, but it pronounces certain childish words very nicely."[80]

Most accounts of the speaking machine noted with some surprise that Kempelen's contrivance did not have a human form. Kempelen suggested that he might make an exterior which would give the machine the appearance of a six-year-old child. Not only would such an innocent exterior fit the machine's approximate verbal age and high voice, but Kempelen thought that it might make the audience more tolerant of the machine's often indistinct pronunciation.[81]

Kempelen never mentioned the efforts of Mical, but a few observers did compare Kempelen's work with that of his French contemporary. Baron Friedrich Melchior Grimm, who also wrote enthusiastically about Kempelen's chess player (he claimed that the chess player was "for the mind & the eyes what the *Flute Player* of M. de Vaucanson is for the ear, but which seems to us in all respects far superior"), regarded Mical's heads as less articulate than Kempelen's device.[82] Rivarol, although his judgment may have been tainted by favoritism, insisted that the chess player was a fake and that Kempelen's speaking machine paled next to Mical's. When Kempelen was asked to reveal the secret of the chess player, Rivarol wrote,

> *When you know it*, he responded, *it will no longer be anything*. Such is in effect the difference between the works of genius and of art, and the simple prestige of skill. If the genius astonishes us by some great effects, it surprises us even more when we allow ourselves to see its causes, and these it shows entirely: and this is why the study of nature is so beautiful; we admire the author more by becoming better acquainted with him; but the confidence man [*le joueur des gobelets*] is lost when he is discovered. M. Kemplein [*sic*] also had a casket [*coffret*] from which escaped a few words, so to speak: but this honest traveler has rendered a genuine tribute to M. l'abbé Mical; as soon as he knew of the speaking heads, he removed his automaton, his casket, and himself.[83]

To a modern student, it certainly seems incongruous to see the speaking machine alongside the chess player. One was a complete sham—worthy of a place in a carnival sideshow (where it eventually landed)—while the other was the result of the most sophisticated research in physiology and phonetics of its day.

Part of the inconsistency that modern readers detect in Kempelen's oeuvre arises from the limitations of applying modern categories to eighteenth-century events. A true object of science appears alongside an amusing illusion. But both of these were used for performance—an aspect of seventeenth- and eighteenth-century science, or "natural philosophy," which has become vastly diminished.[84] Athanasius Kircher and other natural magicians who used their understanding of nature to produce wondrous effects belong in this category. Vaucanson is another figure whose efforts seem entirely playful today, but who earned the respect of the scientific and philosophical thinkers of his era. Electrical investigators, such as Benjamin Franklin and the abbé Nollet—and especially the demonstrators, lecturers, and showmen of natural philosophy, like Benjamin Martin or Franklin's friend Ebenezer Kinnersly—may offer the best analogue for Kempelen.[85] Their researches were full of parlor entertainment: the "electric spider," the "electric kiss," and so on. The modern reader recognizes these exhibitions as the mere "wrapping" for a theory of the electric fluid; nonetheless, the *experiment* and the *amusement* were not distinguishable.

Franklin and Kempelen actually met—in Paris in 1783. Franklin, a chess aficionado, seems to have held a high opinion of both the chess-playing automaton and its maker.[86] Another of Franklin's acquaintances was interested in speaking machines: Erasmus Darwin wrote him in 1772 and asked, "I have heard of somebody that attempted to make a speaking machine, pray was there any Truth in such Reports?"[87] Darwin's letter discussed several scientific subjects, including the pronunciation of consonants among the British dialects. Franklin's reply offered the American's own phonetic considerations, as well as a report of a speaking clock made in Ireland.[88] The previous year, however, Darwin had built a speaking machine of his own.

Darwin came to the subject of speaking machines through his interest in the origin of language.[89] His phonetic observations appeared in the additional notes to the *Temple of Nature* (1803). Darwin asserted that the larynx resembled the trumpet stop of an organ (a type of pipe with a striking reed), as one might demonstrate by "blowing through the wind-pipe of a dead goose."[90] "These sounds," he continued, "would all be nearly similar except in their being an octave or two higher or lower; but they are modulated again, or acquire various tones, in their passage through the mouth; which thus converts them into eight vowels."[91] He subsequently described the positions of the tongue, lips, and teeth in the formation of all the sounds of speech. In his analysis of the mouth's disposition in the formation of the vowels, Darwin had recourse to an ingenious technique. He inserted rolled cylinders of tinfoil into his mouth while uttering the vowels. The points of compression of the misshapen tubes would reveal the configurations of the oral cavity. Since Darwin stammered badly, it is entirely likely that he had made a close study of human articulation in order to understand—or possibly to cure—his own affliction. He was also familiar with John Wilkins's writings and sympathized with Wilkins's views on alphabetic reform.[92]

At the end of his discussion, Darwin described the product of his phonetic curiosity united with his mechanical dexterity.

> I have treated with greater confidence on the formation of articulate sounds, as I many years ago gave considerable attention to this subject for the purpose of improving shorthand; at that time I contrived a wooden mouth with lips of soft leather, and with a valve over the back part of it for nostrils, both which could be quickly opened or closed by the pressure of the fingers, the vocality was given by a silk ribbon about an inch long and a quarter of an inch wide stretched between two bits of smooth wood a little hollowed; so that when a gentle current of air from bellows was blown on the edge of the ribbon, it gave an agreeable tone, as it vibrated between the wooden sides, much like a human voice. This head pronounced the p, b, m, and the vowel a, with so great nicety as to deceive all who heard it unseen, when it pronounced the words mama, papa, map, and pam; and had a most plaintive tone, when the lips were gradually closed. My other occupations prevented me from proceeding in the further construction of this machine; which might have required but thirteen movements . . . unless some variety of musical note was to be added to the vocality produced in the larynx; all of which movements might communicate with the keys of a harpsichord or forte piano, and perform the song as well as the accompaniment; or which if built in gigantic form, might speak so loud as to command an army or instruct a crowd.[93]

The efforts of Wilkins, Mical, Kempelen, Darwin, and several of Vaucanson's contemporaries shared a consistent approach to the imitation of the voice. All of them defined the vowels and other speech sounds in terms of the configurations of the human organs of speech. The devices of Mical, Kempelen, and Darwin contained similar components for the duplication of the lungs, larynx, and mouth. Most of these instrument makers—as well as some of their proponents, such as Rivarol—related the construction of speaking machines to linguistic concerns like the origin of language and the instruction of the deaf. All of these figures also managed to transcend the distinction between entertainment and science.

Among these early figures, Robert Hooke is the exception. His approach to acoustical questions revolved around a different set of concerns from those of his contemporaries who studied the voice.[94] Although only a few sentences of his observations are available, Hooke's work reveals a way of conceiving the problem that would not be taken up again until the nineteenth century.

Imitating Vocal Sounds in the Nineteenth Century

Many artificial glottises, similar in design to those that seem to have activated Mical's heads, were made in the nineteenth century. In his attempt to discern the mechanism of the voice, Ferrein had also experimented with elastic strips covering a tube, and numerous nineteenth-century endeavors followed his example. These physiologists did *not* aim to copy articulate speech, but simply to disclose the operation of the larynx.

One of the most important acoustical physicists of the first half of the nineteenth century, Félix Savart, offered a singular explanation of the voice in his "Mémoire sur la voix humaine" (1825).[95] Savart compared the larynx to a hunter's birdcall—a short cylinder, each end of which was covered by a thin plate with a small hole in the center. Although such birdcalls were known predominantly for their high-pitched sounds, Savart argued that the pyramidal shape of the vocal cavity, along with the softness of its walls, would allow this instrument to achieve the tonal range of the human voice.

To support his suggestion, Savart referred to a plaster cast of the inside of a cadaver's throat. He insisted that the superior and inferior ligaments of the larynx cooperate to form an analogue for his birdcall. (It is the inferior ones that are usually called the vocal cords.) Allegedly, the combination of these features—birdcall-like larynx, pyramidal shape, and fleshy walls—would combine to imitate the sounds of the human voice.

Savart's theory was not universally accepted, but his ideas haunted the vocal researches of the French physicist Charles Cagniard de la Tour. In the late 1830s Cagniard constructed several rubber models of the larynx. As he could never bring himself to contradict his predecessor, he always maintained two sets of membranes. One of Cagniard's demonstrations involved what is possibly the most convenient piece of scientific apparatus ever conceived: his "digito-buccal." He simply blew through two fingers pressed against his lips.[96] (In this example, his lips supplied one set of membranes and his fingers supplied the other.)

The German physiologist Johannes Müller experimented with artificial glottises in the 1830s. By the time of his work, most physiologists agreed on the inconsequential role of the upper ligaments and regarded the mechanism of the larynx as a single pair of "membranous tongues."[97] Édouard Fournié was among the many others who continued this variety of research in the nineteenth century.[98] Fournié discussed his own artificial glottis in his *Physiologie de la voix et de la parole* (1866) (see fig. 8.11).

It is difficult to say when the study of the voice became an acknowledged issue for the physical science of acoustics. The authors surveyed thus far belong in the tradition of physiology and phonetics, although the study of the mechanism of the larynx (and the question of what sort of musical instrument it resembles) seems closely allied to the study of vibrating bodies. This species of problem—for example, the mathematical description of the motion of a vibrating string—had been a proper subject for mathematical physicists since the middle of the eighteenth century. No later than the early nineteenth century, physicists began to think about the mechanism of the larynx. E.F.F. Chladni's *Die Akustik* (1802) briefly described the anatomy of the larynx (he cited anatomical and physiological writers, as well as Kempelen and Kratzenstein) in his section on wind instruments. He considered the larynx to behave like a strip of paper or blade of grass held in the fingers

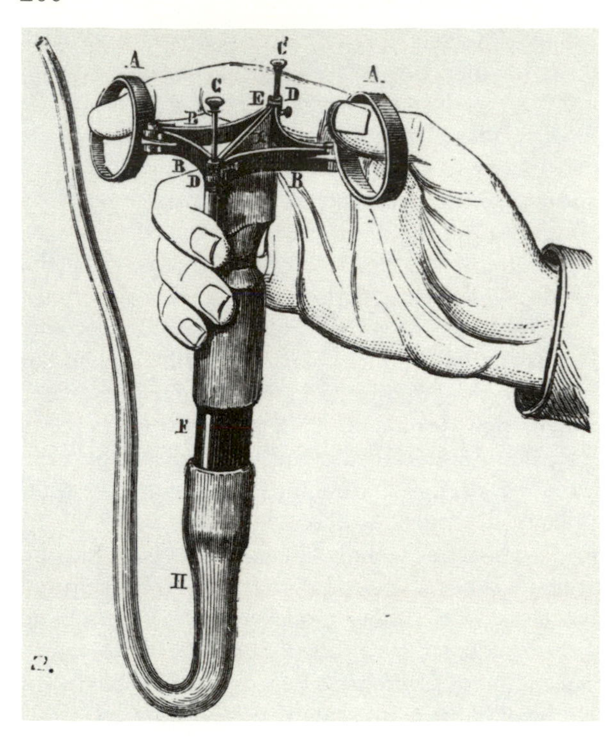

Fig. 8.11. Fournié's artificial glottis. From Fournié, *Physiologie de la voix et de la parole*, p. 397. Courtesy of the University of Washington Libraries.

and blown against. (This is quite similar to Dodart's *chassis bruyant* or Darwin's ribbon.)[99] Jean-Baptiste Biot's *Traité de physique* (1816) compared the vocal organs with a free reed.[100]

Aside from these discussions, most physicists have identified Robert Willis as the pivotal figure who brought the study of the voice into acoustics. Helmholtz's *Sensations of Tone* did not mention Mical, Kratzenstein, Kempelen, or Darwin and claimed instead that the theory of vowel sounds was given first by Wheatstone (discussed below), in his critique of Willis.[101] Rayleigh's *Theory of Sound*—the *Principia* of nineteenth-century acoustics—claimed that "the acoustical treatment of this subject may be considered to date from a remarkable memoir by Willis."[102] Rayleigh made passing references to Kratzenstein and Kempelen.

Willis was a professor of applied mechanics at Cambridge when he wrote his essay "On the Vowel Sounds, and on Reed Organ-Pipes," published in 1830. Ironically, when Willis was an undergraduate, he wrote a pamphlet that exposed the deception of Kempelen's chess-playing automaton, exhibited at that time in London by Maelzel. In his essay, Willis dismissed the efforts of Roger Bacon, Albertus Magnus, Kircher, and Wilkins as "mere deceptions."[103] He mentioned Mical's speaking heads and discussed briefly the labors of Kratzenstein. (He complained that some of Kratzenstein's organ pipes possessed a "most grotesque and complicated figure, for which no reason is offered, save that experience had shewn these forms to be the best adapted to the production of the sounds in question.")[104] Kempelen was discussed rather favorably in the essay. Nevertheless, Willis insisted that "none of these writers . . . have succeeded in deducing any general principles."[105]

At the beginning of his essay, Willis announced the flaw contained in all previous studies: "The generality of writers who have treated on the vowel sounds," he wrote,

> appear never to have looked beyond the vocal organs for their origin. Apparently assuming the actual forms of these organs to be essential to their production, they have contented themselves with describing with minute precision the relative positions of the tongue, palate and teeth, peculiar to each vowel, or with giving accurate measurements of the corresponding separation of the lips, and of the tongue and uvula, considering vowels in fact more in the light of physiological functions of the human body than as a branch of acoustics.[106]

The vowels, he claimed, are "mere affectations of sound, which are not at all beyond the reach of human imitation in many ways, and not inseparably connected with the human organs, although they are most perfectly produced by them."[107] Likewise, Willis argued, although the human voice is the greatest source of musical sounds, "and our best musical instruments offer mere humble imitations of them; . . . who ever dreamed of seeking from the larynx, an explanation of the laws by which musical notes are governed."[108] Willis intended to ignore the anatomy of speech and to employ the familiar tools of acoustics in order to determine what physical conditions contribute to the formation of vowel sounds.

This presented a fascinating and counterintuitive approach; yet one can question Willis's success in this endeavor. His apparatus was a free reed attached to an organ pipe of variable length. He admitted that the free reed was regarded generally as the best mechanical approximation of the larynx.[109] One could argue that he constructed a simple model of the larynx, throat, and mouth. In any case, his desire to reproduce the vowel sounds by purely mechanical means distinct from human morphology was novel.

In his trials, Willis found that as the pipe approached a length at which it would reinforce the sound of the reed, vowel sounds would be produced in the order *U*, *O*, *A*, *E*, *I*. After passing through the resonance point, the vowels would be sounded in the reverse order: *I*, *E*, *A*, *O*, *U*. To explain this phenomenon, Willis looked to Euler's writings. In several essays, Euler proposed that a pulse at the bottom of an organ pipe would reflect off the top and vibrate continually between the ends of the pipe. If, however, pulsations were continuously generated by a reed *and* if the frequency of the reed did *not* coincide with the frequency of the pulses traveling between the ends of the pipe, then this mixture of pitches would create what Willis called a "compound sound."[110]

According to Willis, each vowel possesses a *characteristic pitch*, and when the organ pipe is of a length that will reproduce one of those pitches, a vowel sound will be heard. This theory became known as the "fixed pitch" or "inharmonic" theory of vowel sounds. Willis also suggested another means for producing compound sounds, one that was boldly antimorphological. This technique involved a rotating toothed wheel that would periodically hit a vibrating spring. This combination—the pitch of the teeth beating against the spring, along with the spring's own vibrations—yielded some vowel sounds.[111]

In addition to advancing acoustical theory, Willis believed that his experiments might provide a tool for philologists with which they could quantify the distinctions among vowels and their pronunciation in different nations.[112] William Whewell believed that one of Willis's variable organ pipes could serve that very function. Showing his characteristic passion for nomenclature, he named the hypothetical device a "pthongometer."[113] Willis also thought that his research would enable organ builders to make *vox humana* pipes according to the correct principle of vowel sounds.[114]

Charles Wheatstone championed Willis's vowel theory in his 1837 essay on speaking machines. Wheatstone was heir to several of the traditions discussed in this chapter. He came from a family of musical instrument makers and his earliest scientific papers dealt with acoustics. Communication was another interest that permeated several of his studies. He was one of the inventors of the telegraph and he wrote an essay on cryptography.[115] Wheatstone's name is also associated with scientific toys. One of his earliest productions of philosophical amusement was his "enchanted lyre," which would play music without the aid of a visible performer.[116] The enchanted lyre appeared in Wheatstone's "On the Transmission of Musical Sounds through Solid Linear Conductors" (1831), in which he also called attention to the importance and difficulty of transmitting the sounds of the voice. His essay ended with a suggestion of some sort of machine for reassembling speech sounds.

> The transmission to distant places, and the multiplication of musical performances, are objects of far less importance than the conveyance of the articulations of speech. I have found by experiment that all these articulations, as well as the musical inflexions of the voice, may be perfectly, though feebly, transmitted to any of the previously described reciprocating instruments. . . . [C]ould articulations similar to those enounced by the human organs of speech be produced immediately in solid bodies, their transmission might be effected with any required degree of intensity. Some recent investigations lead us to hope that we are not far from effecting these desiderata; and if all the articulations were once thus obtained, the construction of a machine for the arrangement of them into syllables, words, and sentences, would demand no knowledge beyond that we already possess.[117]

These comments imply that Wheatstone may have been contemplating the role of speaking machines in communication. He had studied Kempelen's book and built a copy of his machine (see fig. 8.10).[118]

Wheatstone's 1837 essay, which discussed Willis's theory, mentioned many mythical and bogus speaking machines of the past, and he summarized the works of Kratzenstein, Kempelen, and Mical. It is clear that Wheatstone hoped to reproduce speech and not simply the vowel sounds. He addressed the problem of artificial consonants—an issue that Willis ignored.[119] He repeated Rivarol's aim (preserving correct pronunciation) as well as that of Erasmus Darwin (directing armies or large crowds), and Wheatstone ended his essay with a sentence from Sir David Brewster's *Natural Magic*: "We have no doubt that, before another century is completed, a talking and a singing machine will be numbered among the conquests of science."[120]

In terms of acoustical theory, Wheatstone suggested a crucial modification to Willis's work. Wheatstone denied that a vowel sound consisted of the mere *coexistence* of a larynx (or reed) tone and a "characteristic" vowel pitch. Rather, Wheatstone asserted that the mouth behaves like a resonant chamber which, according to its size, emphasizes certain partials or overtones of the larynx tone; it is this "multiple resonance" that creates the vowel sound.[121] This proposal became known as the "fixed resonance," "relative pitch," or "harmonic" theory of vowel sounds.

Hermann von Helmholtz's acoustical researches, culminating in his *Sensations of Tone* (first edition 1863), carefully examined the formation of vowel sounds. Helmholtz treated the vowels within the context of his theory of *timbre*—the feature by which tones of the same pitch can be distinguished from one another.[122] His understanding of sound was perceptual in nature, and his instrumental treatment of the problem of timbre was based on the ear's ability to analyze a complex tone into discrete sensual units—the sinusoidal waves of a Fourier series. According to this principle, known as "Ohm's law," the timbre of a complex tone (a fundamental tone accompanied by its partial tones) is recognized only by the relative strengths of its simple partial tones.[123] Helmholtz demonstrated the construction of a complex tone from several simple tones by loudly singing a vowel soud (*A* as in f*a*ther) at the sounding board of a piano, while its damper was raised. His voice sympathetically excited those strings that had frequencies equal to the partials present in the vowel tone. The continued vibration of these strings reassembled the vowel timbre and created a sustained echo.[124] Helmholtz also constructed complex tones artificially with his tuning-fork synthesizer—an array of electromagnetically driven forks combined with resonators—which simultaneously produced several simple partial tones with specific pitches and intensities (see fig. 8.12). Like his demonstration with undamped piano strings, the synthesizer embodied Helmholtz's theory of timbre; it mimicked both the function and the appearance of the tuned elastic anatomical structures in the inner ear.[125]

In order to reproduce vowel sounds with his tuning-fork apparatus, Helmholtz first had to determine the strengths of the partials presenting each vowel tone. Following Wheatstone, Helmholtz regarded the vocal tract as a

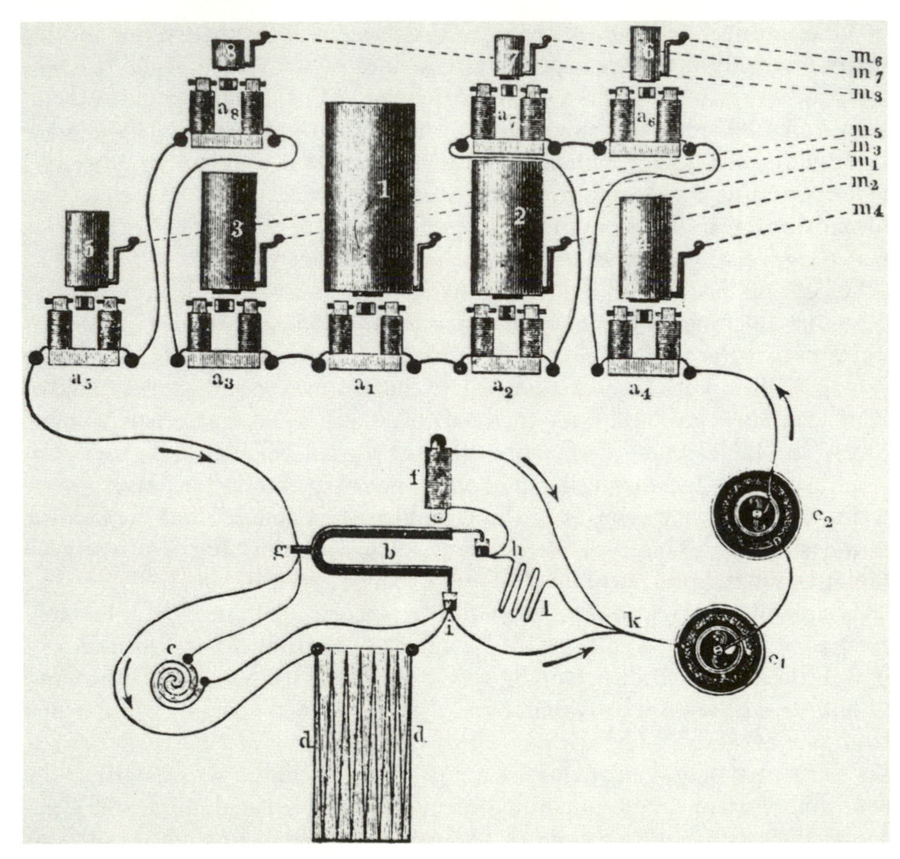

Fig. 8.12. The design of Helmholtz's tuning-fork apparatus for the production of artificial timbres. From Helmholtz, *Sensations*, p. 399. Courtesy of the University of Washington Libraries.

resonator that would either accentuate or diminish the partials of the fundamental (larynx) tone. As Willis had done, Helmholtz identified the particular pitches of the shapes of this cavity that were associated with each vowel. The Dutch physiologist F. C. Donders had also attempted as much, by judging (with his ears alone) the pitches of whispered vowels. According to Helmholtz, Donders's technique was inaccurate. Helmholtz preferred to hold tuning forks in front of the lips while the mouth remained fixed in the position required for the pronunciation of a vowel. Helmholtz identified the characteristic vowel pitches as those of the tuning forks that evoked the strongest resonances from the mouth.[126] Subsequently, Helmholtz determined empirically the strengths of the partials of the larynx tone for each vowel and combined these partials with his tuning-fork synthesizer.[127] Helmholtz also duplicated this result with series of organ pipes, but he judged this method inferior.[128]

Although Helmholtz's theory of timbre became widely accepted and his synthesis of vowel sounds offered a stunning confirmation, the quality of his artificial vowels was still quite poor. Rayleigh claimed that Helmholtz's results were "difficult, and do not appear to have been repeated."[129] Likewise, D. C. Miller considered the relationship of natural vowels to those produced with the tuning-fork apparatus to be "more or less fanciful."[130] (Curiously, Miller devised sets of organ pipes—similar to those of Helmholtz—that he thought produced good reproductions of the vowels.)

The "fixed pitch" and the "relative pitch" theories—conceptions that set the terms for the study of vowels through the early twentieth century—are difficult to distinguish. Rayleigh even concluded that only an apparent difference existed between the two.[131] He explained that the development of Edison's phonograph as a scientific tool should lead to an "*experimentum crucis*."[132] If the fixed pitch theory were correct, increasing the playback speed of a recorded vowel would alter the vowel. If the relative pitch theory were correct, increasing the speed would increase the pitch of the recorded sound, but the vowel would maintain its integrity. This seems straightforward, but the phonograph experiments proved to be inconclusive.[133] The debate continued with figures like L. Hermann and the Yale physiologist E. W. Scripture taking the "fixed pitch" side, and authors such as Louis Bevier at Rutgers and D. C. Miller of Cleveland's Case School of Applied Science selecting the rival theory. By the middle of the twentieth century, the fixed pitch–relative pitch dichotomy had become less meaningful, because the particular theory of vowel sounds was actually inconsequential to the function of some of the more recent acoustical instruments, such as the "sound spectrograph."[134] John Q. Stewart, who contrived an electronic circuit analogue for the vowels, made the same point in *Nature*, in 1922.[135]

In 1879, the British telegraph engineer William H. Preece, along with Augustus Stroh, contrived several means for producing synthetic vowels and for representing them graphically. They based their efforts on Helmholtz's researches, and their instrumentation aimed to combine series of partial tones according to the intensities Helmholtz had assigned. One of their devices—an electromagnetically driven armature with an adjustable spring on the end—could give a prime tone accompanied by one partial (see fig. 8.13). This was extremely similar to Willis's toothed wheel and spring idea, although Preece and Stroh did not mention Willis's paper.[136] This device, they suspected, suffered from the absence of additional partials; therefore, they contrived two other machines. One of these was based on the principle of their "synthetic curve machine" for drawing curves of tones containing eight partials (see fig. 8.14). This device conveyed the vibrations of eight springs against eight toothed wheels (one for each partial) to a diaphragm.[137] Their most refined instrument Preece and Stroh called an "automatic phonograph" (see fig. 8.15).[138] With this machine, synthetically drawn curves that had been cut on the edges of brass disks and mounted on a rotating cylinder

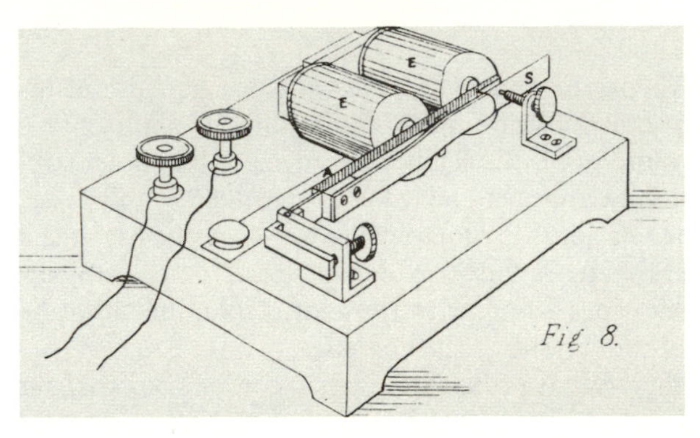

Fig. 8.13. Preece and Stroh's simple device. The spring (S) could be adjusted to give the desired tone relative to that of the electromagnetically driven armature (A). From Preece and Stroh, "Studies in Acoustics," pp. 358–366. Courtesy of the University of Washington Libraries.

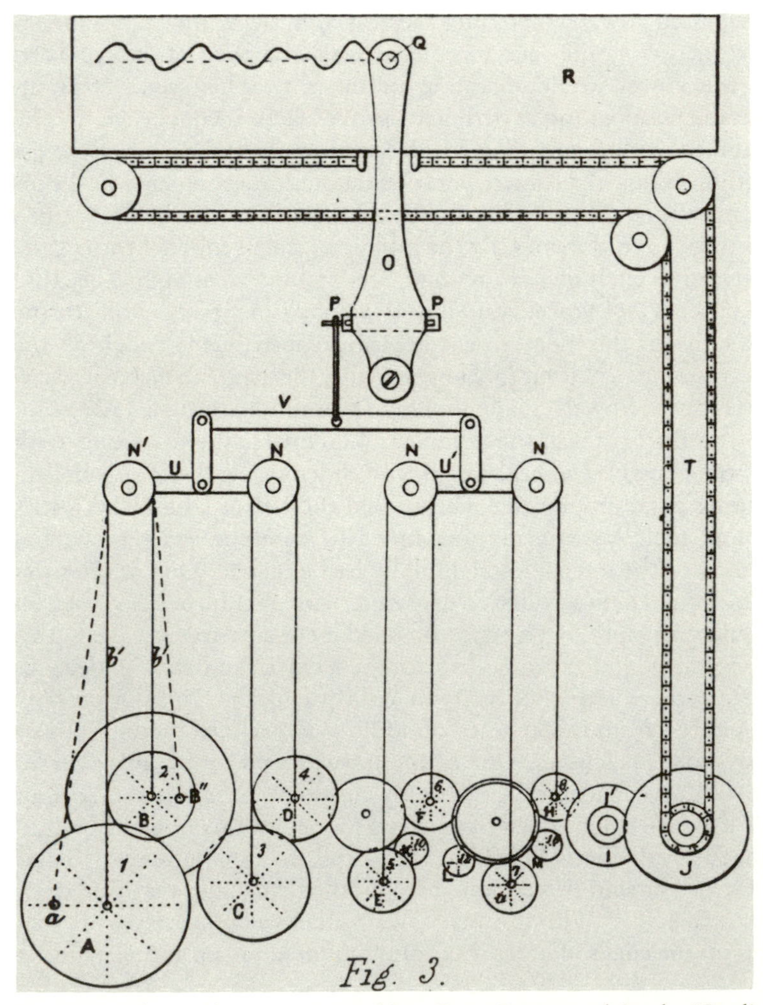

Fig. 8.14. The synthetic curve machine. From Preece and Stroh, "Studies in Acoustics," pp. 358–366. Courtesy of the University of Washington Libraries.

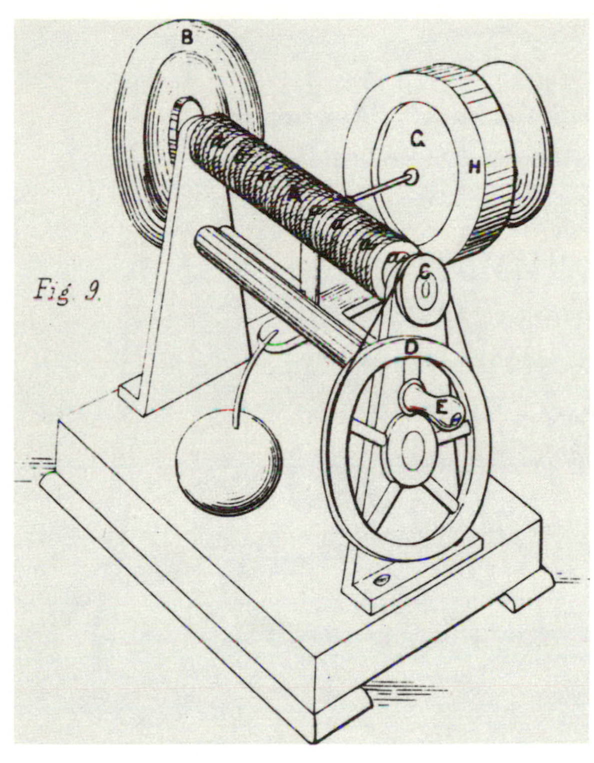

Fig. 8.15. The automatic phonograph. This device differed little from a phonograph, except that the edges of the disks on the cylinder A were carved according to curves made by the synthetic curve machine. From Preece and Stroh, "Studies in Acoustics," pp. 358–366. Courtesy of the University of Washington Libraries.

were played with a phonograph stylus and diaphragm. Preece and Stroh believed that their automatic phonograph made good imitations of vowel sounds. This judgment was not shared by the prominent British phonetician Alexander John Ellis, who witnessed a demonstration by Stroh himself. Ellis complained: "In the artificial vowels just considered I could not recognise any exact form of human vowel with which I was acquainted, although I have made speech sounds an especial study for more than forty years. We have an analogy in the multiform presentment of the human countenance, which is nevertheless unhesitatingly recognised as distinct from that of the anthropoid ape."[139] Similar in concept to Preece and Stroh's machine, the "wave siren" of the great acoustical instrument maker Rudolph König provided another means to replicate vowel sounds. König contrived this device in response to Helmholtz's theory of timbre. Unlike Helmholtz's perceptual and analytic approach, König's conception of sound was fundamentally pictorial. He believed that a tool for studying and re-creating complex tones should be able to disclose readily any audible distinctions among the variations that become apparent in the graphical representations of a single set of partial tones arising from their numerous possible phase relationships.[140] The wave siren was König's way of performing the converse of the graphical and optical techniques for rendering the vibrations of sound visible; from these wave pictures, it created the corresponding sound. König developed

Fig. 8.16. A wave siren for the production of beats. From König, *Quelques expériences d'acoustique*, p. 160. Courtesy of the University of Washington Libraries.

wave sirens in several forms, but they all shared an essential feature: a metal band or disk with an edge cut in an undulatory shape that rotated in front of a blast of air channeled through an elongated slit. In this way, König endeavored to re-create the pattern of compression and rarefaction in the air that corresponded to a specific mathematical curve (see fig. 8.16). Alfred Eichorn devised a *Vocalsirene* based on this principle in 1889.[141] Subsequently, König himself reproduced the complex aerial vibrations of vowels with his instrument.[142]

Many physicists doubted the wave siren's ability to reproduce the tones corresponding to the vibrational curves cut on the edges of its disks. Yet for the present study of scientific instruments, a more significant issue than its success involves the approach to the problem of the voice that the wave siren entailed. Like the other tools of nineteenth-century acoustics discussed here, the wave siren made no attempt to mimic the function of the larynx, tongue, and lips. Instead, these apparatus endeavored to reproduce the vibrational pattern of the air that corresponded to the vowels or to re-create vowel stimuli in the organ of hearing. The labors of these physicists demonstrated the movement that Willis suggested—away from devices of the eighteenth century that copied the shape of the mouth. (Of course many of the eighteenth-century devices duplicated not only the internal anatomy of the organs of speech but the external features of the human visage, as well.)

Nineteenth-century physicists who studied the voice were generally much less interested in reproducing articulate speech than their eighteenth-century counterparts had been. For physicists, the important questions pertained to the acoustic properties of vowel sounds alone. Even though the sounds of a pneumatic speaking machine like Kempelen's may have more closely resembled the voice than one of Preece and Stroh's machines, its form and function, and the theory it promoted, would have rendered Kempelen's device unfit to serve the needs of late-nineteenth-century acoustics.

Despite their theoretical and instrumental differences, the efforts of Willis, Wheatstone, Helmholtz, Preece and Stroh, and König constitute a consistent attempt to duplicate the voice from the vantage of physical acoustics. Yet elocutionists and phoneticians of the nineteenth century approached the problem in a different way. Such investigators found a more meaningful heritage in Wilkins and in the anthropomorphic instrumentation of Kempelen and Erasmus Darwin than in the efforts of researchers in physical and physiological acoustics.

However, no sharp line distinguishes the labors of these two traditions. For example, Alexander Ellis, who provided a splendid translation of Helmholtz's *Tonempfindungen*, had been familiar with techniques for imitating the voice long before he encountered Helmholtz's text. Ellis's *Alphabet of Nature* (1845) continued Wilkins's aim of simplifying written language by introducing an unambiguous phonetic alphabet capable of representing all the sounds of speech.[143] *The Alphabet of Nature* did not ignore physics. It included a discussion of acoustics that drew heavily on John Herschel's article "Sound" in the *Encyclopedia Metropolitana*. Turning to the sounds of speech, Ellis explained the important role of physical apparatus—especially speaking machines—in phonetic studies. "It is impossible," he wrote, "that any person in analysing sounds can do more than analyse his own sensations. In proposing characters for sounds, he proposes characters which represent certain of his sensations, which sensations may never occur in any other individual. ? [*sic*] How then can we hope to render this subjectivity objective. We cannot do it perfectly without the aid of a machine."[144] Ellis claimed that phonetic precision depended upon speaking machines, and he hoped his comments would raise them "far above the grade of simple *curiosities*" and place them "in the ranks of *necessaries for human improvement*."[145] Ellis quoted lengthy passages from Robert Willis's essay. Like William Whewell, Ellis hoped that the studies of Willis and of his successors would provide a device for the accurate comparison of vowels.

In a brief 1873 paper, Richard Potter observed that the problem of producing vowel sounds was an issue of interest to grammarians, physiologists, and natural philosophers.[146] Potter himself imitated vowels with an apparatus that would have seemed familiar to the eighteenth-century investigators mentioned above: a free reed connected to a hollow india rubber sphere that could be deformed to copy the shape of the mouth and produce a variety of vowels.

The Liverpool phonetician R. J. Lloyd made some similar remarks and experiments beginning in 1890. Lloyd recognized a schism between investigators in his field who concerned themselves only with sound—the "acoustic value and affinities" of speech sounds—and others who studied merely the oral features of articulation. Lloyd sought to bridge this gap.

He offered his thoughts concerning what sort of instrumentation would be appropriate for studying vocal sounds. Lloyd complained of the difficulty in obtaining precise results from direct observations of the organs of speech. Measurements of particular characteristic vowel resonances, he noted, could differ by entire octaves.[147] "Such being the equivocal nature of the best evidence derivable from the direct observation of vocal phenomena," he argued,

> it seemed advisable to study the conditions of their artificial reproduction. But the apparatus employed by Helmholtz, Willis, Preece, Stroh, and others for the synthesis of vowels was in no case found to resemble the human vocal organs at all. None of them, therefore, could give any hints respecting that connection between configuration and timbre which is the great object of the present enquiry. Any apparatus designed to throw light upon such a subject would need to possess some resemblance in its nature and form to actual human organs: and yet not too close a resemblance, because it would then be of little more use to observe the effects of this apparatus than those of the vocal organs themselves. The desiderated apparatus seemed to be really one which would be rather a caricature than an exact likeness of the vocal organs: it must be capable of reproducing the broad essential features of every vowel configuration, but *not* the details.[148]

The physically correct yet phonetically abstract devices of nineteenth-century physicists did not satisfy the needs of Lloyd's approach. While Ellis wanted a speaking machine to duplicate accurately the acoustic characteristics of vocal sounds, Lloyd's researches demanded devices that also offered articulatory information. To imitate the vocal cavities, he devised a series of cylindrical glass bottles representing vocal tracts "pretty closely in size but only roughly in shape."[149] Lloyd employed these bottles in his studies of whispered vowels. One of Lloyd's methodological descendants, Sir Richard Paget, contrived plasticine models of the vocal tract in his study of imitation vowel sounds during the 1920s (see fig. 8.17).[150]

While Lloyd's models avoided the precise physiological details of each vowel configuration, copying the manifest appearance of the organs of speech was the ultimate end of the French physiologist Georges René Marie Marage, who worked in the late nineteenth and early twentieth centuries

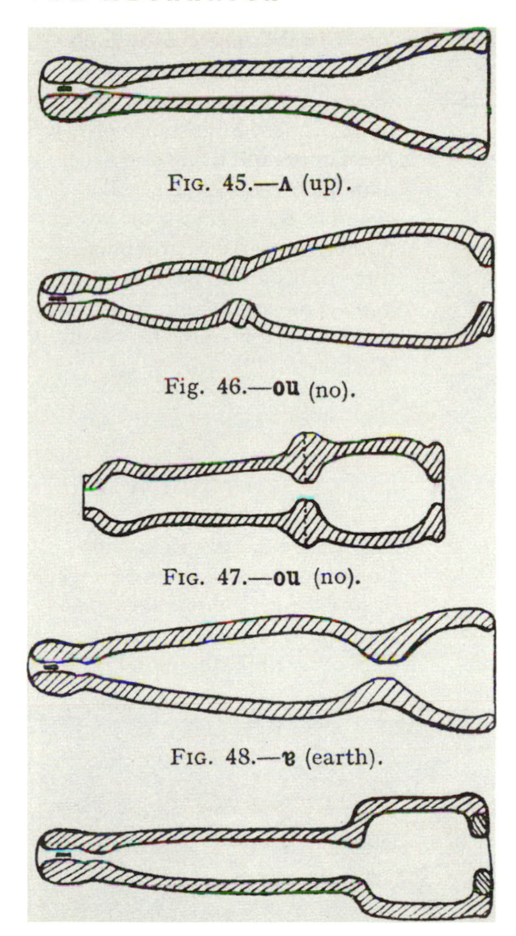

Fig. 8.17. Paget's plasticine vocal tracts. From Paget, *Human Speech*, p. 62. Courtesy of Routledge.

and offered voice instruction at the Sorbonne. Imitating the exact shape of the vocal cavity became the guiding principle of his work. The idea of duplicating human anatomy so completely dominated Marage's thought that he even regarded Helmholtz's vowel synthesizer as an attempt to satisfy this goal: "The tuning forks represented the larynx; the resonators, the supralaryngeal cavities."[151] Information gleaned from techniques for the representation of sound, such as phonautograph traces and manometric flame photographs, aided Marage's studies. He believed that his artificial larynx tone should replicate the phonautograph results, and he crafted sirens for this purpose (see fig. 8.18). Marage's resonant cavities exactly copied the shape of the oral cavity. In fact, they were cast from molds of the mouth, complete with lips and teeth (see fig. 8.19).

In 1905, E. W. Scripture discussed another attempt based on the model of the vocal anatomy—one that did not even bother to make a copy of the human form. His "vowel organ" involved fitting a human skull with artifi-

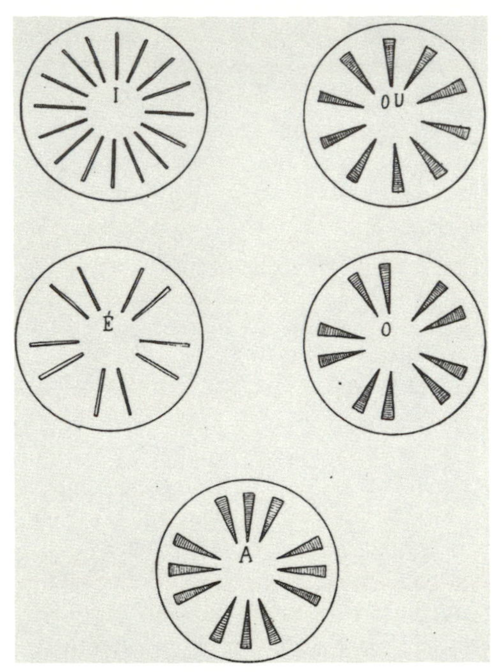

Fig. 8.18. Marage's vowel siren disks. The sounds produced by these disks on a siren were intended to duplicate the apearance of vowel phonautograph traces and manometric flame images. For example, the traces for the vowel *A* (f*a*ther) appeared as groupings of three peaks. From Marage, *Petit manuel de physiologie de la voix*, p. 93. Courtesy of the University of Washington Libraries.

Fig. 8.19. Marage's siren with buccal resonators. From Marage, *Petit manuel de physiologie de la voix*, p. 93. Courtesy of the University of Washington Libraries.

cial cheeks and lips to re-create a resonance chamber. Rubber glottises imitated the larynx. Scripture speculated about the role of such vowel organs in musical performance, mentioning church music in particular.[152]

Phoneticians and physiologists of the late nineteenth and early twentieth centuries were interested in speaking machines, but they did not share the instrumental criteria and the scientific goals that commanded the attention of acoustical physicists. While Helmholtz, Preece and Stroh, and König made devices that reproduced vowels without any connection to the organs of speech, Lloyd, Paget, and others needed information about the positions of the tongue and the shapes of the oral cavity. The rift between these two ways of studying vocal sounds is evident in D. C. Miller's disparaging estimate of Marage's vowel molds. "Such an apparatus," Miller wrote, "like the doll that says 'ma-ma,' is very interesting, but it gives no evidence regarding any particular theory of vowel quality; the vowels so made are not synthetic reproductions scientifically constructed, but are more properly imitations."[153]

Speaking Machines and Entertainment in the Nineteenth Century

It was not the limited vocabulary of these "imitations" that irritated Miller; it was their lack of physical theory. However, for many speaking machines of the nineteenth century, theory was less important than effect. The first patent for a talking doll was awarded to J. N. Maelzel, the inventor of the metronome, in 1824.[154] It consisted of a bellows, reed, and cup-shaped resonator that would be covered and opened mechanically, thus producing the words "mama" and "papa." (Maelzel also exhibited Kempelen's chess player in Europe and America in the early nineteenth century, and he devised a means for the Turk to pronounce "Échec!" after vanquishing an opponent.)[155] Talking baby dolls were only one version of speaking machines as amusement. Most of the devices presented as entertainment, however, were fakes. Ventriloquists, mirror illusions, heads with speaking tubes, and other deceptions that relied on concealed confederates have a long history, and they were staples in the shows and fairs of the eighteenth and nineteenth centuries.[156] Wordsworth's description of the Bartholomew Fair in his *Prelude* (1805 and later versions) provides a sense of the seedy company in which these tricks appeared:

> All moveables of wonder, from all parts,
> Are here—Albinos, painted Indians, Dwarfs,
> The Horse of knowledge, and the learned Pig,
> The Stone-eater, the man that swallows fire,
> Giants, Ventriloquists, the Invisible Girl,
> The Bust that speaks and moves its goggling eyes,
> The Wax-work, Clock-work, all the marvellous craft
> Of modern Merlins, Wild Beasts, Puppet-shows,
> All out-o'-the-way, far-fetched, perverted things,
> All freaks of nature, all Promethean thoughts
> Of man, his dullness, madness, and their feats
> All jumbled up together, to compose
> A Parliament of Monsters. Tents and Booths

> Meanwhile, as if the whole were one vast mill,
> Are vomiting, receiving on all sides,
> Men, Women, three-years Children, Babes in arms.[157]

Among such curiosities and frauds, one could also find Joseph Faber's "Euphonia"—probably the most loquacious pneumatic speaking machine ever made. Faber was a Freiburg native and former astronomer whose failing eyesight led him to take up mechanics and anatomy. Faber began demonstrations of his Euphonia in the early 1840s, and he appeared in London's Egyptian Hall in 1846. The device had the appearance of the head and torso of a man dressed like a Turk (see fig. 8.20).[158] Fourteen keys, laid out like a piano, controlled the disposition of the jaw, lips, and tongue, while a bellows and ivory reed fulfilled the roles of the lungs and larynx. The Euphonia capably pronounced a great variety of words and phrases, although most accounts noted the machine's poor diction and monotonous voice.[159]

Punch used the Euphonia as a tool for satire, pondering how Faber's machine might replace preachers and politicians, since their verbiage was dull and inconsequential anyway. For example: "A clear saving of 10,000 a year might be effected by setting up a machine *en permanence* in the Speaker's chair of the House of Commons. Place the mace before it. Have a large snuff-box on the side, with rappee and Irish for the convenience of Members, and a simple apparatus for crying out 'Order, order,' at intervals of ten minutes, and you have a speaker at the most trifling cost."[160] "By the way," the article suggested, "why should not Lord George Bentinck have one of these machines constructed, with a Benjamin Disraeli figure-head, and play upon it himself at once, and spare the honourable Member for Shrewsbury the bother of being his Lordship's Euphonia?"[161]

The London theater manager John Hollingshead has provided the most complete, as well as the most depressing, portrait of Faber's machine:

> The exhibitor, Professor Faber, was a sad-faced man, dressed in respectable well-worn clothes that were soiled by contact with tools, wood, and machinery. The room looked like a laboratory and workshop, which it was. The Professor was not too clean, and his hair and beard sadly wanted the attention of a barber. I have no doubt that he slept in the same room as his figure—his scientific *Frankenstein* monster—and I felt the secret influence of an idea that the two were destined to live and die together. The Professor, with a slight German accent, put his wonderful toy in motion. He explained its action: it was not necessary to prove the absence of deception. One keyboard, touched by the Professor, produced words which, slowly and deliberately in a hoarse sepulchral voice came from the mouth of the figure, as if from the depths of a tomb. It wanted little imagination to make the very few visitors believe that the figure contained an imprisoned human—or half human—being, bound to speak slowly when tormented by the unseen power outside. No one thought for a moment that they were being fooled by a second edition of the "Invisible Girl" fraud. There were truth, laborious invention, and good faith, in every part of the melancholy room. As a crowning display, the head

Fig. 8.20. Faber's Euphonia. From "The Euphonia," *Illustrated London News* 9 (1846): 96. Courtesy of the University of Washington Libraries.

> sang a sepulchral version of "God save the Queen," which suggested inevitably, God save the inventor. This extraordinary effect was achieved by the Professor working two key-boards—one for the words, and one for the music. Never probably, before or since, has the National Anthem been so sung. Sadder and wiser I, and the few visitors, crept slowly from the place, leaving the Professor with his one and only treasure—his child of infinite labour and unmeasurable sorrow. He disappeared quietly from London, and took his marvel to the provinces, where it was even less appreciated. The end came at last, and not the unexpected end. One day, in a dull matter-of-fact town—a town that could understand nothing but a Circus or a Jack Pudding—he destroyed himself and his figure. The world went on just the same, bestowing as little notice on his memory as it had on his exhibition. As a reward for this brutality, the world, thirty years afterwards, was presented with the phonograph.[162]

Hollingshead's version of the demise of Faber and the Euphonia is apocryphal. Faber may have taken his own life, but the speaking machine, at least, lived on. The husband of Faber's niece exhibited the machine under

Faber's name, and he joined P. T. Barnum's entourage in 1873. According to Barnum's autobiography, the Euphonia was a financial failure.[163]

It seems surprising that Faber's machine—an authentic technical marvel—did not attract crowds, while fraudulent speaking figures endured in shows and carnivals. One might recall that Mical's speaking heads never brought great rewards, even though they earned the praise of Rivarol and others. Similarly, although there were writers who claimed that Kempelen's speaking machine was his true great scientific achievement, the chess-playing Turk (a trick) captured the imagination of his audience.

The speaking machines of Mical, Kempelen, and Faber were ingenious demonstrations of phonetic theory, but they made poor theater. Since ventriloquists and fake talking heads survived, perhaps one can say that in the realm of entertainment, spectacular deceit is worth more than lackluster authenticity. Henri Decremps said as much in his reflection on the demise of Mical's labor. The marvel went unappreciated because it lacked "that taint of charlatanism so necessary in this century to obtain the support of the multitude."[164]

Faber's fate seems similar to that of Mical. Both men were outside of "official" science. Faber's creation received mere page-and-a-half notices in the *Annalen der Physik* (1843) and the *Journal de physique* (1879), but there are no publications written by Faber himself.[165] Unfortunately for him, Faber seems to have come upon the scene at a time when physicists like Willis, Wheatstone, and Helmholtz were pursuing theories of vowel sounds. For them, imitating the details of articulation had limited appeal. Furthermore, acoustical theory guided the design of the physicists' instrumentation. Such tools bore no direct relationship to speech organs, let alone to a man dressed like a Turk!

The more humanoid apparatus of Lloyd, Marage, Scripture, and Paget did not appear until the last years of the nineteenth century, when these investigators approached the problem from a physiological and phonetic point of view. If Faber had demonstrated his machine either fifty years earlier (in Kempelen's time) or fifty years later than its introduction in the 1840s, the Euphonia might have been greeted by an enthusiastic audience.

The reproduction of vocal music and the perfection of *vox humana* organ pipes provided another entertaining application of artificial voices. As we have seen, this was an acknowledged goal of individuals from many differing scientific traditions. Furthermore, comparisons of the voice to musical instruments were extremely common. Oliver Wendell Holmes—who wrote glowingly of the stereoscope and saw human attributes reflected in the "great glassy eyes" of the binocular camera—adopted a similarly enthusiastic tone in his appraisal of the new organ installed in the Boston Music Hall in 1863. "The organ," Holmes wrote,

> as its name implies, is *the instrument*, in distinction from all other and less noble [musical] instruments. We might almost think it was called organ as being part of an unfinished *organism*, a kind of Frankenstein-creation, half framed and half

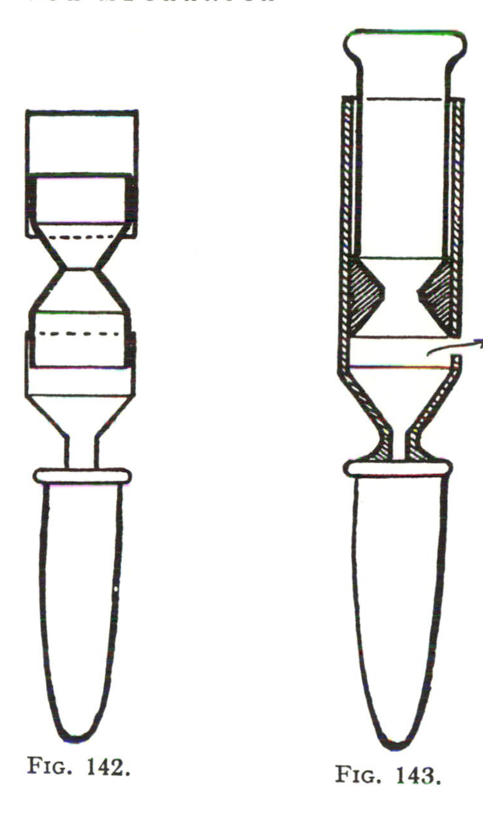

Fig. 8.21. Vowel organ pipes. From Paget, *Human Speech*, p. 235. Courtesy of Routledge.

> vitalized. It breathes like an animal, but its huge lungs must be filled and emptied by an alien force. It has a wilderness of windpipes, each furnished with its own vocal adjustment, or larynx. Thousands of long, delicate tendons govern its varied internal movements, themselves obedient to the human muscles which are commanded by the human brain, which again is guided in its volitions by the voice of the great half-living creature. A strange cross between the form and functions of animated beings, on the one hand, and the passive conditions of inert machinery, on the other![166]

Among all the sonorous capacities of the "Great Instrument," Holmes was most fascinated by the possibility of imitating the voice. "It is the highest triumph of our artificial contrivances to reach a tone like that of the singer, and among a hundred organ-stops none excites such admiration as the *vox humana*."[167] Despite his fascination with the possibility of creating the acoustic illusion, Holmes confessed the poor resemblance to the human model such devices usually exhibited—an observation with which other writers on *vox humana* pipes have concurred.[168] Sir Richard Paget's *Human Speech*, however, indicated that by 1930 some organ builders had incorporated the principles of vowel resonances in *vox humana* pipes and produced vowel pipes that worked quite well (see fig. 8.21). Ironically, these devices strongly resemble some of those made by Kratzenstein.[169]

Conclusion: From Euphonia to the Telephone?

At the beginning of this chapter we argued that efforts to duplicate the human voice had many sources and appeared in many different contexts. This multiplicity of origins shows up strikingly in the invention of the telephone by Alexander Graham Bell. Robert V. Bruce's biography of Bell offers an intriguing suggestion about the relationship between him and Faber. In the opening section, "Prelude: Philadelphia 1846," Bruce speculates on why neither Michael Faraday nor the American physicist Joseph Henry (who encouraged Bell) managed to assemble a telephone before Bell, despite their sufficient understanding of acoustics and electromagnetism. Bruce claims that Faber, Henry, and the city of Philadelphia were three key pieces in the story of Bell's success: "By one of history's numerous ironies, all three of these elements came together in 1846. And to them was added a blurred vision of the great goal itself. The incident shows how loud and close the knock of opportunity can come without awakening the strangely spellbound mind. And it thereby shows the greatness of first conceiving what in retrospect seems obvious."[170] In 1846, Henry was in Philadelphia and saw a demonstration of Faber's machine. He was amazed. "I have seen the speaking figure of Mr. Wheatstone in London," he wrote to a friend, "but it cannot be compared with this, which, instead of uttering a few words, is capable of speaking whole sentences, composed of any words whatever." "The keys," Henry continued,

> could be worked by means of electromagnetic magnets and with a little contrivance, not difficult to execute, words might be spoken at one end of the telegraph line, which had their origin at the other. . . . Thus if an image of the kind were placed in the pulpit of several churches . . . the same sermon might be delivered at the same minute to all. Or—but I will leave your own invention to make other applications.[171]

Bruce ponders the would-be moment of triumph: "The telephone was calling through a mist to the one man in 1846 who could have brought it forth if anyone could have. Yet Joseph Henry never did conceive the simple plan."[172] Bruce, of course, overplays this connection; the function of the telephone depended on an entirely different principle.

Bell, it so happens, was familiar with a surprising number of the themes sketched in this chapter. Like his father and uncle, Bell entered the profession of his grandfather—a Scottish actor who turned his oratorical talents to aiding the diction of others. Bell became an "elocutionist and a corrector of defective utterance."[173] He was acquainted with acoustics, since his work involved representational devices, like the phonautograph and the manometric flame.[174] In an essay on what he termed the "prehistory" of his invention, Bell cited this training as a factor in his development of the telephone. He also identified other influences on his work, such as his personal acquaintance with phoneticians including Henry Sweet (the model for Henry Higgins in Shaw's *Pygmalion*) and Alexander Ellis, who guided the young Bell in his first endeavor in electromagnetism—an attempt to duplicate Helmholtz's tuning-fork vowel synthesizer.[175]

A more intriguing encounter than Henry's meeting with the Euphonia took place in 1863, when the young Alexander Graham Bell and his father attended a performance of Faber's machine in London.[176] After the demonstration, Bell's father—Alexander Melville Bell, who was already well-known for his development of a phonetic alphabet called "visible speech" that could illustrate the mechanics of articulation to deaf students—took his son to meet Charles Wheatstone.[177] Wheatstone honored his guests by performing a few words on his old Kempelen-style machine.

Wheatstone loaned his copy of Kempelen's book to Melville Bell, and the young Alexander read it voraciously. Melville encouraged Alexander and his brother to learn about the physiology of speech by building a speaking machine of their own. The prescribed project was to copy nature exactly, rather than Kempelen's design. The boys used the cast of a human skull, with rubber cheeks and rubber strips for the larynx. The finished product could utter a few humanoid sounds, and it became, as Bell said, an "educational toy"—entertainment as well as a lesson in phonetics.[178]

Although one cannot say that the Euphonia and the other speaking machines either directly or indirectly inspired the telephone, Bell's background in both acoustics and phonetics was significant. But these sciences alone could not have effected the telephone. Bell would never have arrived at his great invention if it were not for his intensive researches on telegraphy. The focus of this labor was his attempt to create a "harmonic telegraph" that distinguished variations of frequency in the electric current and enabled several messages to travel over the same wire simultaneously. This entailed the germ of the telephone.[179]

The telephone and all of the speaking machines discussed above belonged to distinct social and scientific, as well as technological, contexts. The devices possessed different forms, capabilities, and means of operation corresponding to their unique roles: to amuse a crowd, or to discover, prove, or demonstrate a principle in physiology, phonetics, or acoustics. The diversity of apparatus tells no single tale of progress, but rather a history of attempts to satisfy an assortment of scientific goals and ways of understanding and studying the human voice.

It is tempting to look at this history as an evolution from antiquated efforts that merely copied vocal anatomy to modern machines rooted in physical theory. However, Marage's vowel sirens grinningly demonstrate that these developments did not follow any strict chronological order. Nor were these two styles of duplicating human function absolutely disparate. For example, in his attempt to supplant the human model of vowel generation, Helmholtz's tuning-fork synthesizer came to resemble another aspect of human anatomy: the structures of the inner ear. Likewise, the graphical method guided Marey's researches; yet he made functional anatomical models also. Not surprisingly, physiological and phonetic authors who imitated the voice retained instruments close in form to the vocal tract, while acoustical researchers were more apt to depart from human anatomy.

The imitative instruments discussed in this book were linked with a wide array of human activities—music, language and communication, visual art, and entertainment. They hold a special place among the instruments created by human art, because it is through the organs of sense and speech that we know each other and objects in the external world. Duplicating these organs of sense and speech is not only "art imitating nature" but also art imitating the artificer. It is a thoroughly human enterprise. For this reason duplicating the senses has never been limited to what we would call natural science; it inevitably spills out into all forms of discourse, both artistic and practical. It also holds out the possibility—enticing to some, but horrifying to others—that through the knowledge of automata we may come to know ourselves.[180]

CHAPTER NINE

Conclusion

In the previous chapters we have explored instruments on the margins of science. We have seen how they move easily from the realm of science to the realms of literature, philosophy, the fine arts, and entertainment. We find no simple transition from natural magic to natural philosophy, no easy division between science and technology, no single method that can be called "scientific demonstration" except insofar as we choose to define it as such. What we do find are changes in the ways that instruments are regarded, but even here the old ways persist through the changes. The search for "wonder" does not give way entirely to "matter-of-fact," and "matter-of-fact" as event or deed does not yield completely to the enthusiasm for "fact" as precise measure. An instrument of natural magic may reappear as a philosophical instrument, as an instrument of entertainment, or as a practical "invention" in a new guise. To understand actual scientific practice, we have to understand instruments, not only how they are constructed, but also how they are used and, more important, how they are regarded.

The Boundaries of Science

We have argued that natural science has a different appearance if we approach it through the study of instruments rather than through the study of theory or experiment. Part of the reason for the difference is that the territory covered by instruments is not congruent with the territories covered by these other approaches to the natural world. Instruments like the magic lantern or the stereoscope find employment as much outside of science as in it. Some instruments, such as the spectroscope, fit more neatly within the boundaries of their respective sciences. Others, such as the telescope and the microscope, are less confined. These differences in the "territory" covered by instruments raise questions that we need to consider.

First, we can ask whether instruments define their respective sciences or are defined by them. It has been our contention that instruments play a greater defining role than has usually been recognized by historians. When we focus on instrumentation, rather than on theory, science appears to be determined by what instruments can do. The science of acoustics is a good example. Acoustics became recognized as a part of experimental physics—distinct from its origins in music or harmonics—when a battery of new instruments was developed to analyze sound and to record it pictorially. With the graphical trace, a complex sound wave could be described in terms of precise physical quantities, rather than by the subjective evaluation of the human ear. To be sure, the construction of instruments rests on certain physical concepts that may be more or less well formed. Thus, Scott's phonautograph was grounded in the principle that sound is a pressure wave in the air, Kircher constructed the sunflower clock because he believed in a cosmic magnetic force, and Castel's ocular harpsichord assumed an exact analogy between light and sound. Any measuring instrument assumes an understanding of what is being measured, and often what is being measured is quite abstract; examples from experimental physics include force, temperature, charge, and potential difference. Thus, the creator of the instrument must possess a prior understanding of the physical concepts involved. However, once an instrument is granted an established role in a field of inquiry, it may determine the questions that are asked and the answers that may be considered valid. Or, as in the case of the instruments that we have studied, it can find an application outside of natural science altogether.

A second question about the boundaries of science concerns the temporal framework of their setting. Were they set in the past, when natural philosophers and scientists did their work, or are they set by us today? Historians are taught to avoid the sins of anachronism and "Whig" history and to interpret historical events in terms of their contexts. This endeavor is especially challenging in the history of science, where the successes of modern science loom as a potential guide to the "right" and "wrong" answers of past investigators. A critique of the past in terms of present standards of correctness would scarcely tolerate a discussion of most of the persons and devices in the preceding chapters. However, such a critique would also produce a warped view of history. Although modern students will see the instruments discussed in this book as marginal to the main lines of development of modern science, their creators considered them as absolutely central. Let us recall Robert Darnton's anthropological approach to history. The historian in the archive, like the anthropologist in the field, finds the greatest value in exploring that which is *least* comprehensible. It is the process of unraveling the meaning of events that are obscure to a modern mind which enables historians to understand thoughts and actions of the past.

Of course we must realize that historians set their own agendas and that the questions they ask come from their own perspectives. Not only may the boundaries of modern science differ from those of seventeenth-century science, they may *both* differ from the boundaries of the historian's field of inquiry. Rather than a weakness, however, this incongruity is the historian's strength. We understand the past by making it comprehensible to the present. Historians are in a position to choose those parts of history that are most revealing to them and their peers. By investigating some rather odd instruments from the past, we gain insights about the role of instruments throughout the history of modern science.

A third question is whether terms like "margins," "boundaries," and "territories" accurately describe the changes that occur in natural science.

These words imply that there is something called "science" which endures even while its boundaries change. Our ability to write its history rests on the assumption that there exists an identifiable body of knowledge called "science" with a history to be explored. Whether the boundaries of science are set by theory, experiment, or instruments, the assumption remains that something called "natural science" is an identifiable reality. By exploring boundaries we assume a common core, but is there a common core?

In an 1854 lecture, Hermann von Helmholtz reflected on the efforts of eighteenth-century automaton builders. "The marvel of the last century," Helmholtz wrote,

> was Vaucanson's duck, which fed and digested its food; the flute-player of the same artist, which moved all its fingers correctly; the writing-boy of the elder, and the pianoforte-player of the younger Droz; which latter, when performing, followed its hand with its eyes, and at the conclusion of the piece bowed courteously to the audience. That men like those mentioned, whose talent might bear comparison with the most inventive heads of the present age, should spend so much time in the construction of these figures which we at present regard as the merest trifles, would be incomprehensible, if they had not hoped in earnest to solve a great problem.[1]

Helmholtz understood that the mechanical duck had not been a "merest trifle" for Vaucanson. It became so only for Helmholtz's own generation, which regarded it as a misdirected effort toward a mechanical description of nature. Even if one assumes that Helmholtz shared Vaucanson's goal, there remains an obvious qualitative difference between a mechanical duck and his own tuning-fork apparatus for creating artificial timbres. Vaucanson's duck was situated in the realm of public entertainment and performance; Helmholtz's tuning-fork apparatus was situated in a university laboratory. Vaucanson's duck copied animal function as it was observed; Helmholtz's tuning-fork synthesizer copied what he thought to be its mechanism. From our perspective, Helmholtz's approach was more valid—more scientific.

Helmholtz's words confront historians with a "great problem" of their own. Was Helmholtz's inability to understand Vaucanson caused by a shift in the boundaries of acceptable science, or was it caused by a completely different concept of what science should be? The debates over the first philosophical instruments—such as the telescope, microscope, and air pump—would indicate that they were transforming the central core of science, not merely adjusting the boundaries. If this is indeed the case, we have to be careful about assigning the label "scientific" to some instruments and withholding it from others. As modern scientists we can agree with Helmholtz that Vaucanson's duck was a "merest trifle," but as historians we need to understand why it was not a trifle for Vaucanson. The fact that the question can be asked in the first place suggests that the criteria for what counts as "scientific" have never been fixed.

The instruments discussed in this book broaden our understanding of the ways that instruments have been employed in the study of nature—roles that are not as apparent in the better-known tools of past and present science. Rather than merely suggesting that an arbitrary boundary of scientific respectability should be pushed out a little farther, these marginal instruments lead us to inquire about the fundamental nature of the scientific enterprise and the role of instruments in it, as well as the way that attitudes toward instruments have changed over time.[2] In particular, the objects studied in this book disclose the ways that scientific instruments mediate between investigators and the phenomena they study—by expanding the senses, modeling nature, generating visual images, engendering conventions of scientific correctness, or creating a new language of scientific discourse. They also reveal the role of technically functional performance, in teaching, entertainment, communication, or the demonstration of scientific principles. Finally, these instruments provide a historical understanding of the search for objectivity in science.

Mediation

In one of his more desperate moods, Ralph Waldo Emerson bemoaned the tragedy of the human condition.

> It is very unhappy, but too late to be helped, the discovery we have made that we exist. That discovery is called the Fall of Man. Ever afterwards we suspect our instruments. We have learned that we do not see directly, but mediately, and that we have no means of correcting these colored and distorting lenses which we are, or of computing the amount of their errors. Perhaps these subject-lenses have a creative power; perhaps there are no objects. Once we lived in what we saw; now, the rapaciousness of this new power, which threatens to absorb all things, engages us. Nature, art, persons, letters, religions, objects successively tumble in, and God is but one of its ideas. Nature and literature are subjective phenomena; every evil and every good thing is a shadow which we cast.[3]

Emerson's cosmic drama encapsulates the history and key epistemological problems raised by instruments as a way of understanding nature. Instruments, whether they be objects that we create or our own organs of sense, stand between us and the natural world. They are our only connection to that world, assuming that it exists at all, and they are all flawed. In the fall from grace that Emerson described, the divine human instrument was hopelessly fractured; the perfect and instantaneous perception of the world was lost. The deficiencies of our instruments—their dullness, their distortions—gained power over us.

In a more optimistic mood Emerson might have noted that mediate knowledge is better than no knowledge at all, and that our instruments, imperfect as they are, save us from total ignorance. Whether we take Emerson's insight as cause for hope or despair, the recognition that we know nature mediately means that we can know it only through our instruments.

Many of the instruments that we have discussed imitate human senses rather than objects in the external world. When natural philosophers and scientists came to recognize eyes and ears, not as sacred mediators of truth, but as complex tools with their own sets of flaws and imperfections, the replacement of the sense organs with external and mechanical ones became a possibility. Speaking machines and automata have a long history in natural magic, but in the nineteenth century, as we have seen, the efforts to copy human organs of sight and sound took on a different character. Rather than create human likenesses, these instruments duplicated or surpassed human function. Recording instruments "sensed" and inscribed data more accurately than the natural philosopher could do it himself, and speaking machines duplicated the sounds of speech without necessarily copying the anatomy of the speech organs.

Nevertheless, the model of the human body remained an enduring one. Whether the moral guideposts of natural theology pointed the way, or the demands of a specific discipline or approach helped to shape instruments into a human form, the sense organs remained a standard against which artificial instruments were judged.

An extreme perspective in the debate over the human body's status as a device for understanding nature is exemplified in the treatment of automata in literature. Duplicating human function or surpassing it in some spectacular way raises strong emotions, both positive and negative, that have long been exploited by writers. An exemplar of the negative reaction may be found in the tales of E.T.A. Hoffmann, whose literary employment of automata and other instruments may be read as an index of a more "scientific" approach to nature. A speaking automaton terrified Lewis, one of the main characters in Hoffmann's "Automata" (1814).[4] The pulse of romanticism was strong in Lewis; the mechanical imitation of human function disturbed him deeply. Figures such as the automaton, he felt, "can scarcely be said to counterfeit humanity so much as to travesty it," and are "mere images of living death or inanimate life."[5] Lewis had similar complaints about automaton musicians, like Vaucanson's flute player. Cleverly made automata may counterfeit the motions of skilled human fingers, but the mind, soul, and heart of a living performer can never be duplicated. Without these, Lewis explained, the spiritual element in music is lost. Hoffmann was not opposed to all artificial instruments, however. He believed that a more worthy pursuit for skilled mechanists would be an investigation of the physical world dedicated to the "discovery of the marvellous acoustical secrets which lie hidden all around us in nature."[6] Lewis regarded the glass harmonica, the Aeolian harp, and especially the *Wetterharfe* as profound contributions to the aim of "tempt[ing] Nature to give forth her tones."[7] Thus the Aeolian harp, the instrument of romantic science, stood diametrically opposed to the stiff mechanical image of humanity embodied in automata.

Hoffmann's "The Sand-Man" (1816) carries the interaction of human and automaton to a fatal conclusion. In that story, Nathanael, a university student, is driven to madness and to his own demise by Coppelius, a demonic lawyer who reappears thinly disguised as Giuseppe Coppola—a peddler of barometers, telescopes, and thermometers. Coppelius and Nathanael's physics professor conspire to create the beautiful automaton Olimpia with whom the young man falls in love. But his discovery of the truth about her is too awful for his heart to bear.

These two tales employ automata as part of a symbol system in which the harsh scientific analysis of human senses and motions assaults the ineffable core of human experience. The automaton Olimpia formed the link between Coppola's philosophical instruments and the maleficence of his alter ego Coppelius. The casting of Coppola as a merchant of those particular wares, together with the pivotal part played by the physics professor, creates a strong indictment of the methods and tools of modern science. Likewise, the stiffly performed music of automata created a painful dissonance with Hoffmann's romantic conception of music as the most noble and profound form of human expression. A similar motif informed many nineteenth-century critiques of photography. Lady Elizabeth Eastlake and others regarded the sense of vision as an absolutely personal faculty that defied mechanical description and imitation.

Despite this criticism from some quarters, respect for the dignity of the human form and the perfection of its design led many to attempt to imitate it. In this estimate, the human frame constituted the template for a truthful study of nature. Automaton builders and vocal physiologists believed that only by accurately duplicating the body could one discover the principles of its operation; many stereoscopic photographers viewed the dimensions of the human eyes as the only measure of visual fidelity; and Léon Scott conceived his phonautograph as an artificial ear. Oliver Wendell Holmes was utterly thrilled by the artificial duplication of human function in all its forms. He embraced the stereoscopic camera and the *vox humana* pipe (imperfect as it was) as signs of human advancement.[8]

However, it is the attempt to surpass the ordinary capacity of the human sense organs that has been the hallmark of modern science. Like the telescope and microscope, De la Rue's stereoscopic cameras expanded the range of human vision. Helmholtz introduced resonators to assist the ear. Willis departed from mouth-shaped speaking machines. Marey's inscription devices possessed a sensitivity far greater than the most attentive physician. All of these investigators sought to improve or outstrip the capacity of human organs, or to replace them with superior instruments. For figures like these, accuracy must be defined in mechanical terms and not in human ones.

All of these contentions and controversies turn on the ways that our instruments mediate between our conscious minds and the external world. Is this mediation the same in the cases of artificial instruments and the human organs of sense, or is it fundamentally different? If it is the same, then our artificial instruments can give us knowledge about ourselves and power over our environment; if it is fundamentally different, our efforts to duplicate what God has created in his own image are not only misguided but blasphemous.

If we turn from the instruments that imitate and extend the human senses to those that operate on inanimate objects, we see the same arguments as to whether they should imitate or manipulate nature. Those philosophers in the natural magic tradition were not happy with instruments that explained nature by taking it apart and distorting it. The instruments of natural magic modeled nature. If they copied its actions faithfully, then the mechanism that made them work was probably similar to the cause hidden in nature. By analogy the instrument hinted at the occult cause. It took many such hints to give any clear idea of the cause, and therefore natural magicians attempted to encompass all the particular aspects of a phenomenon—-all the examples of sound, for instance, both common and wondrous—in an encyclopedic account of known cases. A single experiment or a single instrument could prove nothing by itself.

The instruments of experimental philosophy, on the other hand, manipulated nature. They were used to establish matters of fact—repeatable events that could then be used to analyze and generalize natural phenomena. These events were not part of common experience, and therefore observers could not validate them by checking them with the unaided senses. The events could, however, lead to new knowledge, and a single experiment could invalidate a theory if the results disagreed with what the theory predicted. In both of these cases—natural magic and the experimental philosophy—the instrument mediated between the observer and nature, but the observer operated from different assumptions in each case.

We have argued that instruments and language have much in common. Both are mediators between ourselves and nature; both are artificial since they are both created by us (except for our organs of sense, if we want to consider them as instruments); both produce signs that we use to reason about nature; and both employ conventions by which the signs are to be understood. Whether the conventions are entirely arbitrary was a subject of debate. For the natural magician there was a hidden connection between the object and the word that named it, just as there was a hidden connection between the operations of his instruments and the operations of nature. For those philosophers in the tradition of John Locke, the conventions were entirely arbitrary; Locke regarded figurative language that hinted at hidden connections between words and things as false and deceptive. It is important to remember that philosophers in the seventeenth century were using new instruments—the telescope, microscope, thermometer, barometer, air pump—in an entirely new way. Through instruments and language they were establishing conventions of scientific practice and discourse. Instruments defined the standards for acceptable experimentation and communication. Not surprisingly, these new tools and new ways of studying nature were looked upon with suspicion by some.

A remarkable feature of the instruments that were part of the natural magic tradition was their capacity to produce visual images, and this feature was carried on into the nineteenth century. Graphs, maps, inscriptions, and other representations that constitute scientific expression follow conventions of style that reflect the assumptions and purposes of their authors. The way that stereographs or phonautograph traces were created and understood embodied a complex set of beliefs about how nature should be studied. Some of these new "languages," such as Linnaeus's binomial system in botany and Lavoisier's chemical nomenclature, were closer to common language. Others, such as graphs, vertical sections in geology, and Chladni figures, were conventions of an entirely new kind. It is significant that their creators often called them languages, indicating that they recognized the connection between linguistic conventions and those of instruments.

Of course the most important language of science, other than written words, is mathematics. Although quantification did not become a major goal of philosophical instruments until the late eighteenth century, it was the natural consequence of an experimental method that sought to establish the existence of specific repeatable events rather than to encompass natural phenomena in their greatest complexity. Both Galileo and the Jesuit experimenters regarded mathematics as the proper language for natural philosophy, but they disagreed vigorously over how that language should be used and what it could tell us about nature. To Jesuits like Kircher and Castel the importance of mathematics lay in its rhetorical form, which allowed one to generalize from particular phenomena, while for Galileo, Descartes, Huygens, and Newton it lay in its ability to give precise formulation to laws of nature.

The extension of mathematics from astronomy, geometrical optics, mechanics, and harmonics into the sciences of heat, light, sound, electricity, and magnetism required the creation of new quantitative instruments, but again, it was not obvious how those instruments were to be used. Most notably, there was ambiguity in what was to count as a "demonstration": whether it was to be a "showing" of a phenomenon with an instrument to establish a matter of fact, or an argument, usually mathematical, that confirmed a theory. As scientific instruments became more quantitative, beginning around 1780, the demonstration lecture diverged from the physics laboratory, not only in its purpose and location, but also in how it "demonstrated" natural science.

If mathematics was the language that mediated between us and the natural world as Galileo claimed, then instruments were the mediators that gave meaning to the "words" or symbols of that language. In the late eighteenth century with the appearance of more quantitative instruments, along with graphs and recording instruments, the languages of instruments and mathematics converged. When figures such as Galileo and Marey claimed that they had at last found a true "language of nature," they assumed that they had hit upon a perfect correspondence between signs and things which would lead them to the very core of phenomena. Emerson could have told them otherwise. None of our mediators, instrumental or linguistic, can ever be perfect. Galileo never imagined that the key of mathematics which he said would rescue him from wandering in a "dark labyrinth" would lead him instead into Peirce's labyrinth of signs, where the Minotaur—that indestructible nugget of reality—would prove to be ever more elusive.

Objectivity and Moral Certainty

In an essay entitled "The Experiment as Mediator between Object and Subject" Goethe argues that the natural way of judging things is to relate the objects perceived by the senses to oneself, as to whether they please or displease, attract or repel, help or harm. He adds that a far more difficult task is to view objects in their own right and in relation to one another. He says that when we attempt to be "objective," that is, to remove from objects their relations to ourselves and judge them only in relation to each other, then "all the inner enemies of man lie in wait [to bias his judgment]: imagination, which sweeps him away on its wings before he knows his feet have left the ground; impatience; haste; self-satisfaction; rigidity; formalistic thought; prejudice; ease; frivolity; fickleness—this whole throng and its retinue." Goethe concludes that "nothing is more dangerous than the desire to prove some thesis directly through experiments."[9] Goethe's doubts about an experiment's ability to mediate between the subject and object come from the obvious fact that the experimenter, who designs the experiment, must in some sense be mediator and subject at the same time. It is impossible for him to be totally "objective" (that is, like the object) if he is also the subject and judge.

One way to make an experiment more objective is to employ an instrument. An instrument cannot be prejudiced or passionate because it does not suffer from the "inner enemies" that worried Goethe. The quantitative and recording instruments introduced into natural philosophy at the end of the eighteenth century represented further steps in the direction of greater objectivity.[10] A graph, especially if it was drawn directly by an instrument, could reveal a quantitative relationship that did not seem to depend on human judgment at all. The same was true of the photograph and the stereograph. They were images "drawn by the sun" and therefore free from subjective error. Of course instruments introduced errors of their own, but somehow these "objective" errors were more easily "delivered and reduced" (in the words of Francis Bacon) than the subjective errors of the experimenter. As a result, experimenters sometimes found it convenient to consider the organs of sense themselves as instruments and to treat the errors that they produced as "instrumental" errors.

It is impossible, however, to remove human judgment entirely from scientific inference. At some point the experimenter has to judge the images that reach his or her consciousness. Even if the eye, for instance, can be regarded as an instrument and the optic nerve as a physical conduit for the image to the brain, at some point the image must become a subjective perception.[11] The effort to make the subject into an object leads us directly to the paradox of scientific "objectivity." An "objective" observation is one that does not make a value judgment, and yet we judge it to be valuable precisely because it does not assign value.[12] The judgment by which we place value on objectivity is both subjective and moral. It is a decision to accept the objective evidence as adequate for the conclusion being "demonstrated," and that decision must ultimately be subjective. It must also be moral in the sense that we are required to reach a judgment without conclusive proof.

When Willem 'sGravesande stated that instruments provide us with "moral evidence" by which we can reach "moral certainty," he meant that they do not give us deductive or mathematical certainty. Nevertheless, they can establish in us a conviction comparable to that raised by a mathematical proof. The *Encyclopédie* defines moral certainty as that kind of certainty which would lead a wise and prudent man to think and act as if he were presented with mathematical certainty.[13] The term comes from law, where mathematical certainty is not possible. According to Robert Boyle, moral certainty is "such a certainty, as may warrant the judge to proceed to the sentence of death against the indicted party," and he sought the same kind of certainty in his experimental philosophy.[14]

It is important to note that Boyle and the *Encyclopédie* define moral certainty in terms of the degree of conviction that it arouses in the judge, not by any criterion drawn from the evidence itself. 'SGravesande's definition is the same. The evidence from an experiment is morally certain when it achieves a very high level of conviction in the minds of witnesses to the experiment. The criterion of validity is essentially subjective—an instrument produces moral certainty if those using it and observing it are convinced by it. If natural philosophers were all convinced equally, they would also agree on the validity of the experiments that produce the evidence, but, as we have seen, they do not. Different philosophers have used instruments in different ways. Kircher's clocks were frauds by our standards, but not by Kircher's standards. For him they illustrated by analogy the cosmic magnetic force that was forever hidden from our senses. According to the assumptions and methods of natural magic, the sunflower clock brought stronger "moral evidence" than the imagined magnetic particles of Descartes's mechanical philosophy.

A similar problem arose in the controversy over whether an instrument should duplicate nature or distort it to the advantage of the philosopher. The telescope, microscope, and air pump were the subjects of controversy in the seventeenth century just as were the photograph and the stereograph in the nineteenth. An enhanced image is not a true rendering of nature, and it is not clear how such an image can bring moral certainty. It is significant that the debates over these instruments often turned on arguments from natural theology and were concerned with aesthetic and moral value.

The realization that instruments give us only moral certainty does not mean that there are no criteria for truth in natural philosophy or that any method is as good as any other. It does suggest, however, that because the criteria are ultimately subjective, they may differ from person to person and will not necessarily be constant through time. It is, perhaps, not surprising that attempts to define a single "scientific method" always fail to do justice to the complexity of scientific practice both in our era and in times past.

Objectivity, however one defines that term, has not always been a universal goal of natural philosophy. The romantics did not want to study nature as something apart from the observer. For them, "moral" evidence could never be objective. The Aeolian harp appealed to them because it was not objective. It was an instrument and a mediator, to be sure, but a mediator that connected the poet more closely to nature rather than one that kept him apart in the service of "objectivity." Also they believed that the effort to make science objective, analytical, and therefore amoral defeated any purpose that science might have. For them, a philosophy of nature that lacked moral content did not deserve to be called a science.

The search for moral certainty enters into the process of mediation whether we like it or not. *We choose* how to represent the natural world to ourselves. In fact, the development and practice of experimental science is contingent upon a long series of such choices; modern science is by no means an obvious or inevitable approach to the study of nature. Granted, the answers that our instruments give us are not entirely our choice, but we do choose how to look at nature, what mediators we want to use, what questions we want to ask, which answers signify and which do not. As we have seen, natural philosophers used instruments in many different ways. What was a proper instrument for Louis-Bertrand Castel or Athanasius Kircher was not a proper one for René Descartes or Johann Heinrich Lambert. What attracted the skill and devotion of Vaucanson was a merest trifle for Helmholtz. The instrument expressed the way that the experimenter wished to question nature. It was the language of his inquiry. Whether it brought certainty was a moral judgment that each scientist had to make, using the criteria of his own particular time and tradition.

NOTES

Chapter One
Instruments and Images: Subjects for the Historiography of Science

1. Francis Bacon, *The Works of Francis Bacon*, ed. and trans. James Spedding, Robert Leslie Ellis, and Douglas Denon Heath, 14 vols. (London: Longman & Co., 1857–1874), 4:47.

2. Deborah Jean Warner, "What Is a Scientific Instrument, When Did It Become One, and Why?" *British Journal for the History of Science* 23 (1990): 83–93; E.G.R. Taylor, *The Mathematical Practitioners in Tudor and Stuart England* (Cambridge: Cambridge University Press for the Institute of Navigation, 1954) and *The Mathematical Practitioners of Hanoverian England* (Cambridge: Cambridge University Press, 1966).

3. Steven Shapin and Simon Schaffer, *Leviathan and the Air-Pump: Hobbes, Boyle, and the Experimental Life* (Princeton: Princeton University Press, 1985), pp. 174–175.

4. Giambattista Della Porta, *Natural Magick* (1st ed. Naples, 1558); reprint of the 1st English ed. (London: Thomas Young and Samuel Speek, 1658), ed. Derek J. Price (New York: Basic Books, 1957), pp. 1–2.

5. On the books of secrets see William Eamon, "Arcana Disclosed: The Book of Secrets Tradition and the Development of Experimental Science," *History of Science* 22 (1984): 111–150, and *Science and the Secrets of Nature: Books of Secrets in Medieval and Early Modern Culture* (Princeton: Princeton University Press, 1994).

6. On the telescope: Della Porta to Federigo Cesi, August 28, 1609, Galileo Galilei, *Le opere di Galileo Galilei*, edizione nazionale, 20 vols. in 21 (Florence: Barbera, 1890–1909), 10:252; also Albert Van Helden, "The Invention of the Telescope," *Transactions of the American Philosophical Society* 67, pt. 4 (1977): 44–45. On the thermometer: W. E. Knowles Middleton, *A History of the Thermometer and Its Use in Meteorology* (Baltimore: The Johns Hopkins University Press, 1966), chap. 1, esp. pp. 21–22. On the air pump: Shapin and Schaffer, *Leviathan and the Air-Pump*, p. 239. On the prism: Simon Schaffer, "Glass Works: Newton's Prisms and the Uses of Experiment," in *The Uses of Experiment: Studies in the Natural Sciences*, ed. David Gooding, Trevor Pinch, and Simon Schaffer (Cambridge: Cambridge University Press, 1989), pp. 67–104, at pp. 73–74 and 78–79. On Hooke as a natural magician, see John Henry, "Robert Hooke, the Incongruous Mechanist," in *Robert Hooke: New Studies*, ed. Michael Hunter and Simon Schaffer (Woodbridge: The Boydell Press, 1989), pp. 149–180.

7. David Brewster, *Letters on Natural Magic Addressed to Sir Walter Scott, Bart.* (London: J. Murray, 1832).

8. On instruments in historiography see Peter Galison, "History, Philosophy, and the Central Metaphor," *Science in Context* 2 (1988): 197–212.

9. Robert Hooke, *Micrographia: or Some Physiological Descriptions of Minute Bodies made by Magnifying Glasses with Observations and Inquiries Thereupon* (London, 1665; reprint, New York: Dover, 1961), preface, unpaginated.

10. Katharine Park, "Bacon's 'Enchanted Glass,'" *Isis* 75 (1984): 290–302, esp. 294–295.

11. Bacon, *Works*, 3:394–395.
12. Ibid., p. 289 and 4:47, 84.
13. Ibid., 4:193–194.
14. Ibid., pp. 60–61.
15. Ibid., p. 84.
16. Bacon insisted that a new method of inquiry was more important than the new instruments. Ibid., p. 26.
17. John Locke, *An Essay Concerning Human Understanding*, ed. Alexander Campbell Fraser, 2 vols. (Oxford: Clarendon Press, 1894), 2:146.
18. Ibid., p. 100. Catherine Wilson, "Visual Surface and Visual Symbol: The Microscope and the Occult in Early Modern Science," *Journal of the History of Ideas* 49 (1988): 85–108, at 103.
19. Historians are beginning to deal with words and images: Martin J. S. Rudwick, "The Emergence of a Visual Language for Geological Science, 1760–1840," *History of Science* 14 (1976): 149–195; David Gooding, " 'Magnetic Curves' and the Magnetic Field: Experimentation and Representation in the History of a Theory," in Gooding, Pinch, and Schaffer, *The Uses of Experiment*, pp. 183–223; Homer E. Le Grand, "Is a Picture Worth a Thousand Experiments?" in *Experimental Inquiries: Historical, Philosophical and Social Studies of Experimentation in Science*, ed. Homer E. Le Grand (Dordrecht: Kluwer, 1990), pp. 241–270; Greg Myers, "Every Picture Tells a Story: Illustrations in E. O. Wilson's *Sociobiology*," in *Representation in Scientific Practice*, ed. Michael Lynch and Steve Woolgar (Cambridge: MIT Press, 1990), pp. 231–265; and Mary G. Winkler and Albert Van Helden, "Representing the Heavens: Galileo and Visual Astronomy," *Isis* 83 (1992): 195–217. A classic study in this vein is Erwin Panofsky, *Galileo as Critic of the Arts* (The Hague: Nijhoff, 1954). On visual communication as opposed to verbal communication see Rudolph Arnheim, *Visual Thinking* (Berkeley and Los Angeles: University of California Press, 1969).
20. Jamie Croy Kassler, "Man—a Musical Instrument: Models of the Brain and Mental Functioning before the Computer," *History of Science* 22 (1984): 59–92.
21. Harold I. Brown, "Galileo on the Telescope and the Eye," *Journal of the History of Ideas* 46 (1985): 487–501.

Chapter Two
Athanasius Kircher's Sunflower Clock

1. John E. Fletcher, "Astronomy in the Life and Correspondence of Athanasius Kircher," *Isis* 61 (1970): 52–54.
2. Peiresc to Gassendi, March 2, 1633; Nicolas Claude Fabri de Peiresc, *Lettres de Peiresc*, ed. Philippe Tamizey de Larroque, 7 vols. (Paris: Imprimerie Nationale, 1888–1898), 4:293–296, esp. pp. 295–296.
3. Peiresc to the brothers Pierre and Jacques DuPuy, May 21, 1633; ibid., 2:525–530.
4. Peiresc to Claude Mydorge in Paris, June 6, 1633; Marin Mersenne, *Correspondance du Père Marin Mersenne, religieux minime*, ed. Paul Tannery and C. de Waard, 13 vols. (Paris: G. Beauchesne et fils, 1932–1977), 3:426–428.
5. Descartes to Mersenne, July 22, 1633; ibid., pp. 457–460, esp. p. 459.
6. Godefroid Wendelin to Mersenne, June 15, 1633; ibid., pp. 430–441, esp. pp. 431–432.

7. Bibliothèque inguimbertine, Archives Communales musée municipal, Carpentras, Peiresc MS 1864, fols. 215ff.

8. Peiresc to Mersenne, October 13–15, 1633; Mersenne, *Correspondence*, 3:497–505, esp. p. 504.

9. J. E. Fletcher, "Claude Fabri de Peiresc and the Other French Correspondents of Athanasius Kircher (1602–1680)," *Australian Journal of French Studies* 9 (1972): 255.

10. Raffaello Magiotti to Galileo, March 18, 1634; Galileo Galilei, *Le opere di Galileo Galilei*, edizione nazionale, 20 vols. in 21 (Florence: Barbera, 1890–1909), 16:65. Quoted from Bern Dibner and Stillman Drake, *A Letter from Galileo*, Burndy Library Publications, 24 (Norwalk, Conn.: Burndy Library, 1967), p. 49.

11. Three years later Magiotti complained to Galileo that Kircher had gone off on a trip without leaving him the sunflower seeds that he had promised, which would indicate that interest in the sunflower clock was still alive in Rome as late as 1637 (Raffaello Magiotti to Galileo, May 16, 1637; Galileo, *Opere*, 17:80–81). Another equally distressed letter from Kircher's friend in Avignon, Antoine Payen, asked Kircher what he should do with the seed supply that Kircher had left behind, and remarked that the seeds were rapidly losing their virtue (Fletcher, "Claude Fabri de Peiresc," pp. 270–271).

12. Silvestro Pietrasancta, *De symbolis heroicis libri IX* (Antwerp: Ex officina Plantiniana B. Moreti, 1634), pp. 145–147, quoted from Martha Baldwin, "Magnetism and the Anti-Copernican Polemic," *Journal for the History of Astronomy* 16 (1985): 162, and Dibner and Drake, *Letter from Galileo*, pp. 48–49.

13. Peter Paul Rubens to Peiresc, December 18, 1634; *The Letters of Peter Paul Rubens*, trans. and ed. Ruth Saunders Magurn (Cambridge: Harvard University Press, 1955), p. 394.

14. Dibner and Drake, *Letter from Galileo*, p. 47n.14; and Georges Monchamps, *Galilée et la Belgique: essai historique sur les vicissitudes du système de Copernic en Belgique* (Saint-Trond: G. Moreaus-Schouberects, 1892), p. 131. For Rubens's contribution to Pietrasancta's book see Julius S. Held, ed., *Rubens and the Book: Title Pages by Peter Paul Rubens*, catalog to an exhibition held at Chapin Library, Williams College, Williamstown, Mass., May 2–31, 1977; and *P. P. Rubens als Boekillustrator*, Stadt Antwerpen Museum Plantin-Moretus, May 7–July 4, 1977.

15. Peiresc to Galileo, April 1 and April 17, 1635; Galileo, *Opere*, 16:245–248, 259–262.

16. Wendelin to Gassendi, May 1, 1635; Mersenne, *Correspondence*, 3:436, Monchamps, *Galilée et la Belgique*, pp. 135–136.

17. Galileo to Peiresc, May 12, 1635, quoted from Dibner and Drake, *Letter from Galileo*, pp. 54–55, with minor changes. Roberto Galilei, who forwarded the letter to Peiresc, congratulated Galileo for having "exposed the stratagem of Father Linus" (p. 56).

18. In a letter to DuPuy, May 29, 1635 (Peiresc, *Lettres*, 3:318–322), Peiresc informed DuPuy that he had heard from Galileo and Pietrasancta, but was waiting for a report from Dormalius.

19. Peiresc to DuPuy, June 19, 1635; ibid., pp. 331–334. With this report from Dormalius, Rubens felt himself relieved of his commission to investigate the clock. He wrote: "I have nothing more to tell you about the marvelous discoveries of Father Linus, since you have had a first-hand report from M. Dormal. (All the Fathers of the Society in our country believe there is some magnetic force in this operation, and

some of them are trying to imitate it.)" Rubens to Peiresc, August 16, 1635; *Letters of Peter Paul Rubens*, pp. 399–402.

20. Peiresc to DuPuy, October 28, 1636; Peiresc, *Lettres*, 3:595–599.

21. Descartes to Constantijn Huygens, January 14, 1643; Mersenne, *Correspondance*, 12:29–30.

22. Monchamps, *Galilée et la Belgique*, p. 137.

23. Martha Baldwin has studied Kircher's magnetic philosophy in great detail in her "Athanasius Kircher and the Magnetic Philosophy" (Ph.D. diss., University of Chicago, 1987). She discusses Linus's clock and the Jesuit writings on magnetism in her "Magnetism and the Anti-Copernican Polemic." Also on Kircher's magnetism see William L. Hine, "Athanasius Kircher and Magnetism," in *Athanasius Kircher und seine Beziehungen zum gelehrten Europa seiner Zeit*, ed. John Fletcher (Wiesbaden: Otto Harrassowitz, 1988), pp. 79–97, and "The Mersenne-Kircher Correspondence on Magnetism," *Proceedings of the Western Society for French History* 10 (1982): 106–117. On the existence of a magnetical philosophy see James A. Bennett, "Cosmology and the Magnetical Philosophy, 1640–1680," *Journal for the History of Astronomy* 12 (1981): 165–177. On other aspects of Kircher's career see Conor Reilly, *Athanasius Kircher, S.J., Master of a Hundred Arts, 1602–1680*, Studia Kircheriana 1 (Wiesbaden: Edizioni del Mondo, 1974), and Jose Alfredo Bach, "Athanasius Kircher and His Method: A Study in the Relations of the Arts and Sciences in the Seventeenth Century," Ph.D. diss., University of Oklahoma, 1985.

24. Johannes Kepler, *Epitome astronomiae copernicae*, in *Gessamelte Werke*, ed. Walther van Dyck and Max Casper, 17 vols. (Munich: C. H. Bech, 1937–1975), 7:298–300. See also Baldwin, "Athanasius Kircher," pp. 197–208.

25. Giambattista Della Porta, *Magia Naturalis* (1st ed. Naples, 1558), reprint of the 1st English ed. (London: Thomas Young and Samuel Speek, 1658), ed. Derek J. Price (New York: Basic Books, 1957), pp. 199–200. Bk. 7 is "Of the wonders of the Load-stone" and contains much solid information about the lodestone and the compass scattered among the magic tricks.

26. Baldwin, "Athanasius Kircher," pp. 333–359. Baldwin discusses the "magnetic" properties of plants in detail.

27. Gilbert suggests this analogy in *De Magnete* (1600). William Gilbert, *On the Magnet*, trans. S. P. Thompson, ed. Derek J. Price (New York: Basic Books, 1958), pp. 130–131.

28. Kircher had access to Peiresc's garden of exotic plants at Aix and to the garden at the Collegium Romanum. The sunflower was first grown in Europe at the Botanical Garden of Padua in 1568. Cortuso, one of the early directors of the garden, claimed to have grown a specimen 120 hands high, and Kircher's engraving shows a dramatic plant. Arturo Paganelli, Giancarlo Cassina, and Fiorella Rocca, "The Botanical Garden of the University of Padua," *The Herbalist* (Concord, Mass.), no. 49 (1983): 72–78, at 75.

29. Peiresc to Gassendi, March 2, 1633; *Lettres de Peiresc*, 4:354. In this letter Peiresc says that Kircher himself admitted that Father Malapert and Clavius, and presumably Kircher himself, "did not believe the opinion of Copernicus could be improved upon, even though they had been pressed and obliged to write in behalf of the common suppositions of Aristotle."

30. Baldwin, "Magnetism and the Anti-Copernican Polemic," p. 160. Baldwin describes the work of other Jesuits after Kircher who also argued for a stationary earth.

31. This story is told in Dibner and Drake, *Letter from Galileo*, in which Drake publishes a previously unknown letter from Galileo to Peiresc about Linus's clock.

32. Baldwin, "Magnetism and the Anti-Copernican Polemic," p. 165; Monchamps, *Galilée et la Belgique*, p. 141.

33. Bibliothèque inguimbertine, Archives Communales musée municipal, Carpentras, Peiresc MS 1864, fols. 215ff. This unpublished manuscript is a rough draft of a report written hurriedly by Peiresc, and some of the words are illegible. Illegible or questionable words are marked by curly brackets {}; explanatory additions by the author are in square brackets [].

34. Kircher's *Magnes* contains many such "magnetic clocks," which differ in the kinds of time scales they contain. One clock gives the time according to different systems of unequal hours, in this case Babylonian, Italian, astronomical, and planetary hours; another is a magnetic astrolabe; another is a planetarium (*Magnes, sive de arte magnetica* [Rome: Sumptibus H. Scheus, 1641], pp. 311–323). He mounts these clocks both horizontally, as in the demonstration at Aix, and in glass globes.

35. Galileo, *Opere*, 16:65. Quoted from Dibner and Drake, *Letter from Galileo*, p. 49.

36. See Peiresc to DuPuy, May 21, 1633; Peiresc, *Lettres*, 2:525–530.

37. Kircher, *Magnes* (1641), p. 737.

38. Ibid.

39. Ibid.

40. Pierre Gassendi, *The Mirrour of True Nobility and Gentility being the Life of the Renowned Nicolaus Claudius Fabricius Lord of Peiresk, Senator of the Parliament at Aix*, trans. W. Rand (London: Printed by J. Streater for Humphrey Moseley, 1657), p. 119.

41. It is a commentary on the weakness of biography that Gassendi makes the relation of Dormalius, who had actually seen Linus's clock, an important argument in favor of it, while in fact the relation increased Peiresc's doubts.

42. Peiresc did acquire versions of these clocks for his *cabinet de physique*, if the inventory of his property at his death is to be trusted. Pierre Humbert, *Un amateur: Peiresc, 1580–1637* (Paris: Desclée de Brouwer, 1933), p. 253.

43. Godefroid Wendelin to Mersenne, June 15, 1633; Mersenne, *Correspondance*, 3:431–432. Collegium Anglorum Societatus Jesu Leodii, *Florus Anglo-Bavaricus* . . . (Liège: Ex officina Typographica Guilielmi Henrici Streel, 1686), pp. 49–50; Dibner and Drake, *Letter from Galileo*, pp. 49–50.

44. Galileo to Peiresc, May 12, 1635, quoted from Dibner and Drake, *Letter from Galileo*, pp. 54–55.

45. Kircher, *Magnes* (1641), pp. 738–739.

46. Augustin de Backer and Carlos Sommervogel, *Bibliothèque de la Compagnie de Jésus*, 12 vols. (Brussels: O. Schepens, 1890–[1932]), 6:736, and *Nouvelle Biographie Générale*, ed. Jean Chrétien Hoefer, 46 vols. (Paris: Firmin Didot Frères, 1852–1866), 40:206.

47. Kircher, *Magnes* (1641), p. 309. In this first edition of the *Magnes* Kircher describes the hidden clepsydra in detail and scarcely mentions the clock itself. In the second edition of 1643, which was revised and edited by Kircher's student Gaspar Schott, Linus's clock is described in much greater detail.

48. Kircher, *Magnes* (1641), p. 739.

49. Kircher's museum at Rome was one of the wonders of the age. In 1656 Queen Christina of Sweden marveled at Kircher's magnetic clocks "which by vertue of the

loadstone turn about with secret force" (Fletcher, "Astronomy in the Life and Correspondence of Athanasius Kircher," p. 57). His failure at Aix obviously did not deter Kircher.

50. John L. Heilbron, *Electricity in the Seventeenth and Eighteenth Centuries: A Study of Early Modern Physics* (Berkeley and Los Angeles: University of California Press, 1979), pp. 101–114, and on Kircher specifically, pp. 183–192.

51. Baldwin, "Athanasius Kircher," p. 23.

52. Keith Hutchison, "What Happened to Occult Qualities in the Scientific Revolution?" *Isis* 73 (1982): 233–253.

53. Peter L. Galison, "Descartes' Comparisons," *Isis* 75 (1984): 311–326.

54. Kircher, *Magnes* (1641), pp. 63–64. See also Baldwin, "Athanasius Kircher," pp. 152–170, and Hine, "The Mersenne-Kircher Correspondence" and "Athanasius Kircher and Magnetism."

55. Johann Stephan Kestler, *Physiologia kircheriana experimentalis* (Amsterdam: Janssonio-Waesbergiana, 1680), describes Kircher's collection of instruments.

56. Georg Philipp Harsdörffer, *Delitiae mathematicae et physicae* (Nuremberg, 1651), pp. 348–349. Quoted from Otto Mayr, *Authority, Liberty, and Automatic Machinery in Early Modern Europe* (Baltimore: The Johns Hopkins University Press, 1986), p. 44.

57. Quoted from Mayr, *Authority, Liberty, and Automatic Machinery*, p. 61.

58. Kircher, *Magnes*, 3d ed. (Rome, 1654), pp. 613–618; Baldwin, "Athanasius Kircher," pp. 453–454.

Chapter Three
The Magic Lantern and the Art of Demonstration

1. Some would want to see a further distinction between the terms "instruments" and "apparatus." Deborah Warner, who has studied this distinction, finds that for most of the early modern period the terms were synonymous and that the first clear definitions were given by James Clerk Maxwell in 1876: "Everything which is required in order to make an experiment is called Apparatus. . . . A piece of apparatus constructed specially for the performance of experiments is called an Instrument" (Deborah Warner, "Mathematical and Philosophical Instruments in Early Modern England," unpublished MS). Because "instrument" and "apparatus" have such a wide variety of meaning, and because they were synonymous during the period that we are discussing, we will not insist on Maxwell's distinction.

2. John L. Heilbron, *Electricity in the Seventeenth and Eighteenth Centuries: A Study of Early Modern Physics* (Berkeley and Los Angeles: University of California Press, 1979), p. 13.

3. See Ernan McMullin, "The Conception of Science in Galileo's Work," in *New Perspectives on Galileo*, ed. Robert E. Butts and Joseph C. Pitt, The University of Western Ontario Series in Philosophy of Science, 14 (Dordrecht: D. Reidel, 1978), pp. 212–215.

4. In law, *demonstratio* meant the pointing out of boundaries before sale, and in rhetoric it was a form of expository speech.

5. Aristotle, *Posterior Analytics*, trans. and ed. Jonathan Barnes (Oxford: Clarendon Press, 1975), chap. 6, pp. 10–13. A good discussion of demonstrative science in the Middle Ages is Eileen Serene, "Demonstrative Science," in *The Cambridge His-*

tory of Later Medieval Philosophy, ed. Norman Kretzmann, Anthony Kenny, and Jon Pinborg (Cambridge: Cambridge University Press, 1982), pp. 496–517.

6. *Posterior Analytics* 1.75b.33–37 (p. 14) and 1.78a.31–40 (p. 21).

7. Abraham Rees, *The Cyclopaedia; Or Universal Dictionary of Arts, Sciences, and Literature*, 39 vols. (London: Longman, Hurst, Rees, Orme, & Brown, 1802–1820), vol. 11, unpaginated. Most of Rees's article "Demonstration" is taken from the French *Encyclopédie, ou dictionnaire raisonné*, ed. Jean d'Alembert and Denis Diderot, 35 vols. (Paris: Briasson, David l'aîné, Le Breton, Durand, 1751–1780), 4:822–824.

8. I have looked at Rudolphus Goclenius, *Lexicon philosophicum quo tanquam clave philosophiae fores aperiuntur* (1613) (reprint, Hildesheim: Georg Olms Verlag, 1980); Johannes Micraelius, *Lexicon philosophicum terminorum philosophis usitatorum* (1662) (reprint, Stern-Verlag Janssen & Co., 1966); John Harris, *Lexicon Technicum or, an Universal English Dictionary of Arts and Sciences Explaining Not Only the Terms of Art but the Arts Themselves* (London: Dan. Brown, Tim. Goodwin . . . , 1704); Stephanus Chauvin, *Lexicon philosophicum* (1713) (reprint, Stern-Verlag Janssen & Co., 1967); *Encyclopédie, ou dictionnaire raisonné*; and Rees, *Cyclopaedia*.

9. Ernan McMullin distinguishes between these three common modes of inference during the Scientific Revolution but goes on to show that actual scientific practice usually employed a combination of them. As a result, the word "demonstration" was used very loosely. McMullin, "Conceptions of Science in the Scientific Revolution," in *Reappraisals of the Scientific Revolution*, ed. David C. Lindberg and Robert S. Westman (Cambridge: Cambridge University Press, 1990), pp. 27–92.

10. Aristotle, *Posterior Analytics*, chap. 13, line 15.

11. William Harvey, *Exercitatio anatomica de motu cordis et sanguinis in animalibus*, with English translation and annotations by Chauncey D. Leake (Baltimore: Charles C. Thomas, 1928), separate paginations, Latin on p. 5, English on p. 4.

12. Ibid., Latin p. 6, English p. 4.

13. Ibid., Latin p. 58, English p. 103.

14. C. D. O'Malley, *Andreas Vesalius of Brussels 1514–1564* (Berkeley and Los Angeles: University of California Press, 1964), p. 16. René Taton, *Enseignement et diffusion des sciences en France au XVIIIe siècle*, Histoire de la pensée 11 (Paris: Hermann, 1964), p. 313.

15. Steven Shapin and Simon Schaffer, *Leviathan and the Air-Pump: Hobbes, Boyle, and the Experimental Life* (Princeton: Princeton University Press, 1985), pp. 22–26.

16. Valeriano Magni, *Demonstratio ocularis. Loci sine locato: Corporis successive moti in vacuo: Luminis nulli corpori inhaerentis* (Warsaw, n.d., approbation 1647). We thank Peter Dear for calling Magni's book to our attention. Unfortunately we have not been able to consult it. See also Dear's "Narratives, Anecdotes, and Experiments: Turning Experience into Science in the Seventeenth Century," in *The Literary Structure of Scientific Argument: Historical Studies*, ed. Peter Dear (Philadelphia: University of Pennsylvania Press, 1991), p. 135.

17. Robert Hooke, *Micrographia: or Some Physiological Descriptions of Minute Bodies made by Magnifying Glasses with Observations and Inquiries Thereupon* (London, 1665; reprint, New York: Dover, 1961), p. 85. Steven Shapin, in his "The

House of Experiment in Seventeenth-Century England," *Isis* 79 (1988): 373–404, at 399–400, explains that Hooke carefully distinguished between "trying" an experiment, "showing" it, and "discoursing" upon it. Shapin argues that the major difference between these kinds of experiments was whether they were performed in private or public spaces. Following this distinction, an "ocular demonstration" was public because it was performed before witnesses, as was the anatomical lecture.

18. Hooke, *Micrographia*, p. 86.

19. Ibid., p. 93. See also Mary B. Hesse, "Hooke's Philosophical Algebra," *Isis* 57 (1966): 67–83, at 70.

20. Robert Hooke, *Posthumous Works*, 2d ed., intro. T. M. Brown ([London]: F. Cass, [1971]), p. 7, quoted from Barbara Shapiro, *Probability and Certainty in Seventeenth-Century England: A Study of the Relationships between Natural Science, Religion, History, Law, and Literature* (Princeton: Princeton University Press, 1983), p. 57n.153.

21. Hooke, *Micrographia*, p. 221.

22. At the very end of the *Micrographia* Hooke argues that the moon has a principle of gravitation, adding, however, "But this, I confess is but a probability, and not yet a demonstration, which (from any Observation yet made) it seems hardly capable of" (ibid., p. 246). Dugald Stewart associated Hooke's "philosophical algebra" with the Greek method of geometrical analysis and with reason, because Hooke believed that the Greeks "had another kind of Analyticks, which went backwards through almost the same Steps by which their Demonstrations went forwards, though of this we have no certain account . . ."(quoted from Hesse, "Hooke's Philosophical Algebra," p. 72).

23. Shapiro, *Probability and Certainty*, p. 56.

24. Thomas Sprat, *History of the Royal Society*, ed. Jackson I. Cope and Harold Whitmore Jones (1667; reprint, St. Louis: Washington University Press, 1958), p. 108.

25. Experiments performed at the Royal Society were "elaborate experiments" and had been invented in an "elaboratory." This placed Hooke in an anomalous social position, because gentlemen, who constituted the membership of the society, did not "labor." That was the duty of mechanics and, in particular, of the curators and their operators, who were paid to perform the "laborious" part of the fellows' investigations. See Stephen Pumfrey, "Ideas above His Station: A Social Study of Hooke's Curatorship of Experiments," *History of Science* 29 (1991): 1–44, at 18 and 36; and Steven Shapin, "Who Was Robert Hooke?" in *Robert Hooke: New Studies*, ed. Michael Hunter and Simon Schaffer (Woodbridge: The Boydell Press, 1989), pp. 253–285.

26. Shapin and Schaffer, *Leviathan and the Air-Pump*, p. 206.

27. Hobbes, *Dialogus physicus*, quoted from ibid., p. 353.

28. Hooke, *Micrographia*, p. 87. Hooke writes that where the matters of fact and the multitude of instances are great, the *inventum* can be so much the less.

29. Huygens to Oldenburg, June 14, 1673, and Brouncker to Oldenburg, ca. June 23, 1673; *The Correspondence of Henry Oldenburg*, ed. A. Rupert Hall and Marie Boas Hall, 13 vols. (Madison: University of Wisconsin Press, 1965–1986), 10:28 and 39; also quoted in Rob Iliffe, "'In the Warehouse': Privacy, Property and Priority in the Early Royal Society," *History of Science* 30 (1992): 29–68, at 30.

30. Aristotle's demonstration *quia* could be understood as a method of discov-

ery (McMullin, "The Conception of Science in Galileo's Work," p. 215), but the *inventum* was more directly associated with the faculty of the imagination and was, therefore, an essentially creative process, while *demonstratio* was associated with the reason.

31. On mirror projection before the magic lantern see Hermann Hecht, "The History of Projecting Phantoms, Ghosts and Apparitions," *New Magic Lantern Journal* 3, no. 1 (1984): 2–6, and no. 2 (1984): 2–6. Because it is part of the history of cinematography, there is an enormous bibliography on the magic lantern. The American Magic Lantern Society has compiled a lengthy bibliography, and Franz Paul Liesegang, *Dates and Sources: A Contribution to the History of the Art of Projection and to Cinematography*, trans. Hermann Hecht (London: The Magic Lantern Society of Great Britain, 1986), gives many early sources. On the invention of the magic lantern, see W. A. Wagenaar, "The True Inventor of the Magic Lantern: Kircher, Walgenstein, or Huygens?" *Janus* 66 (1979): 193–207; S. I. Van Nooten, "Contributions of Dutchmen in the Beginnings of Film Technology," *Journal of the SMPTE* 81 (1972): 116–123; H. Mark Gosser, "Kircher and the Lanterna Magica—a Reexamination," *Journal of the SMPTE* 90 (1981): 972–978. On the later history of the magic lantern see John Barnes, *Catalogue of the Barnes Museum of Cinematography*, vol. 2 (Cornwall, 1970); Martin Quigley, Jr., *Magic Shadows: The Story of the Origin of Motion Pictures* (Washington, D.C.: Georgetown University Press, 1948); and numerous articles in the *New Magic Lantern Journal* published by the Magic Lantern Society of Great Britain.

32. Georgius de Sepibus Valesius, *Romanii collegiis societatus jesu musaeum* (Amsterdam: Janssonio-Waesbergiana, 1678), pp. 38–40, and Johann Stephan Kestler. *Physiologia kircheriana experimentalis* (Amsterdam: Janssonio-Waesbergiana, 1680), pp. 124–128, both claim Kircher as the inventor.

33. As Wagenaar points out, with a point source and a lens of short focal length, it is possible to obtain an upright image with the lens between the light source and the slide, but this would not have been a workable lantern, and with the lamp that Kircher used, the image would have been out of focus. Wagenaar, "The True Inventor of the Magic Lantern," pp. 198–200.

34. Hooke had already published an account of his *camera lanterna*: Robert Hooke, "Contrivance to Make the Picture of any Thing Appear on a Wall," *Philosophical Transactions of the Royal Society of London* 3 (August 17, 1668): 741–743.

35. Athanasius Kircher, *Ars magna lucis et umbrae* (Rome: Sumptibus Hermanni Scheuss, 1646), pp. 768–770.

36. The sketches appear in Manuscript Book A (p. 152) and stand between two pages bearing the dates February 24, 1659 (on p. 85), and October 11, 1659 (on p. 155). Accompanying the drawing is the comment "Pour des représentations par le moyen de verres convexes à la lampe." Christiaan Huygens, *Oeuvres complètes*, 22 vols. (The Hague: Société Hollandaise des Sciences/Nijhoff: 1888–1950), 22:196–197.

37. Ibid., 4:102.

38. Ibid., p. 111.

39. Quoted from Rosalie L. Colie, *"Some Thankfulnesse to Constantine": A Study of English Influence upon the Early Works of Constantijn Huygens* (The Hague: Nijhoff, 1956), p. 93. On Drebbel see Gerrit Tierie, *Cornelis Drebbel (1572–*

1633) (Amsterdam: H. J. Paris, 1932), and Lawrence Ernest Harris, *The Two Netherlanders: Humphrey Bradley and Cornelis Drebbel* (Leiden: E. J. Brill, 1961).

40. Colie, *"Some Thankfulnesse to Constantine,"* p. 98. Drebbel was a "projector," one of those men like Johann Joachim Becher, Martin Elers, Johann Daniel Crafft, and Begt Skytte, who traveled from court to court seeking patronage for their "projects." The projects that seemed to be selling in the seventeenth century were elaborate stage props, perpetual motion machines, magnetic devices, and submarines. Both Drebbel and later Johann Becher pursued these particular projects. See Pamela H. Smith, *The Business of Alchemy: Science and Culture in the Holy Roman Empire* (Princeton: Princeton University Press, 1994), pp. 57–64, 113–114, 247.

41. Colie, *"Some Thankfulnesse to Constantine,"* pp. 106, 110. Svetlana Alpers, *The Art of Describing: Dutch Art in the Seventeenth Century* (Chicago: University of Chicago Press, 1983), chap. 1, "Constantijn Huygens and the New World," reveals how the new optical instruments influenced Dutch painting.

42. Christiaan Huygens, *Treatise on Light*, trans. S. P. Thompson (reprint, New York: Dover, 1952), pp. vi–vii; *Traité de la lumière*, facsimile of the 1690 ed. (London: Dawsons of Pall Mall, 1966), preface (unpaginated).

43. Christiaan Huygens to Pierre Petit, November 28, 1662; Huygens, *Oeuvres*, 4:269.

44. Walgenstein's success annoyed Kircher, who wrote: "As he was going over my discoveries in my descriptions of them, he improved upon the lantern that we described. . . . Afterwards he sold the lamp at great personal profit to various noblemen in Italy, with the result that now the matter is almost too well known at Rome." Kircher, *Ars magna lucis et umbrae* (1671), pp. 768–769.

45. Barnes, *Catalogue*, p. 10.

46. Francesco Eschinardi, *Centuriae opticae* (Rome, 1664), p. 221; Claude François Milliet Dechales, *Cursus seu mundus mathematicus* (Lyons, 1674), p. 696; Johann Christoph Sturm, *Collegium experimentale, sive curiosum*, 2 vols. in 1 (Nuremberg: Wolfgangi Mauritii Endteri & Johannis Andrae Endteri, 1676–1685), p. 164; Johann Zahn, *Oculus artificialis teledioptricus sive telescopium*, 3 vols. (Würzburg: Q. Heyl, 1685–1686), 2:257–259 and 3:253; William Molyneux, *Dioptrica nova; a Treatise of dioptricks* (London: Benj. Tooke, 1692); Jacques Ozanam, *Recréations mathématiques et physiques* (Paris, 1694).

47. *The Diary of Samuel Pepys*, ed. Henry B. Wheatley, 8 vols. (London: G. Bell and Sons, [1892]), 5:40.

48. Heilbron, *Electricity*, pp. 142 and 290. See also Robert E. Schofield, *Mechanism and Materialism: British Natural Philosophy in an Age of Reason* (Princeton: Princeton University Press, 1970), pp. 26, 80, and 140–141.

49. Schofield, *Mechanism and Materialism*, pp. 136–137.

50. Heilbron, *Electricity*, p. 14. Schofield, *Mechanism and Materialism*, p. 141.

51. *Journal de Trévoux*, May and October 1721; from René Taton, ed., *Enseignement et diffusion des sciences en France au xviiie siècle*, Histoire de la pensée, 11 (Paris: Hermann, 1964), p. 622.

52. Willem Jacob van 'sGravesande, *Mathematical Elements of Natural Philosophy Confirmed by Experiments, or an Introduction to Sir Isaac Newton's Philosophy*, 2d ed., 2 vols. (London: J. Senex and W. Taylor, 1721–1726), 2:98–99, and 'sGravesande, *Oeuvres philosophiques et mathématiques*, ed. Jean Nic. Seb. Allamand, 2 pts. in 1 vol. (Amsterdam: M. M. Rey, 1774), p. 81.

53. 'SGravesande, *Mathematical Elements of Natural Philosophy*, 1:xvii. "Ut Physices studium, quantum fieri potest, amoenum & facile reddatur, omnia experimentis esse elucidanda, ipsasque conclusiones mathematicas, hac methodo sub oculos esse ponendas, necessarium duxi." 'SGravesande, *Physices elementa mathematica, experimentis confirmata; sive introductio ad philosophiam Newtonianam* (Geneva: Apud Henricum-Albertum Gosse & Soc., 1721), p. x.

54. 'SGravesande, *Mathematical Elements*, 1:xvii–xviii. "Mathematicus enim circa illa, quae mathematicé demonstrantur, experimenta superflua credit: nos autem mathematicas demonstrationes, semper abstractas, faciliores reddi, si experimentis conclusiones sub oculos ponantur, extra omne dubium habuimus; in hoc imitati Anglos, quorum docendae Philosophiae naturalis methodus nobis occasionem dedit cogitandi de hac, quam in hoc opere fecuti fumus." 'SGravesande, *Physices elementa mathematica*, p. xi.

55. 'SGravesande, "Discours sur l'évidence," *Oeuvres*, pp. 330–333, 339–341. The concept of "moral" certitude comes from British common law and was used by Robert Boyle. See Rose-Mary Sargent, "Scientific Experiment and Legal Expertise: The Way of Experience in Seventeenth-Century England," *Studies in History and Philosophy of Science* 20 (1989): 19–45, esp. 26–28 and 36–39.

56. 'SGravesande, "Discours sur l'évidence," *Oeuvres*, pp. 332–333.

57. Isaac Newton, *Mathematical Principles of Natural Philosophy and His System of the World*, ed. Florian Cajori (Berkeley and Los Angeles: University of California Press, 1960), pp. xvii–xviii; emphasis added.

58. Benjamin Martin, *A New and Compendious System of Optics* (London: James Hodges, 1740), p. 167; emphasis added.

59. 'SGravesande, himself, scorned those who would use mathematical form without mathematical content, and d'Alembert repeated his criticism in the *Encyclopédie*. Jean d'Alembert, "Discours préliminaire," *Encyclopédie*, 1:vii, and article "Expérimentale," 6:301.

60. We have not had space to include two very important aspects of "demonstration" in our discussion. The first is the emphasis in the seventeenth century on mathematical as opposed to logical demonstration. The second is the recognition by natural philosophers that because their demonstrations rest on empirical premises, their conclusions can be known only with a high probability, not with certainty.

61. Pieter van Musschenbroek, *Essai de physique* (Leyden: S. Luchtmans, 1739), pp. 621–623; Jean Antoine Nollet, *Leçons de physique expérimentale*, 6 vols. (Amsterdam and Leipzig: Arkste'e & Merkus, 1754–1765), 5:542–544.

62. 'SGravesande, *Mathematical Elements*, p. xvii.

63. Nollet, *Leçons de physique expérimentale*, 5:547.

64. Antoine Libes, quoted from Heilbron, *Electricity*, p. 16; Nollet, *Leçons de physique expérimentale*, 5:542–543.

65. Martin, *A New and Compendious System*, pp. 292–293.

66. Henry Baker, *The Microscope Made Easy* (London: R. Dodsley, 1743), discusses the origin of the solar microscope. See also his "Account of Mr. Leeuwenhoek's Microscopes," *Philosophical Transactions of the Royal Society* 41, pt. 2 (1740): 516. The instrument maker John Cuff added the external mirror that made it possible to use the microscope when the sun was not directly opposite the hole in the camera obscura. Liesegang, *Dates and Sources*, p. 15.

67. Nollet, *Leçons de physique expérimentale*, 5:549.

68. Ibid., p. 551.

69. Leonhard Euler, "Emendatio laternae magicae ac microscopii solaris," *Nova Commentarii Academiae Scientiarum Imperialis Petropolitanae* 3 (1750/51): 363–380, in *Opera omnia*, ser. 3, vol. 2 (Turici: Orell Fussli, 1962), pp. 22–37.

70. Leonhard Euler, *Lettres à une Princess d'Allemagne*, in *Opera omnia*, ser. 3, vol. 12 (Turici: Orell Fussli, 1960), pp. 170–171. In his letter of January 9, 1762, Euler reminded the princess (von Anhalt) that he had presented her with the instrument six years earlier (p. 169).

71. Maurice Daumas, *Les instruments scientifiques aux XVIIe et XVIIIe siècles* (Paris: Presses Universitaires de France, 1953), pp. 184, 187, and 221. On the *cabinet de physique* see Daumas, "Les instruments scientifiques . . . ," chap. 4 in Taton, *Enseignement et diffusion*, pp. 619–645, and Heilbron, *Electricity*, pp. 147–152.

72. Heilbron, *Electricity*, p. 164; Tore Frängsmyr, J. L. Heilbron, and Robin E. Rider, *The Quantifying Spirit in the Eighteenth Century* (Berkeley and Los Angeles: University of California Press, 1990).

73. Lavoisier claimed that "if in any case the word Demonstration may be employed in natural philosophy and chemistry . . . the proofs which we have given of the decomposition and recomposition of water being of the demonstrative order, it is by experiments of the same order, that is to say by demonstrative experiments, which [*sic*] they ought to be attacked" (quoted and translated by Richard Kirwan in *An Essay on Phlogiston and the Composition of Acids*, ed. William Nicholson, 2d ed. [London: J. Johnson, 1789], pp. 59–61, in Jan Golinski, "Precision Instruments and the Demonstrative Order," *Osiris* 9 [1994]: 30–47, at 42). For Lavoisier a demonstrative experiment had to be quantitative.

74. *Encyclopédie*, 4:821.

75. At Edinburgh, where there was a strong experimental tradition in the medical school, the actual experiments were performed by Andrew Fyfe, the "demonstrator" for the anatomy class, whose task it was to point out the parts of the body while the professor, Alexander Monro, explained their function. Lisa Rosner, "Eighteenth-Century Medical Education and the Didactic Model of Experiment," in *The Literary Structure of Scientific Argument: Historical Studies*, ed. Peter Dear (Philadelphia: University of Pennsylvania Press, 1991), pp. 182–194, at p. 188.

76. Taton, *Enseignement et diffusion*, pp. 630, 638. It appears that Sigaud de la Fond was himself *démonstrateur* at the university. The *Oxford English Dictionary* gives four uses of "demonstrator" from the seventeenth through the nineteenth century. All are from botany, anatomy, and surgery.

77. J. B. Gough, "Sigaud de la Fond," in *Dictionary of Scientific Biography*, 12:427–428.

78. Jean-Paul Marat, *Recherches physiques sur le feu* (Paris: C. A. Jombert, 1780), p. 22. On Marat see Heilbron, *Electricity*, pp. 429–430; Joseph W. Dauben, "Marat: His Science and the French Revolution," *Archives Internationales d'Histoire des Sciences* 22 (1969): 235–261.

79. F. Chevremont, *Jean-Paul Marat: borné de son portrait; esprit politique accompagné de sa vie scientifique, politique et privée*, 2 vols. (Paris: Chez l'auteur, 1880), 2:426–427.

80. Ibid., 1:36.

81. Robert Champeix, *Savants méconnus, inventions oubliées* . . . (Paris: Dunod, 1966), p. 6. Joseph B. J. Fourier, "Éloge de J.A.C. Charles," *Académie des science, Paris, Mémoires* 8 (1829): lxxv and lxxxvi.

82. The visiting Danish scientist Thomas Bugge described Charles's portrait: "In

his hand is a solar microscope, which is a very proper emblem, as he had a remarkably fine apparatus for optical experiments." Bugge also describes how one passed from room to room in order to view the instruments in Charles's elaborate *cabinet* (Thomas Bugge, *Science in France in the Revolutionary Era*, ed. Maurice P. Crosland, Society for the History of Technology, 7 [Cambridge: MIT Press, 1969], pp. 166–167, 181–182). See also Fourier, "Éloge de Charles," p. lxxxiv.

83. The police report is given in Champeix, *Savants méconnus*, pp. 237–238.

84. Anatole France, "L'Elvire de Lamartine," from *Pages d'histoire et de littérature*, in *Oeuvres complètes*, 26 vols. (Paris: Calmann-Lévy, 1934), 24:267–333, at 282. Charles's young wife (née Julie-Françoise Bouchaud des Herettes) was the "Elvire" loved by the poet Lamartine.

85. Robert M. Isherwood, *Farce and Fantasy: Popular Entertainment in Eighteenth-Century Paris* (Oxford: Oxford University Press, 1986), pp. 200–201; Jean Torlais, "Un prestidigitateur célèbre chef de service d'électrothérapie au XVIIIe siècle, Ledru dit Comus (1731–1807)," *Histoire de la médecine* 5 (1953): 13–25, esp. 23.

86. Among his patrons, in addition to the king, were the duc de Chartres, the duc d'Orleans, Lieutenant General Lenoir of the police, the cardinal de Rohan, the duchesse de Villeroi, and the vicomte de Bussy. Isherwood, *Farce and Fantasy*, p. 200.

87. See Charles Coulston Gillispie, *The Montgolfier Brothers and the Invention of Aviation 1783–1784: With a Word on the Importance of Ballooning for the Science of Heat and the Art of Building Railroads* (Princeton: Princeton University Press, 1983), for the early history of ballooning in France.

88. Ibid., pp. 52 and 118.

89. Taton, *Enseignement et diffusion*, p. 634.

90. A model of Robertson's mechanism was in the *cabinet* of Charles. *Biographie universelle et portative . . .* (Paris: F. G. Levrault, 1834), 4:1122.

91. Étienne-Gaspard Robertson, *Mémoires recréatifs, scientifiques et anecdotiques d'un physicien-aeronaute*, reprint of vol. 1 (Langres: Café Clima Editeur, 1985); and David Robinson, "Robinson on Robertson," *New Magic Lantern Journal* 4 (1986): 4–13.

92. F. Marion, *The Wonders of Optics*, trans. Charles W. Quin (London: Sampson Low, Son, and Marston, 1868), p. 176.

93. Robinson, "Robinson on Robertson," p. 13.

94. Benjamin Martin, *The Young Gentleman and Lady's Philosophy*, 3d ed. (London, 1781), pp. 288–293. He also suggested mounting natural history specimens and even the impressions of medals on glass slides for the lantern (John Barnes, "The Projected Image: A Short History of Magic Lantern Slides," *New Magic Lantern Journal* 3, no. 3 [1985]: 4). In his *New and Compendious System* (p. 293) Martin had argued in 1740 that the lantern should be "applied to more useful purposes."

95. Jean-Philippe-Guy Le Gentil, Marquis de Paroy, *Mémoires du Comte de Paroy: Souvenirs d'un défenseur de la famille royale pendant la Révolution (1789–1797)* (Paris: Plon, 1895), pp. 275–283.

96. Laurent Mannoni, "The Magic Lantern Makers of France," *New Magic Lantern Journal* 5, no. 2 (1987): 3–7, at 5.

97. Jacques Perriault, *Mémoires de l'ombre et du son: une archéologie de l'audiovisuel* (Paris: Flammarion, 1981), chap. 3. See also Amalia Toledo, *Contribution à l'histoire de l'enseignement de projections lumineuses, l'abbé Moigno (1872–1880)* (Diplôme de l'Ecole des hautes études en sciences sociales, Paris, 1976).

98. See Derek Greenacre, "Great Heavens: Some Unusual Slides," *New Magic Lantern Journal* 5 (1988): 6.

99. Eileen Yeo and E. P. Thompson, *The Unknown Mayhew* (New York: Pantheon Books, [1971]), pp. 295–296.

100. John Barnes, "Philip Carpenter 1776–1833," *New Magic Lantern Journal* 3, no. 2 (1984): 8–9. Barnes writes, "He was thus instrumental in raising the status of the magic lantern from a mere optical novelty or toy, to that of a scientific instrument which could be used for educational instruction as well as amusement" (p. 8).

101. Lewis Wright, *Optical Projection: A Treatise on the Use of the Lantern in Exhibition and Scientific Demonstration* (London: Longmans, Green, and Co., 1901), pp. 63 and 180.

102. Liesegang, *Dates and Sources*, p. 40.

103. W. F. Ryan, "Limelight on Eastern Europe: The Great Dissolving Views at The Royal Polytechnic," *New Magic Lantern Journal* 4 (1986): 48–55.

104. Adolf Ferdinand Weinhold, *Physikalische Demonstrationen: Anleitung zum Experimentieren im Unterricht an Gymnasium, Realgymnasium, Realschulen und Gewerbschulen* (Leipzig: Quandt & Handel, 1905).

105. John Tyndall, *Lectures on Light Delivered in the United States in 1872–73* (New York: D. Appleton and Company, 1873).

106. Katherine Russell Sopka, "John Tyndall: International Popularizer of Science," in *John Tyndall, Essays on a Natural Philosopher*, ed. W. H. Brock, N. D. McMillan, and R. C. Mollan (Dublin: Royal Dublin Society, 1981), pp. 193–203, at pp. 195–196.

107. Erik Barnouw, "Magicians Created 'Living Pictures' Then Rued the Day," *Smithsonian* 12, no. 4 (1981): 114–116, 118, 120, 122, 124, 126–128.

108. An important question for which I have no immediate answer is why experimental physics so rigorously excluded medicine and physiology (traditional parts of physics) in the eighteenth century. Whatever the demonstration lecturers might have borrowed from the anatomical lecture, they kept the two subjects distinct.

109. William Eamon, "Technology and Magic in the Late Middle Ages and in the Renaissance," *Janus* 70 (1983): 171–212. Robert Hooke boasted that he had considered over thirty ways of flying. John Henry, "Robert Hooke, the Incongruous Mechanist," in *Robert Hooke: New Studies*, ed. Michael Hunter and Simon Schaffer (Woodbridge: The Boydell Press, 1989), pp. 149–180, at p. 174.

Chapter Four
The Ocular Harpsichord of Louis-Bertrand Castel; or, The Instrument That Wasn't

1. Robert Darnton, *The Great Cat Massacre and Other Episodes in French Cultural History* (New York: Basic Books, 1984), pp. 75–104.

2. According to Louis-Bertrand Castel and "Dr. Z. . ." (see below), Athanasius Kircher described the cat piano in his *Musurgia universalis*, 2 vols. (Rome: Francisci Corbelletti, 1650), facsimile ed. (Hildesheim: Olms, 1970), but we have not been able to find it there. His pupil, Gaspar Schott, described it in his *Magia universalis naturae et artis, sive recondita naturalium et artificialium rerum scientia*, 4 vols. (Würzburg, 1657–1659), vol. 2, chap. "Felium musicam exhibere," pp. 372–373. It appeared again in the popular French journal *La Nature* 2 (1883): 519–520, described by a Dr. Z. . . , from which we have taken the account of its invention. Its

greatest success was in Sweden in the eighteenth century. (See Michael Bernhard Valentini, *Museum museorum; oder, Vollständige schau-bühne aller materialien und specereyen, nebst deren natürlichen beschreibung, election, nutzen und gebrauch* . . ., 2d ed., 3 vols. in 2 [Frankfurt am Main: J. D. Zummer und J. A. Jungen, 1714], p. 73 and table 31; and Gunnar Jungmarker, "Kattklaver och voterings-instrument Verklighet och fantasi," *Artes*, no. 5 [1982]: 116–125. We thank Gunnar Broberg for calling our attention to the Swedish cat piano.) The cat piano was not unique. Schott proposed a donkey chorus, and Pierre Bayle tells us that the abbé de Beigne built a pig piano at the order of Louis XI. In every case the animal instrument was created to entertain a noble patron. Pierre Bayle, *The Dictionary Historical and Critical*, 5 vols. (London, 1736), facsimile ed. (New York: Garland, 1984), 3:803; and Isaac Nathan, *Musurgia vocalis*, 2d ed. (London: Fentum, 1836), p. 160. The cat piano occasioned a recent debate in *Experimental Musical Instruments* 5, no. 5 (1989–1990): 6; and—in 6 (1990–1991)—no. 1, p. 4, no. 2, p. 3, and no. 5, p. 2.

3. Louis-Bertrand Castel, "Clavecin pour les yeux, avec l'art de peindre les sons, et toutes sortes de pièces de musique, Lettre écrite de Paris le 20 Février 1725 par le R. P. Castel, Jesuite, à M. Decourt, à Amiens," *Mercure de France*, November 1725, pp. 2552–2577. The best study of Castel is Donald S. Schier, *Louis Bertrand Castel, Anti-Newtonian Scientist* (Cedar Rapids, Iowa: Torch Press, 1941). On Castel's ocular harpsichord see Anne-Marie Chouillet-Roche, "Le clavecin oculaire du Père Castel," *Dix-huitième siècle* 8 (1976): 141–166; Albert Wellek, "Farbenharmonie und Farbenklavier: ihre Entstehungsgeschichte im 18. Jahrhundert," *Archiv für die Gesamte Psychologie* 94 (August–December 1935): 347–375; and, the most recent and most detailed, Maarten Franssen, "The Ocular Harpsichord of Louis-Bertrand Castel: The Science and Aesthetics of an Eighteenth-Century *Cause célèbre*," *Tractrix* 3 (1991): 15–77. Franssen discusses the importance of the ocular harpsichord for theories of aesthetics in the eighteenth century. I am grateful to him for sending me a preprint of his article.

4. Castel credited Kircher, a fellow Jesuit and the author of the cat piano, with the idea of the ocular harpsichord. It is true that Kircher had not made a harpsichord nor had he found the exact correspondence between musical pitches and colors, but he had provided the "seed," the key analogy, from which the theory of the ocular harpsichord could be constructed: Castel, "Clavecin pour les yeux," pp. 2553–2560. Kircher describes the analogy in his *Ars magna lucis et umbrae* (Rome: Sumptibus Hermanni Scheuss, 1646), pp. 131–132. See also Kircher, *Musurgia*, 2:567–568; and Kircher, *Phonurgia nova sive conjugium mechanico-physicum naturae pananymphia phonosophia concinnatum* (Kempten: R. Dreherr, 1673), preamble 1, fol. 6.

5. Louis-Bertrand Castel, "Suite et sixième partie des nouvelles expériences d'optique et d'acoustique: Addressées à M. le Président de Montesquieu, par le Père Castel Jesuite," *Journal de Trévoux*, December 1735, pp. 2642–2768, at p. 2654. (See n. 10 below for the complete series of articles of which this is one. The formal name of this journal is *Mémoires pour l'histoire des sciences et des beaux arts*, but in the eighteenth century it was almost always referred to as the *Journal de Trévoux*.)

6. Voltaire, "Lettre à Mr. Rameau, Mars 1738," *Correspondence*, ed. Theodore Besterman, 135 vols. (Geneva: Institut et Musée Voltaire, 1953–1977), vol. 7, app. 29, pp. 477–480, at p. 480.

7. Penelope Gouk, "The Harmonic Roots of Newtonian Science," in *Let Newton Be! A New Perspective on His Life and Works*, ed. John Fauvel et al. (Oxford: Oxford University Press, 1988), pp. 101–125. See also *The Optical Papers of Isaac*

Newton, ed. Alan E. Shapiro, vol. 1, *The Optical Lectures 1670–1672* (Cambridge: Cambridge University Press, 1984–), pp. 542–545.

8. *The Correspondence of Isaac Newton*, ed. H. W. Turnbull et al., 7 vols. (Cambridge: Cambridge University Press, 1959–1977), 1:376; quoted from Gouk, "Harmonic Roots," p. 118. Newton's statement of the analogy between the spectrum and the octave caused some confusion because he employed an old system of solmization rather than the more recent system of equal temperament. Note on the illustration that Newton repeats "Sol" at the fifth and at the octave. See Wellek, "Farbenharmonie und Farbenklavier," pp. 351–353.

9. Nicolas Malebranche, *Nicolas Malebranche: The Search after Truth*, trans. Thomas M. Lennon and Paul J. Olscamp; and *Elucidations*, trans. Lennon (Columbus: Ohio State University Press, 1980), pp. 686–718.

10. Louis-Bertrand Castel, "Nouvelle expériences d'optique et d'acoustique," *Journal de Trévoux*, August 1735, pp. 1444–1482; the various "suites" appeared in ibid., pp. 1619–1666; September, pp. 1807–1839; October, pp. 2018–2053; November, pp. 2335–2372; and December, pp. 2642–2768.

11. Voltaire, *Oeuvres complètes*, ed. Louis Moland, 52 vols. (Paris: Garnier frères, 1877–1885), 22:503–507, at pp. 503, 505.

12. Voltaire to Maupertuis, June 15, 1738, Voltaire, *Correspondence*, vol. 7, letter 1454; and "Lettre à Mr. Rameau." See also Chouillet-Roche, "Le clavecin oculaire," p. 162.

13. Voltaire to Thieriot, August 7, 1738, *Correspondence*, vol. 7, letter 1509.

14. Jean-Jacques Dortous de Mairan, "Sur la propagation du son dans les différents tons qui le modifient, IV: En quoi l'analogie du son et de la lumière, des tons et des couleurs, de la musique et de la peinture, est imparfaite, ou nulle," *Académie Royale des Sciences, Paris, Memoires*, 1737, pp. 34–45; Georg Philipp Telemann, *Beschreibung der Augen-orgel oder des Augen-clavicimbels* (Hamburg, 1739), reprinted in Lorenz Christoph Mizler von Kolor, *Musikalische Bibliothek, oder gründliche Nachricht von alten und neuen musikalischen Schrifften und Büchern* . . . (Leipzig, 1742), 2:262–266; and Georg Krafft, *Sermones in solemni academiae scientiarum imperialis* . . . (St. Petersburg, 1742). Many of Castel's critics had reasons other than the ocular harpsichord for opposing him. Dortous de Mairan was protecting the Academy of Sciences, some members of which Castel had criticized.

15. Rousseau called the color-sound analogy false in his *Essai sur l'origine des langues* (written 1749, published 1781); and Castel responded by attacking Rousseau's music theory in his *Lettres d'un académicien de Bordeaux sur le fonds de la musique à l'occasion de la lettre de M. R*** contre la musique françoise* (1754), and *L'homme moral opposé à l'homme physique de M. R.**** (1756). In his *Confessions* Rousseau called Castel "fou mais bonhomme au demeurant." See Chouillet-Roche, "Le clavecin oculaire," p. 165.

16. Denis Diderot, *Lettre sur les sourds et muets*, in *Oeuvres complètes*, édition critique et annotée, publiée sous la direction de Herbert Dieckmann, Jean Fabre et Jacques Proust; avec les soins de Jean Varloot, 25 vols. (Paris: Hermann, 1975–), 4:145; and Diderot, "Clavecin oculaire," in *Encyclopédie, ou dictionnaire raisonné des sciences, des arts et des métiers*, 35 vols. (Paris: Briasson, David l'aîné, Le Breton, Durand, 1751–1780), 3:511a–512a. "Le facture de cet instrument est si extraordinaire, qu'il n'y a que le public peu éclaire qui puisse se plaindre qu'il se fasse toujours et qu'il ne s'achève point." Diderot also included Castel and the ocular harpsichord

in his *Bijoux indiscrets* (1747). Castel was "un certaine brame noir, fort original, moitié sensé, moitié fou" whose writings were a "tissu de rêveries." See Chouillet-Roche, "Le clavecin oculaire," p. 164.

17. *Explanation of the Ocular Harpsichord upon Shew to the Public* (London: Printed for S. Hooper and A. Morley, 1757), pp. 2–3; and Chouillet-Roche, "Le clavecin oculaire," pp. 156, 166.

18. Castel, "Suite et sixième partie des nouvelles expériences," *Journal de Trévoux*, December 1735, p. 2722.

19. Chouillet-Roche, "Le clavecin oculaire," p. 158.

20. Schier, *Louis Bertrand Castel*, pp. 179, 183.

21. Louis-Bertrand Castel, "Difficultez sur le clavecin oculaire, avec leurs réponses," *Mercure de France*, March 1726, pp. 455–463, at p. 455.

22. Louis-Bertrand Castel, "Suite et seconde partie des nouvelles expériences d'optique et d'acoustique adressées à M. le Président de Montesquieu," *Journal de Trévoux*, August 1735, pp. 1619–1666, at pp. 1620, 1624.

23. Castel, "Suite et cinquième partie des nouvelles expériences . . . ," *Journal de Trévoux*, November 1735, pp. 2335–2372, at pp. 2350–2351.

24. Castel, "Suite et seconde partie des nouvelles expériences d'optique et d'acoustique adressées à M. le Président de Montesquieu," *Journal de Trévoux*, August 1735, p. 1622.

25. Ibid., p. 1625.

26. Louis-Bertrand Castel, *Le vrai système de physique générale de M. Isaac Newton* . . . (Paris: C.-F. Simon, 1743), p. 6; and Castel, *L'optique des couleurs, fondée sur les simples observations, et tournée sur-tout à la pratique de la peinture, de la teinture et des autres arts coloristes* (Paris: Briasson, 1740), p. 375.

27. Castel illustrates the methods described by Peter Dear, "Jesuit Mathematical Science and the Reconstitution of Experience in the Early Seventeenth Century," *Studies in History and Philosophy of Science* 18 (1987): 133–175.

28. Castel, *L'optique des couleurs*, pp. 376, 393.

29. "Projet d'impression," MS 15747, Bibliotheca Hulthemiana, Royal Library, Brussels, p. 10, quoted from Schier, *Louis Bertrand Castel*, p. 107; and Castel, *L'optique des couleurs*, p. 403.

30. Castel, *Le vrai système de Newton*, pp. 450, 6, 456.

31. Ibid., pp. 10–13. D'Alembert argued just the opposite—that the lack of flexibility in Newton's system was its greatest strength: Jean d'Alembert, "Élémens de philosophie," in *Mélanges de littérature, d'histoire et de philosophie*, 5 vols. (Amsterdam: Zacharie Chatelain & Fils, 1770), 4:231.

32. Castel, *Le vrai système de Newton*, pp. 442, 476; and Castel, *L'optique des couleurs*, pp. 410–414. Castel here registers his particular dislike of the British form of scientific rhetoric that Shapin and Schaffer call "virtual witnessing." See Steven Shapin and Simon Schaffer, *Leviathan and the Air-Pump: Hobbes, Boyle, and the Experimental Life* (Princeton: Princeton University Press, 1985), chap. 5.

33. Castel, *L'optique des couleurs*, p. 488.

34. Dear, "Jesuit Mathematical Science," pp. 145–146.

35. Ibid., p. 142.

36. Louis-Bertrand Castel, "Démonstration géométrique du clavecin pour les yeux et pour tous les sens, avec l'éclaircissement de quelques difficultez, et deux nouvelles observations," *Mercure de France*, February 1726, 277–292, at pp. 277, 287, 291.

37. Louis-Bertrand Castel, "Réflexions sur la nature et la source du sublime dans le discours: sur le vrai philosophie de discours poétique, et sur l'analogie qui est la clef des découvertes," *Mercure de France*, June 1733, pp. 1309–1322, at pp. 1320 (quotation), 1321.

38. Ibid., pp. 1311–1312. Castel raises a significant problem that had challenged natural philosophers since Aristotle. See Henry Guerlac, "Can There Be Colors in the Dark? Physical Color Theory before Newton," *Journal of the History of Ideas* 47 (1986): 3–20.

39. "Journal . . . de la pratique et exécution du clavecin des couleurs," MS 15746, Royal Library, Brussels, p. 50, quoted from Schier, *Louis Bertrand Castel*, p. 100.

40. Castel, "Nouvelles expériences d'optique et d'acoustique," pp. 1458–1459. Castel attributes this method to Kircher.

41. Castel, "Réflexions sur la nature et la source du sublime," pp. 1318–1319.

42. Louis-Bertrand Castel, "Lettre à M. C***, sur l'existence d'un milieu entre le naturel et le surnaturel, qu'il appelle artificiel," *Journal de Trévoux*, December 1722, pp. 2072–2097.

43. Castel, *Le vrai système de Newton*, pp. 447–448.

44. Johann Wolfgang von Goethe, *Scientific Studies*, ed. and trans. Douglas Miller (New York: Suhrkamp, 1988), pp. xvi, 14. See Neil M. Ribe, "Goethe's Critique of Newton: A Reconsideration," *Studies in History and Philosophy of Science* 16 (1985): 315–335, at 324.

45. Goethe, *Scientific Studies*, pp. xvi, 167, 200, 307.

46. Ibid., p. 276.

47. Johann Wolfgang von Goethe, *Die Schriften zur Naturwissenschaft*, hrsg. im Auftrage der Deutschen Akademie der Naturforscher zu Halle, 11 vols. in 18 (Weimar, H. Böhlaus, 1947–), pt. 1, 4:329. On the ocular harpsichord see pt. 2, 6:199–204.

48. Goethe was prepared to describe Castel as an "ingenious man" who, though not one of the first figures of his time, was at least "one of the most distinguished minds of his nation" in spite of the fact that his writing style was "long-winded, nit-picking, and prolix." Goethe, *Schriften zur Naturwissenschaft*, pt. 1, 6:328 and 333.

49. Castel, *Le vrai système de Newton*, p. 499; and Goethe, quoted from Walter D. Wetzels, "Art and Science: Organicism and Goethe's Classical Aesthetics," in *Approaches to Organic Form: Permutations in Science and Culture*, ed. Frederick Burwick, Boston Studies in the Philosophy of Science, 105 (Dordrecht: D. Reidel, 1987), p. 76.

50. Goethe, *Schriften zur Naturwissenschaft*, pt. 1, 3:66. See also Dennis L. Sepper, *Goethe contra Newton: Polemics and the Project for a New Science of Color* (Cambridge: Cambridge University Press, 1988), pp. 88–90; and Michael J. Duck, "Newton and Goethe on Colour: Physical and Physiological Considerations," *Annals of Science* 45 (1988): 512–515.

51. On the phenomenon of synesthesia see Lawrence E. Marks, "On Colored-Hearing Synesthesia: Cross-modal Translations of Sensory Dimensions," *Psychological Bulletin* 82 (1975): 303–331; the article includes an extensive bibliography. Marks (p. 304) claims that the first "scientific" reference to synesthesia was by John Thomas Woolhouse, an English ophthalmologist who lived in Paris, was a friend of Castel, and supported his work on the ocular harpsichord. See also Schier, *Louis*

Bertrand Castel, pp. 10, 20–22, 155. Also valuable are Albert Wellek, "Das Doppelempfinden im abendländischen Altertum und Mittelalter," *Archchiv für die Gesamte Psychologie* 80 (1931): 120–166; Wellek, "Renaissance- und Barock-Synästhesie," *Deutsche Vierteljahrsschrift für Literaturwissenschaft und Geistesgeschichte* 9 (1931): 534–584; and Wellek, "Zur Geschichte und Kritik der Synästhesie-Forschung," *Archiv für die Gesamte Psychologie* 79 (1931): 325–384.

52. For a description of a Carnegie Hall performance of *Prometheus* on March 20, 1915, see H. C. Plummer, "Color Music—A New Art Created with the Aid of Science," *Scientific American* 112 (1915): 343, 350–351.

53. Albert Michelson, *Light Waves and Their Uses* (Chicago, 1903), quoted from Adrian Bernard Klein, *Colour-Music: The Art of Light*, 2d ed. (London: Crosby Lockwood and Son, 1930), p. 223.

54. Ernst H. Gombrich, "Epilogue: Some Musical Analogies," in *The Sense of Order: A Study in the Psychology of Decorative Art* (Oxford: Phaidon Press, 1979), pp. 285–305; and Judith Zilczer, "'Color Music': Synaesthesia and Nineteenth-Century Sources for Abstract Art," *Artibus et historiae: An Art Anthology* 16 (1987): 1101–1126. Most important for disseminating the concept of the musical analogy in abstract painting was Arthur Wesley Dow, whose ideas were picked up by Alfred Steiglitz, Eduard Steichen, Georgia O'Keeffe, Max Weber, and Arthur Dove.

Chapter Five
The Aeolian Harp and the Romantic Quest of Nature

1. Charles Coulston Gillispie, *The Edge of Objectivity* (Princeton: Princeton University Press, 1960), p. 201.

2. Alfred North Whitehead, *Science and the Modern World* (New York: Mentor Books, 1958), p. 84; and Trevor Harvey Levere, *Poetry Realized in Nature: Samuel Taylor Coleridge and Early Nineteenth-Century Science* (Cambridge: Cambridge University Press, 1981), p. 27.

3. A recent collection of essays exploring this problem is Andrew Cunningham and Nicholas Jardine, eds., *Romanticism and the Sciences* (Cambridge: Cambridge University Press, 1990).

4. Keith Hutchison pursues this subject in "Idiosyncrasy, Achromatic Lenses, and Early Romanticism," *Centaurus* 34 (1991): 125–171, at 157.

5. Novalis (Friedrich von Hardenberg), *Werke in einem Band* (selections), ed. Uwe Lassen, 3d ed. (Hamburg, 1966), pp. 376–377. Quoted in Hutchison, "Idiosyncracy, Achromatic Lenses, and Early Romanticism," p. 157.

6. Marjorie Hope Nicolson, *Newton Demands the Muse: Newton's Opticks and the Eighteenth-Century Poets* (Westport, Conn.: Greenwood Press, 1946), and Meyer Howard Abrams, *The Mirror and the Lamp: Romantic Theory and the Critical Tradition* (Oxford: Oxford University Press, 1953).

7. Abrams, *The Mirror and the Lamp*, p. viii.

8. Percy Bysshe Shelley, "Defense of Poetry," in *Shelley's Literary and Philosophical Criticism*, ed. John Shawcross (Oxford: Oxford University Press, 1909), p. 121. According to Shelley, the poet exceeds the harp because he or she adds harmony to melody "by an internal adjustment of the sounds or motions thus excited to the impressions which excite them."

9. William Jones, *Physiological Disquisitions; or, Discourses on the Natural Philosophy of the Elements . . .* (London: J. Rivington and Sons, 1781), p. 341.

10. Coleridge, *The Eolian Harp*, in *The Best of Coleridge*, ed. Earl Leslie Griggs (New York: The Ronald Press, 1938), p. 15:

> . . . And that simplest Lute,
> Placed length-ways in the clasping casement, hark!
> How by the desultory breeze caress'd,
> Like some coy maid half yielding to her lover,
> It pours such sweet upbraiding, as must needs
> Tempt to repeat the wrong! And now, its strings
> Boldlier swept, the long sequacious notes
> Over delicious surges sink and rise,
> Such a soft floating witchery of sound
> As twilight Elfins make, when they at eve
> Voyage on gentle gales from Fairy-Land,
> Where Melodies round honey-dropping flowers,
> Footless and wild, like birds of Paradise,
> Nor pause, nor perch, hovering on untam'd wing!

William Wordsworth, *Prelude* (1850 text, bk. 1, lines 101–105), in *The Prelude or Growth of a Poet's Mind*, ed. Ernest De Selincourt, 2d ed. (Oxford: Clarendon Press, 1959), p. 8:

> It was a splendid evening; and my soul
> Did once again make trial of the strength
> Restored to her afresh; nor did she want
> Eolian visitation—but the harp
> Was soon defrauded

Melville, *The Æolian Harp: At the Surf Inn* (1888), in *The Works of Herman Melville. Standard Edition*, 16 vols. (London: Constable and Co., 1922–1924), 16: 232:

> List the harp in window wailing
> Stirred by fitful gales from sea:
> Shrieking up in mad crescendo—
> Dying down in plaintive key!
> Listen: less as strain ideal
> Than Ariel's rendering of the Real.
> What that real is, let hint
> A picture stamped in memory's mint.

11. Meyer Howard Abrams, "The Correspondent Breeze: A Romantic Metaphor," in his *The Correspondent Breeze: Essays on English Romanticism* (New York: W. W. Norton and Company, 1984), pp. 25–43, at p. 26. On the history of the Aeolian harp see Stephen Bonner and M. G. Davies, eds., *Aeolian Harp*, 4 vols. (Duxford, Cambridge: Bois de Boulogne, 1970–1974); Geoffrey Grigson, *The Harp of Aeolus and Other Essays on Art, Literature and Nature* (London: George Routledge & Sons, 1947), chap. 3; and Georges Kastner, *La harpe d'Éole et la musique cosmique: études sur les rapports des phénomènes sonores de la nature avec la science et l'art* (Paris: G. Brandus, Dufour et Cie, 1856).

12. Giambattista Della Porta, *Natural Magick* (1st ed. Naples, 1558), reprint of the 1st English ed. (London: Thomas Young and Samuel Speek, 1658), ed. Derek J.

Price (New York: Basic Books, 1957), chap. 7, "The Chaos," p. 405. The subheading is "To make a Harp or other instrument be played on by the Winde."

13. Gaspar Schott, *Magica hydraulico-pneumatica* (Frankfurt: J. G. Schonwetteri, 1657), pp. 348f.

14. Athanasius Kircher, *Musurgia universalis*, 2 vols. (Rome: Francisci Corbelletti, 1650), facsimile ed. (Hildesheim: Olms, 1970), 2:352–355, and *Phonurgia nova sive conjugium mechanico-physicum naturae pananymphia phonosophia concinnatum* (Kempten: R. Dreherr, 1673), lib. 2, sec. 7, pp. 144–147.

15. Kircher, *Musurgia*, 2:353.

16. Schott, *Magica hydraulico-pneumatica*, pp. 348f.

17. Kircher, *Musurgia*, 2:353, 355.

18. It is possible that Kircher got the idea for the Aeolian harp from Jesuit missionaries in Japan or Java. The Javanese have long made singing kites that employ a fiber of bamboo for a vibrating string, and Kircher describes in corollary 2 how to make "amazing music coming from way up in the air" from a flying fish or serpent (or a flying angel to make it even more miraculous; p. 354). His picture of the flying fish looks strikingly like a Javanese fish-kite. Kircher carried on an extensive correspondence with Jesuit missionaries all over the globe. On the Javanese Aeolian kites see Anthony Reid, *Southeast Asia in the Age of Commerce 1450–1680*, vol. 1, *The Lands below the Winds* (New Haven: Yale University Press, 1988), pp. 195–196.

19. Johann Jacob Hofmann (Joh. Jacobi Hofmanni), *Lexicon universale historiam sacram et profanum*, 4 vols. (Leyden: Jacob. Hackium, 1698), 1:88–89.

20. James Thomson, *The Castle of Indolence*, lines 343–369, in *Liberty, The Castle of Indolence and Other Poems*, ed. James Sambrook (Oxford: Clarendon Press, 1986), pp. 186–187. Marilyn Butler points out that the English "romantics" did not call themselves by that name. She argues that the roots of romanticism were in the work of the "country" or "patriot" poets such as James Thomson and Thomas Gray. These were also the first to use the Aeolian harp (Marilyn Butler, "Romanticism in England," in *Romanticism in National Context*, ed. Roy Porter and Mikulas Teich [Cambridge: Cambridge University Press, 1988], pp. 37–67, esp. 37–43). Grey opens his poem "The Progress of Poesy" (1754) with the Aeolian harp.

21. Thomson, *Liberty, The Castle of Indolence*, p. 187.

22. Ibid., p. 314. Thomson also wrote "An Ode, on the Winter Solstice" at about this time, which also includes the Aeolian harp with a footnote, but it was unpublished until 1955 (p. 436).

23. From Roger H. Lonsdale, *Dr. Charles Burney: A Literary Biography* (Oxford: Clarendon Press, 1965), p. 29.

24. Quoted from Lewis Mansfield Knapp, *Tobias Smollett: Doctor of Men and Manners* (Princeton: Princeton University Press, 1949), p. 60.

25. On Oswald see Mary Anne Alburger, *Scottish Fiddlers and Their Music* (London: Victor Gollancz, 1983), pp. 42–48.

26. Lonsdale, *Dr. Charles Burney*, p. 31.

27. Ibid., pp. 34–35.

28. Jones, *Physiological Disquisitions*, pp. 338–340.

29. Charles Burney, "Voice," in Abraham Rees, *The Cyclopaedia: or Universal Dictionary of Arts, Sciences, and Literature*, 39 vols. (London: Longman, Hurst, Rees, Orme, & Brown, 1802–1820), 37:n.p.

30. *Gentleman's Magazine*, February 1754, p. 74. A reply to this letter appeared in the April issue of the same year and included extracts from Kircher and J. J.

Hofmann (pp. 174–175). Tobias Smollett also felt the need to explain the Aeolian harp to his readers (*The Adventures of Ferdinand Count Fathom*, ed. Jerry C. Beasley and O. M. Brack, Jr. [Athens: University of Georgia Press, 1988], p. 159).

31. Frederick Hintz, "Guitar-maker to Her Majesty and the Royal Family: makers of Guitars, Mandolins, Viols de l'Amour, Viola de Gamba, Dulcimers, Solitaires, Lutes, Harps, Cymbals, the Trumpet marine, and the Aeolian Harp." Bonner and Davies, *Aeolian Harp*, 2:30.

32. Bonner, who has amassed many details about Aeolian harp history, writes, "From page 338 of Jones's work dates the Aeolian harp's more intense period of popularity." Ibid., p. 18.

33. Jones, *Physiological Disquisitions*, p. 345n.

34. H. Lichtenberg (*Göttingen Taschen-kalendar* [Göttingen: Johann Christian Dietrich, 1792]), Christian Friedrich Quandt ("Versuche und Vermutungen über die Aeolsharfe," *Lausizische Monatschrift*, 1795, pp. 277f.), and Robert Bloomfield, ("Nature's Music" [1801], in *Works*, 8 vols. in 3 [London: Longman, Hurst, Rees, Orme and Brown, 1819–1826], vol. 3) all depended heavily on Jones.

35. Jones, *Physiological Disquisitions*, p. 340.

36. Jones writes, "I know not how to account for the compass of its notes on the principles of the harmonics, but by admitting a new species of sounds, which I call *harmonics of the harmonics*, or *secondary harmonics*. The sharp seventh is very commonly heard, which if deduced as an harmonic, must be of the second species, as the 17th of the 12th; as also the 9th, which is as frequently heard, may be taken for the 12th of the 12th: and thus perhaps we may account for all its varieties." Ibid., p. 341.

37. Kircher, *Musurgia*, pp. 354–355.

38. Daniello Bartoli, a fellow Jesuit of Kircher's at Rome, attacked Kircher's theory as absurd in his *Del Suono de'tremori armonici e dell'udito* (Rome, 1679), p. 108. See Bonner and Davies, *Aeolian Harp*, 2:17.

39. Jones, *Physiological Disquisitions*, p. 341.

40. Ibid., pp. 344–345.

41. Thomas L. Hankins, *Jean d'Alembert: Science and the Enlightenment* (Oxford: Clarendon Press, 1970), pp. 47–48, 218–220; J. Ravetz, "The Representation of Physical Quantities in Eighteenth-Century Mathematical Physics," *Isis* 52 (1961): 7–20. Mathematical treatments of this history are Clifford A. Truesdell, "The Rational Mechanics of Flexible or Elastic Bodies, 1638–1788," in Leonhard Euler, *Opera Omnia*, ser. 2, vol. 11, pt. 2 (Zurich: Orell Fussli, 1960), and "The Theory of Aerial Sound, 1687–1788," ser. 2, vol. 13, pt. 2, pp. XIX–LXXII.

42. Jones, *Physiological Disquisitions*, p. xx.

43. Ibid., pp. 334–337. Tartini tones, or combination tones as they were later called, were drawn to the attention of musicians through d'Alembert's discussion of them in the *Encyclopédie* and in his *Élémens de musique*. V. Carlton Maley, Jr., *The Theory of Beats and Combination Tones, 1700–1863* (New York: Garland, 1990), pp. 48–53.

44. Helmholtz showed that combination tones are not beats but are caused by sounds loud enough to drive the vibrating string or membrane into a nonlinear range where the restorative force is no longer proportional to the displacement. He called them "sum and difference tones" and showed how they produced additional terms in the Fourier series expansion describing the sound wave. See Maley, *The Theory of Beats*, chap. 9.

45. The argument that the nonharmonic tones from the Aeolian harp may be combination tones appears in "Aeolusharfe," in Johann Traugott Gehler, *Physikalisches Worterbuch* (Leipzig: E. B. Schwickert, 1825), 1:210.

46. Matthew Young, *An Enquiry into the Principal Phaenomena of Sounds and Musical Strings* (Dublin: G. Robinson, 1784), pp. 170–182. He compared the phenomenon to wind blowing across a field of ripening grain. When the wind is light the stalks of grain sway back and forth, but when it is strong the stalks bend over and only the heads vibrate rapidly in the gust (pp. 173–174).

47. Charles-Émile Pellisov, "Andeutungen zur Begrundung einer Theorie der Aeolsharfe," *Annalen der Physik und Chemie* 19 (1830): 237ff. Pellisov's work is described in detail in Kastner, *La harpe d'Éole*, pp. 151–162.

48. Quoted from Kastner, *La harpe d'Éole*, p. 158.

49. V. Strouhal, "Über eine besondere Art der Tonerregung," *Annalen der Physik und Chemie* 5 (1878): 216, translated and quoted at length in Bonner and Davies, *Aeolian Harp*, 4:27–46.

50. "Using a thin elastic wire at a slowly increasing speed, at certain speeds an especially loud and clear sound may be noted, which gains considerable intensity if this speed is maintained for some time. If this speed is then further increased, the sound becomes quieter—but soon the next overtone (i.e. harmonic) appears. . . . Such effects can, of course, be produced if the air blows on stationary wires. This is the true reason for the fractioning of Aeolian Harp tones—i.e. mainly from the resonance effects." Bonner and Davies, *Aeolian Harp*, 4:35–36.

51. Ibid., pp. 47–81.

52. Louis-Hector Berlioz, *Voyage musical en Allemagne et en Italie* (1844), quoted from ibid., 3:73.

53. G. N. Cantor, "Revelation and the Cyclical Cosmos of John Hutchinson," in *Images of the Earth: Essays in the History of the Environmental Sciences*, ed. L. J. Jordanova and Roy S. Porter, BSHS Monographs (Chalfont St. Giles: The British Society for the History of Science, 1979), pp. 3–22.

54. Ibid., pp. 7–10.

55. Albert J. Kuhn, "Glory or Gravity: Hutchinson vs. Newton," *Journal of the History of Ideas* 22 (1961): 319.

56. Hutchinson challenged Samuel Clarke to a debate over natural philosophy, but Clarke apparently found Leibniz to be a more worthy opponent.

57. Jones, *Physiological Disquisitions*, pp. iii–iv.

58. Kuhn, "Glory or Gravity," p. 320.

59. Quoted from ibid., p. 318. On the divine analogy see Earl R. Wasserman, "Nature Moralized: The Divine Analogy in the Eighteenth Century," *ELH* 20 (1953): 39–76.

60. Jones had studied at Oxford with Nathan Alcock, who, in turn, had been a student of Boerhaave; therein may be a source of his interest in fire, but it most certainly comes from Hutchinson as well. Jones's ideas are more in accord with Boerhaave's opinions regarding fire and air, but he agrees that it is often difficult to distinguish between the two. Air supports fire as a pabulum. Stephen Hales, who first recognized that air and fire may be "fixed" in solids, talked of "aerial particles of fire" as if air and fire were in some sense the same thing. Thus Jones concludes that air and fire may have the same elementary nature. See Robert E. Schofield, *Mechanism and Materialism: British Natural Philosophy in an Age of Reason* (Princeton: Princeton University Press, 1970), pp. 127–128.

61. Jones, *Physiological Disquisitions*, p. 298.

62. Ibid., p. 328. In 1784 Jones published *A Treatise on the Art of Music; in which the Elements of Harmony and Air are Practically Considered* (Colchester, 1784). He claimed that music has its foundation in harmony but its superstructure in "air." We have not seen this work and do not know the extent to which Jones identifies musical "air" with elemental air. A description of Jones's musical theory is Jamie Croy Kassler, "The Systematic Writings on Music of William Jones (1726–1800)," *Journal of the American Musicological Society* 26 (1973): 92–107.

63. William Law, *A Serious Call to a Devout and Holy Life*, introd. Norman Sykes (London, 1967), p. 204. Quoted from Harriet Guest, *A Form of Sound Words: The Religious Poetry of Christopher Smart* (Oxford: Clarendon Press, 1989), p. 144.

64. Christopher Smart, *The Poetical Works*, vol. 1., *Jubilate Agno*, ed. Karina Williamson (Oxford: Clarendon Press, 1980), verses 246–250, p. 53, and app., "Smart and the Hutchinsonians," p. 131. Also Christopher Devlin, *Poor Kit Smart* (London: Rupert Hart-Davis, 1961), pp. 73, 169. Samuel Johnson argued that Smart should never have been confined, and said that he would "as lief pray with Kit Smart as any one else." Christopher Smart, *Poems*, ed. Robert Brittain (Princeton: Princeton University Press, 1950), p. 39.

65. Smart, *Poems*, p. 225.

66. Acts 2:2–4. The story of Pentecost: "And suddenly there came a sound from heaven as of a rushing mighty wind, and it filled all the house where they were sitting. And there appeared unto them cloven tongues like as of fire, and it sat upon each of them. And they were all filled with the Holy Ghost, and began to speak with other tongues, as the Spirit gave them utterance."

67. Anthony John Harding, *Coleridge and the Inspired Word* (Kingston and Montreal: McGill-Queens University Press, 1985), pp. 5–8.

68. Smart, *Poems*, p. 117.

69. Lichtenberg, *Göttingen Taschen-kalendar*, pp. 137f.

70. *Opusculi Scelti di Milano* (1785), pp. 298–309, in Bonner and Davies, *Aeolian Harp*, 2:75–76 and 4:82–95. The latter is a translation of Gattoni's letter.

71. Bonner and Davies, *Aeolian Harp*, 2:67–68, and Kastner, *La harpe d'Éole*, p. 87.

72. E.T.A. Hoffmann, *Sämtliche poetische Werke*, ed. Hannsludwig Geiger, 4 vols. (Weisbaden: Emil Vollmer Verlag, 1972), 2:342–343 and 3:27; *The Best Tales of Hoffmann*, ed. E. F. Bleiler (New York: Dover, 1967), p. 95; Martin Bidney, "The Aeolian Harp Reconsidered: Music of Unfulfilled Longing in Tjutchev, Morike, Thoreau, and Others," *Comparative Literature Studies* 22 (1985): 335.

73. Schiller, *Würde der Frauen* (1796), quoted from Friedrich Schiller, *Poems of Schiller* (Boston: S. E. Cassino and Co., 1884), pp. 237–238.

74. Goethe, *Zeuginung* to *Faust*, pt. 1 (1797), lines 25–32, quoted from *Faust*, trans. Walter Kaufman (New York: Doubleday/Anchor, 1961), p. 67. Goethe wrote a poem titled *Aeolsharfen*, but not until 1822. The most important early poetic effort in German was Baron J. F. von Dalberg's *Die Aeolsharfe, ein allegorischer Traum* (1801).

75. This difference was noticed by Bidney, "The Aeolian Harp Reconsidered," pp. 329–330.

76. Meyer Howard Abrams, "Coleridge's 'A Light in Sound': Science, Metascience, and Poetic Imagination," *Proceedings of the American Philosophical Society* 116, no. 6 (December 1972): 458–476.

77. Erasmus Darwin, *Loves of the Plants*, 1:101–102, quoted from Desmond King-Hele, *Erasmus Darwin and the Romantic Poets* (London: Macmillan, 1986), p. 96.

78. Coleridge borrowed Ralph Cudworth's *True Intellectual System of the Universe* (1678) from the Bristol Library between November 9 and December 13, 1796; see David Jasper, *Coleridge as Poet and Religious Thinker: Inspiration and Revelation* (London: Macmillan, 1985), p. 20. Cudworth uses the harp as a symbol for his "plastick natures" (*True Intellectual System of the Universe, Part I* [For Richard Royston, 1678], bk. 1, chap. 3, sec. 37, pp. 157–158). Denis Diderot's *D'Alembert's Dream* contains a metaphor wherein the association of ideas is explained by the sympathetic vibration of the strings in a harpsichord (Denis Diderot, *Oeuvres philosophiques*, ed. Paul Vernière [Paris: Garnier Frères, 1961], pp. 271–274).

79. Some of these themes are more obvious than others. For instance, the Aeolian harp "warbles." It warbled for Kircher ("tremulum"), it warbled for Thomson, and in 1817 it warbled for Coleridge. It did not warble for Jones, but then Jones was not a poet.

80. Samuel Taylor Coleridge, *Collected Letters of Samuel Taylor Coleridge*, ed. Earl Leslie Griggs, 6 vols. (Oxford: Oxford University Press, 1956–1971), 4:771 and 773.

81. See Duane B. Schneider, "Coleridge's Light-Sound Theory," *Notes and Queries*, May 1963, pp. 182–183, and J. B. Beer, "Coleridge and Boehme's 'Aurora,'" *Notes and Queries*, May 1963, pp. 183–187.

82. Samuel Taylor Coleridge, *Marginalia*, pt. 1, ed. George Whalley, vol. 12 of *Collected Works* (Princeton: Princeton University Press, 1980), p. 572.

83. See Robin C. Dix, "The Harps of Memnon and Aeolus: A Study in the Propagation of an Error," *Modern Philology* 85 (1988): 288–293; and the anonymous article in *Quarterly Review* (London) 138 (1875): 529–540.

84. S. T. Coleridge, *Confessions of an Inquiring Spirit and Some Miscellaneous Pieces*, ed. H. N. Coleridge (1849), p. 51, quoted from Coleridge, *Marginalia*, pt. 2, p. 787. In *The Friend* he wrote: "We are far from being Hutchinsonians, nor have we found much to respect in the twelve volumes of Hutchinson's works, either as biblical comment or natural philosophy: though we give him credit for orthodoxy and good intentions. . . . Those who would wish to learn the most important points of the Hutchinsonian doctrine in the most favorable form, and in the shortest possible space, we can refer to Duncan Forbes's Letter to a Bishop" (Coleridge, *The Friend*, ed. Barbara E. Rooke, in *The Collected Works of Samuel Taylor Coleridge* [London: Routledge & Kegan Paul, 1969], vol. 4, pt. 1, p. 502). Duncan Forbes was a Hutchinsonian and a close friend and patron of James Thomson. Possibly he knew William Jones, although he died in 1747 and would have had little chance to meet Jones in London. In 1817 Coleridge made extensive marginal comments in Forbes's *Whole Works*, and it is probably through Forbes's *Letter to a Bishop* that he learned about Hutchinsonianism (Coleridge, *Marginalia*, pt. 2, pp. 784–790).

85. Kuhn, "Glory or Gravity," p. 316. Levere, in his *Poetry Realized in Nature* (p. 235n.10), says Coleridge rejected Hutchinsonianism out of hand because of its mechanism.

86. Percy Bysshe Shelley, *The Complete Poetical Works*, ed. Neville Rogers, 4 vols. (Oxford: Clarendon Press, 1975), 2:48. Glenn O'Malley, "Shelley's 'Air-Prism': The Synesthetic Scheme of *Alastor*," *Modern Philology* 55 (1958): 178–187. O'Malley claims that Shelley used Jones's theory of the air-prism, although Shelley

never mentioned the air-prism explicitly. If Shelley did read Jones, it was undoubtedly in the same place that O'Malley read him, in Robert Bloomfield's "Nature's Music."

87. Stephen Prickett, *Romanticism and Religion: The Tradition of Coleridge and Wordsworth in the Victorian Church* (Cambridge: Cambridge University Press, 1976), pp. 19 and 189.

88. Samuel Taylor Coleridge, *The Complete Works*, 7 vols. (New York: Harper & Brothers, 1853), 5:593. Also Harding, *Coleridge and the Inspired Word*, pp. 1–18.

89. The harp remained an ambiguous image, however, because Coleridge also condemned the idea that the "sweet *Psalmist of Israel* was himself as mere an instrument as his harp, an *automaton* poet" and that the Bible could be only a "colossal Memnon's head, a hollow passage for a voice" (Coleridge, *Complete Works*, 5:593 and 591). In these passages, intermingled with the other passages quoted in the text, the harp is a mere automaton, a mere conduit for God's word.

90. Emerson's harp is still in his study at the Emerson House in Concord. Thoreau's harp is in the Concord Museum.

91. F. O. Matthiessen, *American Renaissance: Art and Expression in the Age of Emerson and Whitman* (New York: Oxford University Press, 1941), p. 50.

92. Ibid., pp. 47–48.

93. Ralph Waldo Emerson, *The Complete Works*, Concord Edition, ed. Edward Waldo Emerson (Boston: Houghton Mifflin, 1903–1904), 10:129–130.

94. Ibid., 9:203–207. In 1868 Emerson gave an Aeolian harp to his daughter Edith and her husband, and accompanied it with his poem "Maiden Speech of the Aeolian Harp"; see Jeanetta Boswell, "Three Poets and the Aeolian Harp: Emerson, Thoreau, and Melville," *University Forum, Fort Hays State University*, no. 31 (September 1983): 5. Although this poem seems to be better known than "The Harp," it is merely a presentation poem and lacks the yearning for transcendence that one finds in "The Harp."

95. Ralph Waldo Emerson, *Journals*, ed. Edward Waldo Emerson and Waldo Emerson Forbes, 10 vols. (Boston: Houghton Mifflin, 1909–1914), 9:311; Robert S. Matteson, "Emerson and the Aeolian Harp," *South Central Bulletin* (South Central Modern Language Association) 24, no. 1 (February 1964): 7.

96. Matthiessen, *American Renaissance*, p. 48.

97. Ibid., p. 84.

98. Henry David Thoreau, *The Journal of Henry D. Thoreau*, ed. Bradford Torrey and Francis H. Allen, 14 vols. in 2 (New York: Dover, 1962), 2:450.

99. Henry David Thoreau, *A Week on the Concord and Merrimack Rivers* (New York: New American Library, 1961), p. 154. Thoreau's river trip took place in 1839, but his account of it appeared only ten years later.

100. Writing about the Aeolian harp also inspired Thoreau to quote his poem "Rumors from an Aeolian Harp." The poem is a utopian vision of a valley "Where foot of man has never been." It is a place where "love is warm, and youth is young / And poetry is yet unsung." The harp appears only in the title, but there is no doubt that it inspired this transcendant vision. *Week*, pp. 153–154.

101. Thoreau, *Journal*, 4:206 and 515. There are many more references to the telegraph harp in Thoreau's journal. See Paul Sherman, "The Wise Silence: Sound as the Agency of Correspondence in Thoreau," *New England Quarterly* 22 (1949): 511–527.

102. Thoreau, *Journal*, 2:268.

103. Thoreau, *Week*, p. 155.

104. Ibid., p. 154.

105. "Mr Gray carried usually with him on these tours a Plano-convex Mirror of about four inches diameter on a black foil, and bound up like a pocket book. A glass of this sort is perhaps the best and most convenient substitute for a camera obscura, of anything that has hitherto been invented, and may be had of any optician." Thomas Gray, *Journal in the Lakes* (1775), quoted from Deborah Warner, "The Landscape Mirror and Glass," *Antiques* 105 (1974): 158–159.

106. Samuel Taylor Coleridge, *The Notebooks*, ed. Kathleen Coburn, 4 vols. (New York: Pantheon Books, 1957–), 2: no. 3159. See also Abrams, "Coleridge's 'A Light in Sound,'" p. 464. Abrams discusses this theme at length in "The Correspondent Breeze," p. 42, and *The Mirror and the Lamp*, pp. 50–51.

107. Nicolson opens her *Newton Demands the Muse* with an account of this dinner (pp. 1–2). The best source on instruments and romanticism is Erika von Erhardt-Siebold, "Some Inventions of the Pre-Romantic Period and Their Influence upon Literature," *Englische Studien* 66 (1931–1932): 347–363, and "Harmony of the Senses in English, German, and French Romanticism," *PMLA* 47 (1932): 577–592.

108. Samuel Taylor Coleridge, *Aids to Reflection, and the Confessions of an Inquiring Spirit* (London: George Bell and Sons, 1901), p. 224.

109. Ibid., p. 161.

110. Ibid., p. xlvi.

111. These comments were recorded by Eliza Hamilton. See Robert P. Graves, *Life of Sir William Rowan Hamilton*, 3 vols. (Dublin: Hodges, Figgis and Co., 1882–1891), 1:313.

112. William Wordsworth, *The Complete Poetical Works*, ed. Andrew J. George (Boston: Houghton Mifflin, 1932), *Excursion*, bk. 4, lines 1251–1263.

113. Ibid., lines 1144–1145.

Chapter Six
Science since Babel: Graphs, Automatic Recording Devices, and the Universal Language of Instruments

1. John Locke, *An Essay Concerning Human Understanding*, ed. Alexander Campbell Fraser, 2 vols. (Oxford: Clarendon Press, 1894), 2:146. Hans Aarsleff, *From Locke to Saussure: Essays on the Study of Language and Intellectual History* (Minneapolis: University of Minnesota Press, 1982), pp. 24–26, 42–45, 63–67.

2. Locke, *Essay Concerning Human Understanding*, 2:100. Hobbes said we misuse language, "First, when men register their thoughts wrong, by the inconstancy of the signification of their words; by which they register for their conception, that which they never conceived; and so deceive themselves. Secondly, when they use words metaphorically; that is, in other sense than that they are ordained for; and thereby deceive others." *Leviathan or The Matter, Forme, and Power of a Commonwealth, Ecclesiasticall and Civill* (London, 1651), p. 13.

3. See William B. Ashworth, "Natural History and the Emblematic World View," in *Reappraisals of the Scientific Revolution*, ed. David C. Lindberg and Robert S. Westman (Cambridge: Cambridge University Press, 1990), pp. 303–332, and Martin Elsky, "Bacon's Hieroglyphs and the Separation of Words and Things," *Philological*

Quarterly 63 (1984): 449–460. A typical emblem consists of a visual image, a short motto, and a slightly longer epigram. The debate over whether universal languages came from the tradition of cryptography or from emblem is probably unimportant, since they were not clearly distinguished in the seventeenth century. James Knowlson, *Universal Language Schemes in England and France, 1600–1800* (Toronto: University of Toronto Press, 1975), pp. 17–18.

4. William Playfair's graphs of trade and commerce are the obvious exception.

5. Robin E. Rider, "Measure of Ideas, Rule of Language: Mathematics and Language in the Eighteenth Century," in *The Quantifying Spirit in the Eighteenth Century*, ed. Tore Frängsmyr, J. L. Heilbron, and Robin E. Rider (Berkeley and Los Angeles: University of California Press, 1990), pp. 113–140.

6. On the history of graphs see Laura Tilling, "Early Experimental Graphs," *British Journal for the History of Science* 8 (1975): 193–213; Margaret C. Shields, "The Early History of Graphs in Physical Literature," *American Physics Teacher* 5 (1937): 68–71; Hebbel E. Hoff and L. A. Geddes, "Graphic Recording before Carl Ludwig: An Historical Summary," *Archives internationales d'histoire des sciences* 12 (1959): 1–25; Hoff and Geddes, "The Beginnings of Graphic Recording," *Isis* 53 (1962): 287–310; H. Gray Funkhouser, "Historical Development of the Graphical Representation of Statistical Data," *Osiris*, 1st ser., 3 (1937): 269–404; and Erica Royston, "A Note on the History of the Graphical Presentation of Data," *Biometrika* 43 (1956): 241–247.

7. Many of the authors in the previous note make some effort at a survey.

8. Johann Heinrich Lambert, *Semiotik*, no. 1, in *Philosophische Schriften*, facsimile ed., 7 vols. (Hildesheim: Georg Olms Verlagsbuchhandlung, 1965–), 2:6.

9. Ibid., pp. 8–15.

10. Ibid., p. 15. Lambert credits Francis Bacon with this argument.

11. Lambert, *Dianoiologie, oder Lehre von den Gesetzen des Denkens*, no. 700, in *Philosophische Schriften*, 1:450.

12. Lambert, *Semiotik*, p. 16.

13. There were several attempts to create universal languages employing musical notation. See Knowlson, *Universal Language Schemes*, pp. 119–122.

14. Lambert notes that it is possible to compare abstract concepts with perceptions and with objects by using figurative language. By giving extension to concepts we can make them "*figurlich*," the best example being music, where the notes represent degrees of pitch. Here we are close to graphs. Lambert, *Alethiologie oder Lehre von der Wahrheit*, no. 52, in *Philosophische Schriften*, 1:487.

15. Lambert, *Dianoiologie*, no. 177, in *Philosophische Schriften*, 1:110.

16. On Lambert's "line theory" see Gereon Wolters, *Basis und Deduktion. Studien zur Entstehung und Bedeutung der Theorie der axiomatischen Methode bei J. H. Lambert (1728–1777)* (Berlin: Walter de Gruyter, 1980), pp. 120–166.

17. Lambert carried on a lengthy correspondence with Brander in which he discussed the manufacture and use of scientific instruments. This correspondence is in *Johann Heinrich Lamberts deutscher gelehrter Briefwechsel*, ed. Johann Bernoulli, 5 vols. in 7 (Berlin: Bey dem Herausgeber, 1781–1787). See also Alto Brachner et al., *G. F. Brander 1713–1783: Wissenschaftliche Instrumente aus seiner Werkstatt* (Munich: Deutches Museum, 1983). On Lambert's method see J. J. Gray and Laura Tilling, "Johann Heinrich Lambert, Mathematician and Scientist, 1728–1777," *Historia Mathematica* 5 (1978): 13–41.

18. Lambert, *Pyrometrie oder vom Maasse des Feuers und der Wärme* (Berlin: Bey Hande und Spener, 1779), p. 350. The data were collected by René Réaumur.

19. Ibid., p. 352.

20. "Theorie der Zuverlässigkeit . . . ," in Lambert, *Beyträge zum Gebrauch der Mathematik und derren Anwendungen*, 3 vols. in 4 (Berlin: Im Verlage des Buchladens der Realschule, 1765–1772), vol. 1 (1765), pp. 424–488. Lambert considers cases where the form of the graph is known from theory (variation of the length of the seconds pendulum with latitude) and where the form can be found only from the data (variation of the compass declination with time and mortality tables for London between 1753 and 1785). Laura Tilling has found more than thirty graphs distributed throughout Lambert's work. Tilling, "Early Experimental Graphs," p. 201. On Lambert's theory of errors see also O. B. Sheynin, "J. H. Lambert's Work on Probability," *Archive for History of Exact Sciences* 7 (1971): 244–256; Sheynin, "Origin of the Theory of Errors," *Nature*, no. 5052 (August 27, 1966): 1003–1004.

21. These comments were collected by Funkhouser, "Historical Development," p. 295.

22. Mary Slaughter, *Universal Languages and Scientific Taxonomy in the Seventeenth Century* (Cambridge: Cambridge University Press, 1982), p. 47.

23. Johann Heinrich Lambert, "Essai de taxéometrie, ou sur la mesure de l'ordre," *Akademie der Wissenschaften, Berlin, Nouveau mémoires*, 1770, 327–342, and 1773, 347–368. See also John Lesch, "Systematics and the Geometrical Spirit," in *The Quantifying Spirit in the Eighteenth Century*, ed. Frängsmyr, Heilbron, and Rider, pp. 85–86.

24. Slaughter argues that the interest in universal languages declined around 1700 as the taxonomic approach in science was replaced by the mathematical approach (*Universal Languages*, pp. 182–219). Lesch argues that mathematics *promoted* taxonomy and systematics ("Systematics and the Geometrical Spirit," pp. 73–111, esp. pp. 83–84). The history is not simple. Often philosophers such as Condillac and d'Alembert condemned the "spirit of systems" (that is, the taxonomic approach) but created systems themselves, most notably the classification of the sciences that appeared in d'Alembert's *Discours préliminaire* to the *Encyclopédie*. Buffon, on the other hand, criticized Linnaeus's botanical taxonomy because he thought it was *too much* like mathematics. The relationships between systematics, language, and mathematics in the eighteenth century remain a historical puzzle.

25. William Playfair, *The Commercial and Political Atlas*, 3d ed. (London: J. Wallis, 1801), pp. ix–xii.

26. Ibid., pp. v–vi.

27. Knowlson, *Universal Language Schemes*, pp. 21, 61.

28. Rider, "Mathematics and Language," pp. 125–132.

29. "Playfair, William," in *Dictionary of National Biography*, 15:1300.

30. Playfair, *An Inquiry into the Permanent Causes of the Decline and Fall of Powerful and Wealthy Nations, Illustrated by Four Engraved Charts*, 2d ed. (London, 1807), p. xvi, from Funkhouser, "Historical Development," p. 289.

31. M. C. Shields ("James Watt and Graphs," *American Physics Teacher* 6 [1938]: 162) notes that Watt drew a pressure-volume curve in 1782, which Playfair might have seen. Presumably the curve was an estimate, because Watt invented his first pressure gauge, the "indicator," only in the 1790s and his "indicator card" in 1796 (H. W. Dickinson and Rhys Jenkins, *James Watt and the Steam Engine* [Ox-

ford: Clarendon Press, 1927], pp. 228–233; and Shields, "James Watt and Graphs," p. 162).

32. The earliest indicator diagrams in the Boulton and Watt Collection are from 1803, although there are references to such diagrams from 1796. Dickinson and Jenkins, *James Watt*, pp. 229–230.

33. Shields, "Early History of Graphs," p. 68.

34. See the essays by Hebbel E. Hoff and L. A. Geddes, "Graphic Recording before Carl Ludwig"; "Graphic Registration before Ludwig: The Antecedents of the Kymograph," *Isis* 50 (1959): 5–21; "The Technological Background of Physiological Discovery: Ballistics and the Graphic Method," *Journal of the History of Medicine* 15 (1960): 345–363; and "The Beginnings of Graphic Recording." Benjamin Thompson also used graphs, obviously derived from geometrical diagrams, in describing his ballistic experiments. He gave his data in the form of tables, and presented them graphically to show the agreement between experiment and theory ("New Experiments upon Gun-powder, with Occasional Observations and Practical Inferences . . . ," *Royal Society of London, Philosophical Transactions* 71, pt. 2 [1781]: 229–328). These graphs were less striking than those of Playfair, and Thompson did not introduce them as a new method.

35. Galileo Galilei, *The Assayer* (1623), in *Discoveries and Opinions of Galileo*, ed. and trans. Stillman Drake (New York: Anchor/Doubleday, 1957), pp. 237–238.

36. Galileo Galilei, *Discourses Concerning Two New Sciences*, trans. Henry Crew and Alfonso de Salvio (New York: Dover, 1954), p. 99. V. Carlton Maley has called this "the first technique of 'recording' a sound for later study" (V. Carlton Maley, Jr., *The Theory of Beats and Combination Tones: 1700–1863* [New York: Garland, 1990], p. 5). However, both D. P. Walker and Sigalia Dostrovsky have shown that Galileo fabricated the precise data from these experiments, or "thought experiments." See D. P. Walker, *Studies in Musical Science in the Late Renaissance* (London: Warburg Institute; Leiden: E. J. Brill, 1978), p. 30; and Sigalia Dostrovsky, "Early Vibration Theory: Physics and Music in the Seventeenth Century," *Archive for History of Exact Sciences* 14 (1975): 169–218.

37. On Chladni, see Dieter Ullmann, "Chladni und die Entwicklung der experimentellen Akustik um 1800," *Archive for History of Exact Sciences* 31 (1984): 35–52. For a full list of studies on Chladni figures, see Mary D. Waller, *Chladni Figures: A Study in Symmetry* (London: D. Bell, 1961). Franz Josef Pisko's *Die neuren Apparate der Akustik* (Vienna: Carl Gerold's Sohn, 1865) displays the varieties of this sort of acoustic demonstration in the nineteenth century, as well as a wealth of information on the representational instruments and other devices employed in nineteenth-century acoustics. These included Auguste Seebeck's siren, Hermann von Helmholtz's resonators, and Jules Antoine Lissajous's apparatus for creating his acoustical figures.

38. Ernst Florens Friedrich Chladni, *Traité d'acoustique* (Paris: Courcier, 1809) (this is Chladni's translation of his *Die Akustik* [1802] with a biographical preface added), p. vii. Olexa Myron Bilaniuk, "Lichtenberg, Georg Christoph," in *Dictionary of Scientific Biography*, 8:320–323.

39. See H. C. Oersted, *The Soul of Nature*, trans. Leonora and Joanna B. Horner (London: Dawsons, 1966), pp. 325–351. Related comments may be found in J. B. Stallo, *General Principles of the Philosophy of Nature* (Boston: W. M. Crosby and H. P. Nichols, 1848), pp. 75–76.

40. Walter D. Wetzels, *Johann Wilhelm Ritter: Physik im Wirkungsfeld der deutschen Romantik* (Berlin: Walter de Gruyter, 1973), pp. 87–97.

41. Ritter to Oersted, March 31, 1809, in H. C. Oersted, *Correspondance avec divers savants*, ed. H. C. Harding, 2 vols. in 1 (Copenhagen: H. Aschehoug & Co., 1920), 2:224. See also Stuart Walter Strickland, "Circumscribing Science: Johann Wilhelm Ritter and the Physics of Sidereal Man" (Ph.D. diss., Harvard University, 1992).

42. Thomas Young, *Miscellaneous Works of the Late Thomas Young*, ed. George Peacock, 3 vols. (London: J. Murray, 1855; reprint, New York: Johnson, 1972), 1:87.

43. Thomas Young, *A Course of Lectures on Natural Philosophy and the Mechanical Arts*, 2 vols. (London: J. Johnson, 1807), 1:369.

44. See Pisko, *Die neuren Apparate der Akustik*, chap. 3, "Die Tonschreibkunst, Phono- oder Vibrographie."

45. Charles Wheatstone, "Description of the Kaleidophone, or Phonic Kaleidoscope; a new Philosophical Toy, for the Illustration of several Interesting and Amusing Acoustical Phenomena," *Quarterly Journal of Science* 1 (1827); reprinted in *The Scientific Papers of Sir Charles Wheatstone* (London: Taylor and Francis, 1879), pp. 21–29, at pp. 22, 21.

46. A portion of Young's Göttingen dissertation was devoted to the human voice and to a phonetic alphabet he devised to represent speech sounds. Alexander Wood, *Thomas Young, Natural Philosopher* (Cambridge: Cambridge University Press, 1954), pp. 49–50; George Peacock, *Life of Thomas Young* (London: J. Murray, 1855), p. 90.

47. Édouard-Léon Scott de Martinville, *Histoire de la sténographie depuis les temps anciens jusqu'à nos jours; ou précis historique et critique des divers moyens qui ont été proposés ou employés pour rendre l'écriture aussi rapide que la parole* (Paris: Charles Tondeur, 1849), pp. 7–8.

48. Ibid., pp. 149, 150.

49. E.-L. Scott, *Les noms de baptême et les prénoms* (Paris: Alexandre Houssiaux, 1857, 2d ed. 1858), p. 53.

50. E.-L. Scott de Martinville, "Principes de phonautographie," in *Le problème de la parole s'écrivant elle-même* (Paris: Scott, 1878), pp. 29–33, at pp. 29–30.

51. Ibid., p. 31.

52. Edward Wheeler Scripture, *The Elements of Experimental Phonetics* (New York: Charles Scribner's Sons; London: Edward Arnold, 1902), p. 17.

53. Scott's notion of modeling an instrument on the ear later came full circle in the "Ear Phonautograph" of Clarence Blake, a Boston otologist. In this device, the eardrum and ossicles of a human cadaver were connected to the stylus. "It is readily comprehended," Blake wrote, "that a structure so admirably fitted by nature for the office which it has to fulfill, the reception and transmission of sonorous vibrations, should better answer the purposes of experimentation than any purely mechanical device." Clarence J. Blake, "The Use of the Membrana Tympani as a Phonautograph and Logograph," *Archives of Ophthalmology and Otology* 5 (1876): 108–113, at 110. The visually intriguing intricacy of phonautographic traces was not lost on Scott. In his patent, he also suggested that the inscriptions might be valuable for the creation of ornamental designs. See his 1857 patent, which is reprinted in *Le problème*, pp. 34–37, at p. 35.

54. The communication is reprinted in Scott, *Le problème*, pp. 38–48. The quotation is from p. 39.

55. Ibid., p. 47.

56. Ibid., p. 46.

57. See F. C. Donders, "Zur Klangfarbe der Vocale," *Annalen der Physik* 199 (1864): 527–528; Heinrich Schneebeli, "Expériences avec le phonautographe," *Archives des sciences physiques et naturelles*, n.s., 4 (1878): 78–83; Georges René Marie Marage, "La méthode graphique dans l'étude des voyelles," *Comptes rendus, Académie des sciences, Paris* 128 (1899): 425–427.

58. See Robert J. Silverman, "Instrumentation, Representation, and Perception in Modern Science: Imitating Human Function in the Nineteenth Century" (Ph.D. diss., University of Washington, 1992). These issues are raised in chap. 3, "The Torch of Acoustics: Instrumentation, Representation, and the Science of Sound in the Nineteenth Century."

59. André Millard, *Edison and the Business of Invention* (Baltimore: The Johns Hopkins University Press, 1990), p. 63.

60. Scott, *Le problème*, p. 10.

61. Ibid., p. 10. The *phonéglyphes* "are holes analogous to those of Gruyère cheese and arranged in a straight line," he added.

62. Carl Ludwig, "Beiträge zur Kenntniss des Einflusses der Respirationsbewegungen auf den Blutlauf im Aortensysteme," *Archiv für Anatomie, Physiologie, und wissenschaftliche Medizin*, 1847, 242–302. See the articles by Hoff and Geddes cited above, as well as their "A Historical Perspective on Physiological Monitoring: Sherrington's Mammalian Laboratory and Its Antecedents," *Cardiovascular Research Center Bulletin* 13 (1974): 19–39 and their "A Historical Perspective on Physiological Monitoring: Chaveau's Projecting Kymograph and the Projecting Physiograph," *Cardiovascular Research Center Bulletin* 14 (1975): 3–35. Also see Nancy Roth, "'First Stammerings of the Heart': Ludwig's Kymograph," *Medical Instrumentation* 12 (1978): 348; Merriley Borell, "Extending the Senses: The Graphic Method," *Medical Heritage* 2 (1986): 114–121; Christopher Lawrence, "Physiological Apparatus in the Wellcome Museum. 1. The Marey Sphygmograph," *Medical History* 22 (1978): 196–200; Audrey B. Davis, *Medicine and Its Technology* (Westport, Conn.: Greenwood, 1981); Hughes Evans, "Losing Touch: The Controversy over the Introduction of Blood Pressure Instruments into Medicine," *Technology and Culture* 34 (1993): 784–807; Soraya de Chadarevian, "Graphical Method and Discipline: Self-Recording Instruments in Nineteenth-century Physiology," *Studies in History and Philosophy of Science* 24 (1993): 267–291; Robert G. Frank, Jr., "The Telltale Heart: Physiological Instruments, Graphic Methods, and Clinical Hopes, 1854–1914," in *The Investigative Enterprise: Experimental Physiology in Nineteenth Century Medicine*, ed. William Coleman and Frederic L. Holmes (Berkeley and Los Angeles: University of California, 1988), pp. 211–290; and Lorraine Daston and Peter Galison, "The Image of Objectivity," *Representations* 40 (1992): 81–128.

63. Ludwig admitted that his innovation relied upon the work of predecessors. In the eighteenth century, Stephen Hales connected a long vertical glass tube to the femoral artery of a horse; the height to which the column of blood would rise indicated the blood pressure. A century later, J. M. Poiseuille attached a mercury manometer directly to an artery—a development to which Ludwig added the capacity of graphic registration (Lawrence, "Physiological Apparatus," pp. 196–197). In this

respect, Ludwig considered himself to be a beneficiary of the labors of James Watt (Hoff and Geddes, "Graphic Registration before Ludwig," p. 18).

64. Lawrence, "Physiological Apparatus," p. 197.

65. Ibid., p. 197. Also see H. A. Snellen, ed., *E. J. Marey and Cardiology* (Rotterdam: Kooyker, 1980).

66. Frank, "The Telltale Heart," pp. 211–290.

67. Hoff and Geddes, "Graphic Registration before Ludwig."

68. J. Burdon-Sanderson and Francis E. Anstie, "On the Application of Physical Methods to the Exploration of the Movements of the Heart and Pulse in Disease," *Lancet* (American ed.), 1867, p. 103. This quotation can also be found in Frank, "Telltale Heart," p. 222.

69. On Marey, see François Dagognet, *Étienne-Jules Marey: A Passion for the Trace*, trans. Robert Galeta and Jeanine Herman (New York: Zone Books, 1992); Chadarevian, "Graphical Method and Discipline"; Michel Frizot, *Avant la cinématographie, la chronophotographie: temps, photographie et mouvement autour de E.-J. Marey* (Beaune: Association des amis de Marey, 1984); Daston and Galison, "The Image of Objectivity"; Lisa Cartwright, "'Experiments of Destruction': Cinematic Inscriptions of Physiology," *Representations* 40 (1992): 129–152; and Marta Braun, *Picturing Time: The Work of Étienne-Jules Marey (1830–1904)* (Chicago: University of Chicago Press, 1992).

70. Étienne-Jules Marey, *La méthode graphique dans les sciences expérimentales et principalement en physiologie et en médecine* (Paris: G. Masson, 1878; 2d ed., 1885), p. I.

71. Ibid., p. III.

72. Ibid.

73. Ibid., pp. III–IV.

74. Ibid., p. III. Marey did believe, however, that in order to perfect the universal language of the graphical method, inscription devices would have to be improved—they would have to become more precise and their techniques would have to become standardized. See his "Mesures à prendre pour l'uniformisation des méthodes et le contrôle des instruments employés en physiologie," *Comptes rendus, Académie des sciences, Paris* 127 (1898): 375–381.

75. Marey, *La méthode graphique*, p. V.

76. Ibid., p. VI.

77. Charles Sanders Peirce, *Collected Papers of Charles Sanders Peirce*, ed. Charles Hartshorne and Paul Weiss, 7 vols. (Cambridge: Harvard University Press, 1933), vol. 2, para. 2.219.

78. Ibid., para. 2.227.

79. Ibid., vol. 4, bk. 2, title page.

80. Ibid., para. 4.353. Peirce notes that others have claimed that Christian Weise created logical diagrams before Lambert, but that this claim was based on Lambert's own statement (Lambert, *Anlage zur Architektonik*, in *Philosophische Schriften*, 1:28). See also Thomas A. Sebeok, "'Semiotics' and Its Congeners," in *Frontiers in Semiotics*, ed. John Deely, Brooke Williams, and Felicia E. Kruse (Bloomington: Indiana University Press, 1986), pp. 255–263, at p. 256. Locke used the term "semeiotikė" in his *Essay Concerning Human Understanding*, bk. 4, chap. 20, p. 361 (London: Printed by Elizabeth Holt for Thomas Basset, 1690).

81. Peirce's brilliance and stormy career are treated with care in Joseph Brent, *Charles Sanders Peirce: A Life* (Bloomington: Indiana University Press, 1993).

82. James Joseph Sylvester, "On an Application of the New Atomic Theory to the Graphical Representation of the Invariants and Covariants of Binary Quantics—with Three Appendices," *American Journal of Mathematics* 1 (1878): 64–128, and Sylvester, "Chemistry and Algebra," in *Collected Mathematical Papers*, 4 vols. (Cambridge: Cambridge University Press, 1904–1912), vol. 3, no. 14; William Kingdon Clifford, "Remarks on the Chemico-Algebraic Theory,"in *Mathematical Papers*, ed. Robert Tucker (London: Macmillan and Co., 1882), no. 28.

83. Peirce, *Collected Papers*, vol. 4, para. 4.535. Peirce defines a graph more formally as "a superficial diagram [meaning on a surface] composed of the sheet upon which it is written or drawn, of spots or their equivalents, of lines of connection, and (if need be) of enclosures. The type, which it is supposed more or less to resemble, is the structural formula of the chemist" (para. 4.418–419). This is close to the first definition given by the *Oxford English Dictionary*.

84. The best history of graph theory is Norman L. Biggs, E. Keith Lloyd,and Robin J. Wilson, *Graph Theory, 1736–1936* (Oxford: Clarendon Press, 1976).

85. Peirce, *Collected Papers*, para. 2.778. See also 2.170, 2.444, and 2.782.

86. Ibid., para. 2.385.

87. Ibid., para 4.530.

88. Peirce classified signs in a complex triadic arrangement, each part of a triad generating a new triad. The "second trichotomy" or the "trichotomy of performance" separates signs into icons, indices, and symbols. See Terence Hawkes, *Structuralism and Semiotics* (Berkeley and Los Angeles: University of California Press, 1977), p. 126.

89. See chap. 4.

90. It is doubtful that Peirce would have agreed with our argument. He claimed that in spite of initial controversy, natural philosophers always come to agreement eventually on what is the correct conclusion to draw from an experiment. This agreement is their "truth" and it is the only truth that they could hope to achieve. Charles Sanders Peirce, "A Critical Review of Berkeley's Idealism" (1871), in *Values in a Universe of Chance: Selected Writings of C. S. Peirce*, ed. Philip Wiener (Garden City, N.Y.: Doubleday, 1958), pp. 81–83. Joseph Brent informs us that the figure at the center of the labyrinth is not the Minotaur, but Peirce's dog! However, since it does seem to have cloven hooves and a Minotaur-like tail, we conclude that Peirce's sign is, at the least, ambiguous.

91. See the group of essays in *October*, no. 55 (1990): Thomas Y. Levin, "For the Record: Adorno on Music in the Age of Its Technological Reproducibility," pp. 23–47; Theodor Adorno, "The Curves of the Needle," trans. Levin, pp. 49–55; Adorno, "The Form of the Phonograph Record," trans. Levin, pp. 56–61; Adorno, "Opera and the Long-Playing Record," trans. Levin, pp. 62–66; and Douglas Kahn, "Track Organology," pp. 68–78. Friedrich Nietzsche, "On Truth and Falsity in Their Ultramoral Sense" (1873), in *The Complete Works of Friedrich Nietzsche*, trans. Maximilian A. Mügge, ed. Oscar Levy, 18 vols. (New York: Macmillan, 1924), 2:178.

92. Adorno, "Form of the Phonograph Record," p. 59. Adorno's emphasis.

93. Ibid. The inside quotations are from Walter Benjamin, "Ursprung des deutschen Trauerspiels," in *Gesammelte Schriften*, unter Mitwirkung von Theodor W. Adorno und Gershom Scholem; herausgegeben von Rolf Tiedemann und Hermann Schweppenhauser, 7 vols. (Frankfurt am Main: Suhrkamp Verlag, ca. 1972–), 1:387.

94. Adorno, "Form of the Phonograph Record," p. 60. Levin's notes to his translation of Adorno.

95. See Jonathan Crary, *Techniques of the Observer: On Vision and Modernity in the Nineteenth Century* (Cambridge: MIT Press, 1990), chap. 1, "Modernity and the Problem of the Observer."

96. Marey, *La méthode graphique*, pp. VIII–IX.

Chapter Seven
The Giant Eyes of Science: The Stereoscope and Photographic Depiction in the Nineteenth Century

1. Charles Wheatstone, "Contributions to the Physiology of Vision—Part the first. On some Remarkable, and hitherto Unobserved, Phenomena of Binocular Vision," *Philosophical Transactions of the Royal Society* 128 (1838): 371–394. Also in Nicholas J. Wade, ed., *Brewster and Wheatstone on Vision* (London: Academic Press, 1983), sec. 2.4. Many of the important papers on this subject by Brewster and Wheatstone have been conveniently collected in Wade's book. Wheatstone's papers cited in this chapter may also be found in *The Scientific Papers of Sir Charles Wheatstone* (London: Taylor and Francis, 1879). On Wheatstone, see Brian Bowers, *Sir Charles Wheatstone, F.R.S., 1802–1875* (London: Her Majesty's Stationery Office, 1975).

2. Antecedents who are often cited include Euclid, Galen, Leonardo da Vinci, Giambattista Della Porta, Aguilonius, Joseph Harris, and William Wells. See Edwin G. Boring, *Sensation and Perception in the History of Experimental Psychology* (New York: Appleton-Century-Crofts, 1942), chap. 8.

3. A. C. Crombie, "The Mechanistic Hypothesis and the Scientific Study of Vision: Some Optical Ideas as a Background to the Invention of the Microscope," in *Historical Aspects of Microscopy*, ed. S. Bradbury and G. L'E. Turner (Cambridge: Royal Microscopical Society, 1967), pp. 3–112.

4. Wheatstone, "Contributions" (1838), in Wade, *Brewster and Wheatstone*, p. 67.

5. Wade, *Brewster and Wheatstone*, p. 70.

6. See ibid., p. 322; Boring, *Sensation and Perception*, chap. 8; R. Steven Turner, "Consensus and Controversy: Helmholtz on the Visual Perception of Space," in *Hermann von Helmholtz and the Foundations of Nineteenth-Century Science*, ed. David Cahan (Berkeley and Los Angeles: University of California Press, 1993), pp. 154–204.

7. On "philosophical toys," see Gerard L'E. Turner, *Nineteenth-Century Scientific Instruments* (London and Berkeley: Sotheby/University of California Press, 1983), chap. 16. Additional material on the stereoscope can be found in Moritz von Rohr, *Die Binocularen Instrumente* (Berlin: Springer, 1920); Robert Taft, *Photography and the American Scene: A Social History, 1839–1889* (1938; New York: Dover, 1964), chap. 10; William C. Darrah, *The World of Stereographs* (Gettysburg: W. C. Darrah, 1977); and Edward W. Earle, ed., *Points of View: The Stereograph in America—A Cultural History* (Rochester: Visual Studies Workshop, 1979). Both von Rohr and Earle have excellent bibliographies. Jonathan Crary offers a Foucauldian interpretation of the stereoscope and other nineteenth-century optical devices in his *Techniques of the Observer: On Vision and Modernity in the Nineteenth Century* (Cambridge: MIT Press, 1990).

8. Robert Hunt from *Art Journal*, March 1856, p. 11. Quoted in Earle, *Points of View*, p. 28.

9. Darrah, *World of Stereographs*, p. 3.

10. From *Anthony's Photographic Bulletin*, December 1872, p. 766. Quoted in Earle, *Points of View*, p. 50.

11. William Paley, *Natural Theology: or, Evidences of the existence and attributes of the Deity, collected from the appearances of nature* (London: R. Faulder, 1802).

12. Peter Mark Roget, *Animal and Vegetable Physiology Considered with Reference to Natural Theology*, 2 vols. (London: William Pickering, 1834), 2:445–446.

13. Michael Baxandall, *Painting and Experience in Fifteenth Century Italy; A Primer in the Social History of Pictorial Style*, 2d ed. (1972; Oxford: Oxford University Press, 1988). On p. 152, Baxandall writes: "A society develops its distinctive skills and habits, which have a visual aspect, since the visual sense is the main organ of experience, and these visual skills and habits become part of the medium of the painter: correspondingly, a pictorial style gives access to the visual skills and habits and, through these, to the distinctive social experience. An old picture is the record of visual activity. One has to learn to read it, just as one has to learn to read a text from a different culture, even when one knows, in a limited sense, the language: both language and pictorial representation are conventional activities." Also see Alan Trachtenberg, *Reading American Photographs: Images as History, Mathew Brady to Walker Evans* (New York: Hill and Wang, 1989).

14. See G. Ten Doesschate, *Perspective: Fundamentals, Controversials, History* (Nieuwkoop: B. De Graaf, 1964); M. H. Pirenne, *Optics, Painting, and Photography* (Cambridge: Cambridge University Press, 1970); David C. Lindberg, *Theories of Vision from Al-Kindi to Kepler* (Chicago: University of Chicago Press, 1976), chap. 8; Joel Snyder, "Picturing Vision," *Critical Inquiry* 6 (1980): 499–526.

15. Svetlana Alpers, *The Art of Describing; Dutch Art in the Seventeenth Century* (Chicago: University of Chicago Press, 1983), chap. 2.

16. Michael Baxandall, *Patterns of Intention: On the Historical Explanation of Pictures* (New Haven: Yale University Press, 1985).

17. See E. H. Gombrich, "Standards of Truth: The Arrested Image and the Moving Eye," *Critical Inquiry* 7 (1980): 237–273.

18. Emerson referred to his technique as "naturalistic photography," and he tried to incorporate what he learned from his reading of Helmholtz's researches on physiological optics. See Gombrich, "Standards," p. 260; Peter Turner and Richard Wood, *Peter Emerson: Photographer of Norfolk* (Boston: David R. Goodine, 1974).

19. On Talbot, see John Ward and Sara Stevenson, *Printed Light: The Scientific Art of William Henry Fox Talbot and David Octavius Hill with Robert Adamson* (Edinburgh: Her Majesty's Stationery Office, 1986).

20. See Wade, *Brewster and Wheatsone*, pp. 33–39. The following also discuss the origins of stereoscopic photography: R. S. Clay, "The Stereoscope," *Transactions of the Optical Society* 29 (1927–1928): 149–166; A. T. Gill, "Early Stereoscopes," *Photographic Journal* 109 (1969): 546–559, 606–614, 641–651; and two articles by Steven F. Joseph—"Wheatstone's Double Vision," *History of Photography* 8 (1984): 329–332, and "Wheatstone and Fenton: A Vision Shared," *History of Photography* 9 (1985): 305–309.

21. On Brewster's intellectual development and his commitment to natural theol-

ogy, see Edgar W. Morse, "Natural Philosophy, Hypotheses, and Impiety: Sir David Brewster Confronts the Undulatory Theory of Light" (Ph.D. diss., University of California, Berkeley, 1972).

22. Gill, "Early Stereoscopes," p. 557; A. D. Morrison-Low, "Brewster and Scientific Instruments," in *'Martyr of Science': Sir David Brewster 1781–1868*, ed. A. D. Morrison-Low and J.R.R. Christie (Edinburgh: Royal Scottish Museum, 1984), at p. 62; Wheatstone privately developed a stereoscope quite similar to the lenticular model well before Brewster's announcement. Brewster's claim for the originality of his stereoscope—as well as his jealousy and scientific conflicts with Wheatstone—fueled their rivalry, which is examined throughout Wade, *Brewster and Wheatstone*.

23. "The Stereoscope, Pseudoscope, and Solid Daguerreotypes," *Illustrated London News* 20 (1852): 77–78, at 78.

24. "Mascher's Stereoscopic Books," *Scientific American* 11 (1856): 228.

25. Ibid.

26. "The Stereoscope, Pseudoscope, and Solid Daguerreotypes," p. 78.

27. On Holmes and his stereoscope, see Eleanor M. Tilton, *Amiable Autocrat: A Biography of Dr. Oliver Wendell Holmes* (New York: Henry Schuman, 1947); Thomas F. Currier and Eleanor M. Tilton, *A Bibliography of Oliver Wendell Holmes* (New York: New York University, 1953); "The 'Holmes' Stereoscope," *Philadelphia Photographer* 6 (1869): 23–25; Oliver Wendell Holmes, "History of the 'American Stereoscope,'" *Philadelphia Photographer* 6 (1869): 1–3; Walter LeConte Stevens, "The Stereoscope: Its History," *Popular Science Monthly* 21 (1882): 37–53; and William and Estelle Marder, *Anthony: The Man, the Company, the Cameras; An American Photographic Pioneer; 140 Year History of a Company from Anthony to Ansco to GAF* (Plantation, Fla.: Pine Ridge Publishing Co., 1982).

28. [Oliver Wendell Holmes], "The Stereoscope and the Stereograph," *Atlantic* 3 (1859): 738–748, at 748.

29. [Holmes], "Sun-Painting and Sun-Sculpture," *Atlantic* 8 (1861): 13–29, at 16. On Anthony's "instantaneous views," see Marder and Marder, *Anthony*, p. 119.

30. [Holmes], "The Stereoscope and the Stereograph," p. 744.

31. Ibid., p. 747.

32. Ibid., p. 748.

33. [Holmes], "Sun-Painting and Sun-Sculpture," p. 28.

34. "The Stereoscope, Pseudoscope, and Solid Daguerreotypes," p. 78.

35. On Kepler, see Crombie, "Mechanistic Hypothesis"; Stephen Straker, "The Eye Made 'Other': Dürer, Kepler and the Mechanization of Light and Vision," in *Science, Technology and Culture in Historical Perspective*, ed. L. A. Knafla, M. Staum, and T.H.E. Travers, University of Calgary Studies in History, no. 1 (Calgary: University of Calgary Press, 1976), pp. 7–25, as well as his "What Is the History of Theories of Perception the History Of?" in *Religion, Science, and Worldview: Essays in Honor of Richard S. Westfall*, ed. M. J. Osler and P. L. Farber (Cambridge: Cambridge University Press, 1985), pp. 245–273; Lindberg, *Theories of Vision*, chap. 9.

36. David Brewster, *The Stereoscope: Its History, Theory, and Construction with Its Application to the Fine and Useful Arts and to Education* (London: J. Murray, 1856), pp. 145–147. Other works by Brewster on his camera include "Description of a Binocular Camera," *Report of the British Association. Transactions of the Sections*, 1849, p. 5; "Account of a Binocular Camera, and of a Method of Obtaining

Drawings of Full Length and Colossal Statues, and of Living Bodies, Which Can Be Exhibited as Solids by the Stereoscope," *Transactions of the Royal Scottish Society of Arts* 3 (1851): 259–264, reprinted in Wade, *Brewster and Wheatstone*, sec. 3.5.

37. [David Brewster], "Binocular Vision and the Stereoscope," *North British Review* 17 (1852): 165–204, at 183.

38. David Brewster, "On the Form of Images Produced by Lenses and Mirrors of Different Sizes," *Report of the British Association. Transactions of the Sections*, 1852, pp. 3–7, at p. 4.

39. [Brewster], "Binocular Vision," p. 183.

40. Brewster, "On the Form of Images," p. 5.

41. Ibid.

42. J. F. Mascher, "On Taking Daguerreotypes without a Camera," *Journal of the Franklin Institute* 59 (1855): 344–347. On Mascher's stereoscopic case, see Darrah, *World of Stereographs*, p. 15.

43. J. F. Mascher, "On Taking Daguerreotypes without a Camera," p. 344.

44. Ibid., pp. 345–346.

45. Ibid., p. 346.

46. Ibid., p. 345. When Mascher became aware of Brewster's researches on the subject, he praised the Scottish physicist and recognized his priority. See Mascher's "On the Cause of Distortions in Photographic Pictures—A Disclaimer," *Journal of the Franklin Institute* 60 (1855): 65–66.

47. Wheatstone, for example, cited Bacon's *Sylva Sylvarum* in his "Contributions" (1838) in Wade, *Brewster and Wheatstone*, p. 78. The key passage reads: "We see more exquisitely with *One Eye Shut*, than with *Both Open*. The *Cause* is, for that the *Spirits Visual* unite themselves more, and so become the Stronger." From Francis Bacon, *Sylva Sylvarum or Natural History*, 7th ed. (London: William Lee, 1658), p. 188 (this page is incorrectly numbered as "184"). Needless to say, neither Wheatstone nor the majority of his contemporaries agreed with Bacon's physiological reasoning.

48. "The Stereoscope," *National Magazine* 12 (1858): 49–54, at 52.

49. [Holmes], "Sun-Painting and Sun-Sculpture," p. 15.

50. Ibid.

51. Joseph LeConte, *Sight: An Exposition of the Principles of Monocular and Binocular Vision*, rev. ed. (1881; New York: D. Appleton, 1897), p. 166.

52. Austin Abbot, "The Eye and the Camera," *Harper's Magazine* 39 (1869): 476–482, at 480, 481.

53. *Anthony's Photographic Bulletin* 1 (1870): cover; also see Marder and Marder, *Anthony*.

54. *Punch* 44 (1863): 249.

55. Gaston Tissandier, *A History and Handbook of Photography*, ed. J. Thomson, 2d rev. ed. (New York: Arno, 1973; reprint of 1878 ed.), following p. 312.

56. [Holmes], "The Stereoscope and the Stereograph," p. 743. By "Mokanna," Holmes was referring to Hakim, who led a revolt in eighth-century Persia. He was called *Al-Mokanna* ("the veiled one") because he was never seen in public without a mask. He allegedly possessed magical powers, especially the ability to cause an intense moonlike light to rise from a pit near Nakshab. *Encyclopedia Britannica*, Twentieth Century Edition, s.v. "Mohammedism" and "Mokanna."

57. "The Stereoscope," *Illustrated London News* 20 (1852): 229–230, at 229.

58. From *Art Journal*, 1858, p. 375, quoted in Earle, *Points of View*, p. 30.

59. [Lady Elizabeth Eastlake], "Photography," *Quarterly Review* 101 (1857): 442–468, at 460.

60. "The Photographic Portrait," *The Crayon* 4 (1857): 154–155, at 155.

61. Ibid.

62. Ibid.

63. [Eastlake], "Photography," p. 461.

64. "The Photographic Portrait," p. 155.

65. This account of Claudet's demonstration is given by David Brewster, in his anonymously published "Photography," *North British Review* 7 (1847): 465–504, at 494.

66. [Eastlake], "Photography," pp. 461–462.

67. John William Draper, "On the Process of Daguerreotype, and Its Application to Taking Portraits from Life," *Philosophical Magazine*, 3d ser., 17 (1840): 217–225, at 225.

68. [Eastlake], "Photography," p. 461.

69. "The Photographic Portrait," p. 155.

70. [Brewster], "Photography," p. 504.

71. [Eastlake], "Photography," p. 453. On collodion, see William Crawford, *The Keepers of Light: A History and Working Guide to Early Photographic Processes* (Dobbs Ferry, N.Y.: Morgan and Morgan, 1979), pp. 7–8, 42.

72. On Muybridge, see *Eadweard Muybridge: The Stanford Years, 1872–1882* (Stanford: Stanford Museum of Art, 1972, rev. ed., 1973); Gordon Hendricks, *Eadweard Muybridge: The Father of the Motion Picture* (New York: Grossman, 1975). On Marey, see François Dagognet, *Étienne-Jules Marey: A Passion for the Trace*, trans. Robert Galeta and Jeanine Herman (New York: Zone Books, 1992); Marta Braun, *Picturing Time: The Work of Étienne-Jules Marey (1830–1904)* (Chicago: University of Chicago Press, 1992); as well as the works cited in chap. 6 above.

73. Charles Baudelaire, "The Modern Public and Photography," pt. 2 of "The Salon of 1859," pp. 149–155 in *Art in Paris, 1845–1862*, trans. and ed. Jonathan Mayne (London: Phaidon, 1965). According to Mayne, the "Salon" was originally published in four installments between June 10 and July 20 in the *Revue française*. It is reprinted in Charles Baudelaire, *Variétés critiques*, 2 vols. (Paris: G. Crès, 1924), 1:111–196. The quotation is from the Mayne edition, pp. 152–153.

74. Baudelaire, *Art in Paris*, p. 153. The translation here differs slightly from Mayne's.

75. See Meyer Howard Abrams, *The Mirror and the Lamp: Romantic Theory and the Critical Tradition* (Oxford: Oxford University Press, 1953); and his *Natural Supernaturalism: Tradition and Revolution in Romantic Literature* (New York: Norton, 1971).

76. John Ruskin, *The Eagle's Nest: Ten Lectures on the Relation of Natural Science to Art*, in *The Complete Works of John Ruskin*, ed. E. T. Cook and Alexander Wedderburn, 39 vols. (London: G. Allen, 1903–1912), 22:194. There is a wealth of secondary literature on Ruskin. In particular, see John D. Rosenberg, *The Darkening Glass: A Portrait of Ruskin's Genius* (New York: Columbia University Press, 1961); Robert Hewison, *John Ruskin: The Argument of the Eye* (Princeton: Princeton University Press, 1976); and Elizabeth K. Helsinger, *Ruskin and the Art of the Beholder* (Cambridge: Harvard University Press, 1982).

77. [Eastlake], "Photography," p. 465. See Baudelaire's similar comment in *Art in Paris*, p. 154.

78. See [Eastlake], "Photography," pp. 461–462; "The Photographic Portrait," p. 155.

79. Brewster, "Account of a Binocular Camera" (1851), in Wade, *Brewster and Wheatstone*, p. 220.

80. Ibid., p. 221.

81. Ibid.

82. Ibid.

83. Latimer Clark, who held such a view, wrote, "I know of no good reason why the natural distance of the eyes, viz. 2½ inches should be much exceeded." His comments appear in his "On an Arrangement for Taking Stereoscopic Pictures with a Single Camera," *Photographic Journal* 1 (1853–1854): 57–59, at 59. The problem of stereoscopic angles was also debated feverishly in *Notes and Queries*, although most correspondents were satisfied with the judgment of T. L. Merritt, who claimed that surpassing the 2½-inch barrier would be "false to nature," or "outraging nature." T. L. Merritt, "Stereoscopic Angles," *Notes and Queries*, 1st ser., 8 (1853): 109–110, at 109.

84. "The Stereoscope, Pseudoscope, and Solid Daguerreotypes," p. 78.

85. Ibid.

86. John Leighton, "Binocular Photographs," *Photographic Journal* 1 (1853–1854): 211–212, at 212.

87. Wheatstone, "Contributions . . . Part the Second," *Phil. Trans. Roy. Soc.* 142 (1852): 1–17, in Wade, *Brewster and Wheatstone*, sec. 2.10, at pp. 157 and 158.

88. H. Helmholtz, "On the Telestereoscope," *Philosophical Magazine*, 4th ser., 15 (1858): 19–24, from *Annalen der Physik* no. 9 (1857). See also Helmholtz's *Physiological Optics* (vol. 1, 1856; vol. 2, 1860; vol. 3, 1866), ed. and trans. James P. C. Southall, 3 vols. (New York: Dover, 1962; 1st English ed., Rochester: Optical Society of America, 1924); 3:310–312. Walter Hardie of Edinburgh independently contrived an essentially identical instrument several years earlier. It is found in his "Description of a New Pseudoscope," *Philosophical Magazine*, 4th ser., 5 (1853): 442–446. Hardie pointed out his priority in "The Telestereoscope," *Philosophical Magazine*, 4th ser., 15 (1858): 156–157.

89. Helmholtz, "On the Telestereoscope," p. 20.

90. Ibid.

91. Ibid., p. 22.

92. Darrah, *World of Stereographs*, p. 147.

93. Warren De la Rue, "Report on the Present State of Celestial Photography in England," *Report of the British Association*, 1859, pp. 130–153, at p. 143. Herschel's comment may be found in Note (I) on Article (473) of his *Outlines of Astronomy* (London: Longmans, Green, and Co., 1875), at p. 700. Also see the letter "The Stereoscopic Angle," *Photographic News* 1 (1858): 110. This letter is signed "J.F.W.H." and it says that the lunar stereograph entailed "a step out of and beyond nature." It also says that the moon appears in the stereograph as it would to a "giant" whose eyes were fifty-two thousand miles apart. On De la Rue, see L. Pearce Williams, "De La Rue, Warren," in *Dictionary of Scientific Biography*, 4:18–19. John Darius, *Beyond Vision* (New York: Oxford University Press, 1984), pp. 28–29.

94. [Holmes], "Sun-Painting," pp. 28–29. Étienne-Jules Marey included the stereoscope among the instruments that could reveal the function of the human sense organs, as well as improve upon them. When one employs stereoscopic photographs taken from widely separated cameras, he wrote, "a village seems like a cluster of

those little toy houses that children play with and one will see them at a very short distance. In an English scientific exposition, I have seen stereoscopic photographs of Saturn with its ring. . . . [I]ts ring was detached with such relief that it seemed one could grab it with one's hand." *La méthode graphique dans les sciences expérimentales* (Paris: G. Masson, 1878; 2d ed., 1885), p. 10n.1.

95. "Stereoscope," in *Chambers's Encyclopaedia*, American Revised Edition (Philadelphia: J. B. Lippincott, 1883), 9:117.

96. Ibid., p. 116.

97. "Binocular Vision," *Edinburgh Review* 108 (1858): 437–473, at 469.

98. Brewster, *The Stereoscope*, p. 147.

99. Ibid.

100. Ibid., pp. 147–148.

101. Ibid., p. 157.

102. David Brewster, *A Treatise on the Kaleidoscope* (Edinburgh: Archibald Constable, 1819). Morrison-Low, *'Martyr of Science,'* pp. 60–62.

103. See Wade, *Brewster and Wheatstone*, sec. 3.1; Richard D. Altick, *The Shows of London* (Cambridge: Harvard University Press, Belknap Press, 1978); Ralph Hyde, *Panoramania!* (London: Trefoil Publications in association with Barbican Art Gallery, 1988); Turner, *Nineteenth-Century Scientific Instruments*.

104. LeConte, *Sight*, p. 151. Like most nineteenth-century investigators of binocular vision, LeConte developed an impressive mastery of his ocular muscles. The skill is currently becoming widespread among those who wish to perceive the stereoscopic quality of computer-generated "Magic Eye" drawings. See N. E. Thing Enterprises, *Magic Eye III—Visions: A New Dimension in Art* (Kansas City: Andrews and McMeel, 1994).

105. Deborah Jean Warner, "The Landscape Mirror and Glass," *Antiques* 105 (1974): 158–159.

106. Deborah Jean Warner, ed., *Pike's Illustrated Catalogue of Scientific and Medical Instruments* (1848; 2d ed., New York: Benjamin Pike, 1856), facsimile of the 2d ed., 2 vols. (Dracut, Mass.: The Antiquarian Scientist; San Francisco: Jeremy Norman, 1984),, vol. 1, supp. p. 32; see also 2:373.

107. Holmes, "History of the 'American Stereoscope,'" p. 3. See also Walter LeConte Stevens, "The Stereoscope," p. 52.

108. Holmes, "History of the 'American Stereoscope,'" p. 3.

109. Brewster, *The Stereoscope*, pp. 205–207.

110. Wheatstone introduced the pseudoscope in his 1852 paper, "Contributions . . . Part the Second," in Wade, *Brewster and Wheatstone*, p. 162.

111. Ibid., 164.

112. Ibid., 165.

113. "The Stereoscope, Pseudoscope, and Solid Daguerreotypes," p. 78.

114. "The Stereoscope," *Illustrated London News* 20 (1852): 229.

115. [Holmes], "The Stereoscope and the Stereograph," p. 748.

116. Ibid.

117. Ibid.

118. Brewster, *The Stereoscope*, p. 164.

119. "Stereoscopic Journeys" (originally in *London Literary Journal*) *Eclectic Magazine* 40 (1857): 560–561, at 560.

120. Ibid.

121. Ibid.

122. A. Claudet, "Photography in Its Relation to the Fine Arts," *Photographic Journal* 6 (June 15, 1860), quoted in Helmut and Alison Gernsheim, *The History of Photography from the Earliest Use of the Camera Obscura in the Eleventh Century up to 1914* (London: Oxford University Press, 1955), p. 191.

123. Charles F. Himes, "Contributions to the Subject of Binocular Vision," *Journal of the Franklin Institute* 62 (1871): 263–270, 340–348, 413–419; and 63 (1872): 141–144; 350–356, at 263.

124. See Harvey Green, " 'Pasteboard Masks': The Stereograph in American Culture 1865–1910," in Earle, *Points of View*, pp. 109–115; Richard N. Masteller, "Western Views in Eastern Parlors: The Contribution of the Stereograph Photographer to the Conquest of the West," *Prospects* 6 (1981): 55–71.

125. Much more could be added concerning the stereoscope in education. On the "visual education" trend, see Paul Saettler, *A History of Instructional Technology* (New York: McGraw-Hill, 1968). The role of the stereoscope is examined more closely in Harold A. Layer, "Stereoscopy: An Analysis of Its History and Its Import to Education and the Communication Process" (Ed.D. diss., Indiana University, 1970). Layer's dissertation includes an impressive bibliography regarding the pedagogical applications of the stereoscope.

126. See Nelson Goodman, *Languages of Art* (1968; 2d ed., Indianapolis: Hackett, 1976). E. H. Gombrich has written several pieces on this theme: *Art and Illusion* (1960; Princeton: Princeton University Press, 1984), esp. chap. 11; "The 'What' and the 'How': Perspective Representation and the Phenomenal World," in *Logic and Art: Essays in Honor of Nelson Goodman*, ed. Richard Rudner and Israel Scheffler (Indianapolis: Bobbs-Merrill, 1972), pp. 129–149; "Mirror and Map: Theories of Pictorial Representation," *Philosophical Transactions of the Royal Society* (B) 270 (1975): 119–149; "Image and Code: Scope and Limits of Conventionalism in Pictorial Representation," in *Image and Code*, ed. Wendy Steiner (Ann Arbor: Michigan Studies in the Humanities, 1981). Joel Snyder has explored photography's relation to modes of picture making: Snyder, "Picturing Vision"; Joel Snyder and Neil Walsh Allen, "Photography, Vision, and Representation," *Critical Inquiry* 2 (1975): 143–169.

127. Alpers, *Art of Describing*, p. 240n.4.

128. Harold I. Brown, "Galileo on the Telescope and the Eye," *Journal of the History of Ideas* 46 (1985): 487–501.

129. Steven Shapin and Simon Schaffer, *Leviathan and the Air-Pump: Hobbes, Boyle, and the Experimental Life* (Princeton: Princeton University Press, 1985); Simon Schaffer, "Glass Works: Newton's Prisms and the Uses of Experiment," in *The Uses of Experiment*, ed. David Gooding, Trevor Pinch, and Simon Schaffer (Cambridge: Cambridge University Press, 1989), pp. 67–104. Other relevant sources may be found in chap. 1 above.

Chapter Eight
Vox Mechanica: The History of Speaking Machines

1. The most famous utterance of the head of Orpheus was its prediction of the untimely death of Cyrus the Great, which ended his campaign against the Scythians. David Brewster, *Letters on Natural Magic Addressed to Sir Walter Scott, Bart.* (London: J. Murray, 1832), esp. letters 1 and 7.

2. See W. Niemann, "Sprechenden Figuren. Ein Beitrag zur Vorgeschichte der Phonographen," *Geschichtsblätter für Technik und Industrie* 7 (1920): 2–30; J. Voskuil, "The Speaking Machine through the Ages," *Transactions of the Newcomen Society* 26 (1947–1949): 259–267.

3. Francis Bacon, *The Advancement of Learning and New Atlantis*, ed. Arthur Johnston (Oxford: Clarendon Press, 1974), p. 239.

4. Ibid., pp. 239–244.

5. Ibid., p. 244.

6. Athanasius Kircher, *Musurgia universalis*, bk. 9, quoted in Niemann, "Sprechenden Figuren," pp. 14–15.

7. Gaston Tissandier, *Popular Scientific Recreations* (London: Ward, Lock, 1883), pp. 135–137.

8. On Wilkins, see Barbara J. Shapiro, *John Wilkins, 1614–1672: An Intellectual Biography* (Berkeley and Los Angeles: University of California Press, 1969). John Wilkins, *An Essay Towards a Real Character, and a Philosophical Language* (1668; Menston, England: Scolar, 1968).

9. John Wilkins, *Mathematicall Magick. Or, the Wonders that may be Performed by Mechanicall Geometry* (London: Printed by M. F. for S. Gellibrand, 1648), p. 176.

10. Ibid., p. 177.

11. Ibid. The legend of the frozen words may be found in Rabelais's *Pantagruel*, chaps. 55 and 56.

12. Wilkins, *Mathematicall Magick*, pp. 177–178. The argument that the shape of the letters of the Hebrew alphabet provides a phonetic description of the position of the tongue was proposed by Franciscus Mercurius van Helmont in his *Alphabeti vere naturalis Hebraica Brevissima Delineato* of 1667. Although his thesis claimed that Hebrew was a "natural" language for humans, it is nearly impossible to contort the tongue into the positions he prescribed.

13. John Evelyn, *The Diary of John Evelyn*, ed. E. S. de Beer, 6 vols. (Oxford: Clarendon Press, 1955), 3:110.

14. Shapiro, *John Wilkins*, pp. 120–121.

15. J. G. Crowther, *Founders of British Science* (London: Crescent, 1960), p. 144.

16. Penelope Gouk, "The Role of Acoustics and Music Theory in the Scientific Work of Robert Hooke," *Annals of Science* 37 (1980): 573–605.

17. Robert Hooke, *The Diary of Robert Hooke, 1672–1680*, ed. Henry W. Robinson and Walter Adams (London: Taylor and Francis, 1935), p. 223; Thomas Birch, *The History of the Royal Society of London*, 4 vols. (London: A. Millar, 1756–1757), 4:96. Savart used a rotating toothed wheel in his studies on the limits of human audibility: "Notes sur la sensibilité de l'organe de l'ouïe," *Annales de chimie* 44 (1830): 337–352, and "Notes sur la limite de la perception des sons graves," *Annales de chimie* 47 (1831): 69–74.

18. Richard Waller, "The Life of Dr. Robert Hooke," in *The Posthumous Works of Robert Hooke* (London: S. Smith and B. Walford, 1705), p. xxiii.

19. Shapiro, *John Wilkins*, p. 120.

20. Salomon de Caus, *Les raisons des forces mouvantes avec diverses machines tant utiles que plaisantes* (1615; Amsterdam: Frits Knuf, 1973). On automata see Alfred Chapuis and Édouard Gélis, *Le monde des automates*, 2 vols. (Paris, 1928);

Alfred Chapuis and Edmond Droz, *Automata: A Historical and Technical Study*, trans. Alec Reid (Neuchâtel: Éditions du Griffon, 1958); Silvio A. Bedini, "The Role of Automata in the History of Technology," *Technology and Culture* 5 (1965): 24–42; Derek J. de Solla Price, "Automata and the Origins of Mechanism and Mechanistic Philosophy," *Technology and Culture* 5 (1964): 9–24.

21. André Doyon and Lucien Liaigre, *Jacques Vaucanson, mécanicien de génie* (Paris: Presses Universitaires de France, 1966), chap. 5. Also see Doyon and Liaigre, "Méthodologie comparée du biomécanisme et de la mécanique comparée," *Dialectica* 40 (1956): 292–323; and David M. Fryer and John C. Marshall, "The Motives of Jacques de Vaucanson," *Technology and Culture* 20 (1979): 257–269.

22. Doyon and Liaigre, *Vaucanson*, p. 110.

23. Bacon, *The Advancement of Learning and New Atlantis*, p. 245.

24. René Descartes, *Treatise of Man*, French text with trans. by Thomas Steele Hall (Cambridge: Harvard University Press, 1972), pp. 2n, 3n, 4, 22, and 113. Also see Price's essay ("Automata"), which claims that automata influenced the development of the mechanical philosophy.

25. Doyon and Liaigre, *Vaucanson*, pp. 117–118.

26. Ibid., p. 118.

27. Ibid., p. 18.

28. Ibid., p. 148.

29. Théodore Vetter, "Le Cat, Claude-Nicolas," in *Dictionary of Scientific Biography*, 8:114–116, at 115. See Doyon and Liaigre, *Vaucanson*, pp. 121–123, and p. 149. The principal works on bleeding include J.-B. Silva's *Traité de l'usage des différentes sortes de saignées* (1727), Le Cat's *Traité de la saignée* (1729), Quesnay's *Observations sur les effets de la saignée* (1730), and his *Traité des effets et de l'usage de la saignée* (1750).

30. Doyon and Liaigre, *Vaucanson*, p. 138.

31. Ibid., pp. 150–161.

32. Ibid., pp. 118, 163.

33. Ibid., p. 150.

34. Ibid., p. 162.

35. Ibid., p. 148; Julien Offray de La Mettrie, *L'homme machine* (1748), in his *Oeuvres philosophiques*, 2 vols., new ed. (Berlin, 1774), 1:345.

36. Jean Blanchet, *L'art du chant* (Paris: Dessaint et Saillant, 1755), pp. 47–48; Doyon and Liaigre, *Vaucanson*, pp. 166–167.

37. M. D. Grmek, "Dodart, Denis," in *Dictionary of Scientific Biography*, 4:135–136. Dodart, "Mémoire sur les causes de la voix de l'homme, et de ses différens tons," *Mémoires de l'Académie royale des sciences*, 1700, pp. 238–287. Grmek, "Ferrein, Antoine," in *Dictionary of Scientific Biography*, 4:589–590. Ferrein, "De la formation de la voix de l'homme," *Mémoires de l'Académie royale des sciences*, 1741, pp. 409–432.

38. H.-J.-B. Montagnat, *Eclaircissement en forme de lettre à M. Bertin, médecin, sur la découverte que M. Ferrein a faite du mécanisme de la voix de l'homme par M. Montagnat, médecin* (1746), cited in Doyon and Liaigre, *Vaucanson*, p. 164.

39. Doyon and Liaigre, *Vaucanson*, p. 173.

40. An essay by Peter Galison and Alexi Assmus on C.T.R. Wilson's artificial clouds that led to his development of the cloud chamber shows how imitation—what the authors call "mimetic science"—has been an enduring, if often unrecognized, theme in the history of science. Peter Galison and Alexi Assmus, "Artificial Clouds,

Real Particles," in *The Uses of Experiment: Studies in the Natural Sciences*, ed. David Gooding, Trevor Pinch, and Simon Schaffer (Cambridge: Cambridge University Press, 1989), pp. 225–274. Also see W. D. Hackmann, "The Relationship between Concept and Instrument Design in Eighteenth-Century Experimental Science," *Annals of Science* 26 (1979): 205–224. Georges Canguilhem has pointed out that this has been especially true in the history of biology. Georges Canguilhem, "The Role of Analogies and Models in Biological Discovery," in *Scientific Change*, ed. A. C. Crombie (New York: Basic Books, 1963), pp. 507–520.

41. Étienne-Jules Marey, *La circulation du sang à l'état physiologique et dans les maladies* (Paris: G. Masson, 1881). Marey discussed artificial wings of birds and insects in his *Animal Mechanism* (New York: Appleton, 1873).

42. Chapuis and Gélis, *Le monde des automates*, 2:202; "Mical," in *Biographie universelle, ancienne et moderne*, 52 vols. (Paris: Michaud, 1811–1828), 28:517–519. On entertainment in France during this era, see Robert M. Isherwood, *Farce and Fantasy: Popular Entertainment in Eighteenth-Century Paris* (New York: Oxford University Press, 1986).

43. Félix Vicq d'Azyr's report, cited in Chapuis and Gélis, *Le monde des automates*, p. 204.

44. Antoine Rivarol, "Lettre à M. le Président de ***, Sur le Globe aérostatique, sur les Têtes-parlantes, et sur l'état présent de l'opinion publique à Paris," in *Oeuvres complètes*, 5 vols. (1818; Geneva: Slatkine Reprints, 1968), 2:207–246, at p. 230. His "De l'universalité de la langue française" is in the same volume, and the talking heads are discussed in the notes on pp. 91–94.

45. Rivarol, "Lettre à M. le Président de ***," pp. 236–237.

46. Chapuis and Gélis, *Le monde des automates*, 2:206.

47. Rivarol, "De l'universalité," p. 91; "Lettre à M. le Président de ***," pp. 230–231.

48. *Acta Academiae Scientiarum Imperialis Petropolitanae*, pt. 2, *Histoire*, 1780, pp. 9–10. A summary of Kratzenstein's essay is contained on pp. 13–15 of the *Acta*, and the essay is published in its entirety as "Sur la naissance & la formation des voyelles," *Journal de Physique*, 1782, pp. 358–380. On Kratzenstein, see E. Snorrason, *C. G. Kratzenstein: professor physices experimentalis Petropol. et Havn. and His Studies on Electricity during the Eighteenth Century*, vol. 29 (1974) of *Acta Historica Scientiarum Naturalium et Medicinalium*.

49. Leonhard Euler, "The Wonders of the Human Voice," in *Letters of Euler on Different Subjects in Natural Philosophy Addressed to a German Princess*, 3 vols. (1768, 1772), trans. Henry Hunter, notes by David Brewster, reprint of 1833 ed., 2 vols. in 1 (New York: Arno, 1975), 2:76–79, at p. 78.

50. Ibid., p. 79.

51. Kratzenstein, "Sur la naissance," p. 361.

52. Ibid.

53. Jamie Croy Kassler, "Man—A Musical Instrument: Models of the Brain and Mental Functioning before the Computer," *History of Science* 22 (1984): 59–92.

54. Kratzenstein, "Sur la naissance," p. 362.

55. Ibid., p. 363.

56. Ibid., p. 365.

57. Ibid., pp. 367, 368.

58. Ibid., p. 367.

59. Ibid., pp. 369–379, 373–376.

60. Ibid., pp. 379–380.

61. On Kempelen and his chess player, see Charles Michael Carroll, *The Great Chess Automaton* (New York: Dover, 1975). This book details the career of the chess player, both during Kempelen's time and afterward. It also recounts the many attempts to debunk the automaton, one of which was written by Edgar Allan Poe. (See W. K. Wimsatt, Jr., "Poe and the Chess Automaton," *American Literature* 11 [1939]: 138–151; Henry Ridgely Evans, *Edgar Allan Poe and Baron von Kempelen's Chess-Playing Automaton* [Kenton, Ohio: International Brotherhood of Magicians, 1939]. Poe's essay, "Maelzel's Chess Player," was originally published in the *Southern Literary Messenger* 2 [1836]: 318–326.) Charles Gottlieb de Windisch, *Lettres sur le joueur d'échecs de M. de Kempelen*, trans. Chrétien de Mechel (Basel: Mechel, 1783)—this book was originally published in German and was also translated into English with the title *Inanimate Reason*; all citations here will be from the French edition. Windisch's account also has a letter on Kempelen's speaking machine. The story of the chess player was told cinematically in the 1930 French film *Le joueur d'échecs*.

62. Windisch, *Lettres sur le joueur d'échecs*, p. 41; the translation given here is from Carroll, *The Great Chess Automaton*, p. x.

63. The *Monthly Review* for 1784 reported that "many were simple enough to affirm, both in conversation and in print, that the little wooden man played *really* and *by himself*, (like certain politicians at a deeper game) without any communication with his *Constituent*. It appears indeed, as yet, unaccountable to the spectators, how the artist imparts his influence to the automaton at the time of his playing, and all the hypotheses, which have been invented, by ingenious and learned men, to unfold this mystery, are but vague and inadequate; but were they even otherwise, they rather increase than diminish the admiration that is due to the surprising talents and dexterity of Mr. de Kempelen." "Inanimate Reason," *Monthly Review*, 1st ser., 70 (1784): 307–308.

64. Wolfgang von Kempelen, *Le mécanisme de la parole, suivi de la description d'une machine parlante* and *Mechanismus der menschlichen Sprache nebst der Beschreibung seiner sprechenden Maschine*, both versions (Vienna: J. V. Degen, 1791). The German edition has been reprinted and includes biographical and bibliographical material (Stuttgart-Bad Cannstatt: Friedrich Frommann, 1970). All quotations below will be given from the French edition. On Kempelen, see Homer Dudley and T. H. Tarnoczy, "The Speaking Machine of Wolfgang von Kempelen," *Journal of the Acoustical Society of America* 22 (1950): 151–166; Bolla Kálálman, ed., "In Memoriam Farkas Kempelen: Articles Written in Commemoration of the 250th Anniversary of Farkas Kempelen's Birth," *Magyar Fonetikai Füzetek—Hungarian Papers in Phonetics* 13 (1984)—this volume is in Hungarian with English summaries. Kempelen's given name was "Farkas," although he adopted a moniker more fitting for his position in Vienna.

65. See Allan Dickson Megill, "The Enlightenment Debate on the Origin of Language and Its Historical Background" (Ph.D. diss., Columbia University, 1975).

66. Kempelen, *Le mécanisme de la parole*, chap. 2.

67. Ibid., pp. 395–396.

68. Ibid., p. 396.

69. Ibid., pp. 399–400.

70. Ibid., pp. 400–401.

71. Ibid., pp. 401, 402.

72. Ibid., p. 402.
73. Ibid., p. 403.
74. Ibid.
75. Ibid., p. 259.
76. Ibid., p. 404.
77. Ibid., pp. 403–404.
78. Ibid., p. 394.
79. Ibid., pp. 413–414.
80. Quoted in Chapuis and Gélis, *Le monde des automates*, 2:208.
81. Windisch, *Lettres sur le joueur d'échecs*, p. 48.
82. *Correspondance littéraire, philosophique et critique, par Grimm, Diderot, Raynal, Meister, etc.*, ed. Maurice Tourneux, 16 vols. (Paris: Garnier, 1877–1882), 13:358, 359 (entry for September 1783). Grimm seems to have based his account of the chess player on Windisch's book; see the similarity to Windisch, *Lettres sur le joueur d'échecs*, p. 38. Windisch, however, did not mention Mical's heads.
83. Rivarol, "Lettre à M. le Président de ***," p. 243.
84. See Simon Schaffer, "Natural Philosophy and Public Spectacle in the Eighteenth Century," *History of Science* 21 (1983): 1–43.
85. John R. Millburn, *Benjamin Martin: Author, Instrument Maker, and "Country Showman"* (Leyden: Noordhoff, 1976); J. A. Leo Lemay, *Ebenezer Kinnersley: Franklin's Friend* (Philadelphia: University of Pennsylvania Press, 1964).
86. *The Works of Benjamin Franklin*, ed. Jared Sparks, rev. ed., 10 vols. (Philadelphia: Childs & Peterson, 1840), 10:23–24.
87. *The Letters of Erasmus Darwin*, ed. Desmond King-Hele (Cambridge: Cambridge University Press, 1981), p. 63. Of course, Franklin would not have known about Kempelen's efforts by 1772. On Erasmus Darwin, see Desmond King-Hele, *Erasmus Darwin* (New York: Scribner, 1963) and *Doctor of Revolution: The Life and Genius of Erasmus Darwin* (London: Faber & Faber, 1977); and Robert E. Schofield, *The Lunar Society of Birmingham: A Social History of Provincial Science in Eighteenth-Century England* (Oxford: Clarendon Press, 1963).
88. Darwin, *Letters*, p. 64.
89. The "first language," he explained, consisted of facial expressions and other superficial marks of sentiment. Spoken language, however, arose from the linkage of these ideas with specific vocal sounds. Erasmus Darwin, *The Temple of Nature; or, The Origin of Society* (London: J. Johnson, 1803), canto 3, lines 335–356, pp. 112–113; and lines 363–370, p. 114.
90. Ibid., additional note 15, "Analysis of Articulate Sounds," p. 109.
91. Ibid.
92. Ibid., additional note 6, "Hieroglyphic Characters."
93. Ibid., "Analysis," pp. 119–120.
94. See Gouk, "The Role of Acoustics." She notes that Hooke's toothed wheel, as well as his attempts to exhibit the behavior of vibrating surfaces by covering them with flour, were the very topics taken up much later by Chladni and Savart.
95. Félix Savart, "Mémoires sur la voix humaine," *Annales de chimie*, 2d ser., 30 (1825): 64–87.
96. Charles Cagniard de la Tour, "Voix humaine," *L'Institut* 5, no. 212 (1837): 179–180.
97. Johannes Müller, *Elements of Physiology*, trans. William Baly, 2d ed., 2 vols. (London: Taylor and Walton, 1839), vol. 2, sec. 3.

98. Édouard Fournié, *Physiologie de la voix et de la parole* (Paris: Adrien Delhaye, 1866), pp. 392–398. This book contains a lengthy discussion of the history of vocal physiology.

99. Ernst Florens Friedrich Chladni, *Traité d'acoustique* (Paris: Courcier, 1809), pp. 67–71.

100. Jean-Baptiste Biot, *Traité de physique expérimentale et mathématique*, 4 vols. (Paris: Deterville, 1816), 2:195.

101. Hermann L. F. von Helmholtz, *On the Sensations of Tone as a Physiological Basis for the Theory of Music* (*Die Lehre von den Tonempfindungen als physiologische Grundlage für die Theorie der Musik* [1863]), trans. A. J. Ellis, 2d English ed., based on the 4th German ed. of 1877 (1885; New York: Dover, 1954), p. 103.

102. John William Strutt, 3d Baron Rayleigh, *The Theory of Sound* (vol. 1, 1877, and vol. 2, 1878; 2d ed., vol. 1, 1894, and vol. 2, 1896), 2d rev. ed. (New York: Dover, 1945), 2:470.

103. Robert Willis, "On the Vowel Sounds, and on Reed Organ-Pipes," read Nov. 24, 1828, and March 16, 1829; published in *Transactions of the Cambridge Philosophical Society* 3 (1830): 231–268.

104. Ibid., p. 232.

105. Ibid.

106. Ibid., p. 231.

107. Ibid., p. 233. Although Willis believed that vowel sounds were not the results of physiology exclusively, he also treated physiological matters elsewhere: Willis, "On the Mechanism of the Larynx," *Transactions of the Cambridge Philosophical Society* 4 (1833): 323–352. Many writers had discussed the perfection of the human voice and its ability to mimic the sounds of all animals. See Euler, "The Wonders of the Human Voice," p. 77–78; A. Richerand, *Elements of Physiology*, trans. G.L.M. De Lys, 3d ed. (London: Thomas and George Underwood, 1819), p. 426; Charles Bell, "Of the Organs of the Human Voice," *Philosophical Transactions of the Royal Society*, 1832, pp. 299–320, at pp. 299–300; John Bishop, "Experimental Researches into the Physiology of the Human Voice," *Philosophical Magazine*, 3d ser., 9 (1836): 201–209, 269–277, and 342–348, at 343–344.

108. Willis, "On the Vowel Sounds," pp. 233.

109. Ibid., pp. 235–236.

110. Ibid., p. 244.

111. Ibid., pp. 249–250.

112. Ibid., p. 243.

113. William Whewell, *The Philosophy of the Inductive Sciences, Founded upon their History*, intro. John Herivel, 2d ed., 2 vols., facsimile of the 1847 ed. (New York: Johnson, 1967), 1:343–344.

114. Willis, "On the Vowel Sounds," 261–262.

115. Charles Wheatstone, "Instructions for the Employment of Wheatstone's Cryptograph," in *The Scientific Papers of Sir Charles Wheatstone* (London: Taylor and Francis, 1879), pp. 342–347. He also wrote an essay, "Interpretation of an Important Historical Document in Cipher," in *Papers*, pp. 321–341. Wheatstone thought ciphers would help preserve the privacy of telegraphic transmissions. See Brian Bowers, *Sir Charles Wheatstone. F.R.S., 1802–1875* (London: Her Majesty's Stationery Office, 1975), pp. 181–184.

116. Bowers, *Sir Charles Wheatstone*, pp. 7–8.

117. He presented his version of Kempelen's machine to the British Association in 1835. Wheatstone, "On the Transmission of Musical Sounds through Solid Linear Conductors, and on Their Subsequent Reciprocation," *Journal of the Royal Institution* 2 (1831), reprinted in his *Papers*, pp. 47–63, at pp. 62–63.

118. Charles Wheatstone, "On the Various Attempts Which Have Been Made to Imitate Human Speech by Mechanical Means," *Report of the British Association. Transactions of the Sections*, 1835, p. 14. Wheatstone's essay "Reed Organ-Pipes, Speaking Machines, etc." appeared in the *London and Westminster Review* (1837) and was signed "C.W."; it is reprinted in his *Papers*, pp. 348–367.

119. Wheatstone, "Reed Organ-Pipes, Speaking Machines," pp. 360–364.

120. Ibid., p. 367. Brewster's comment may be found in his *Natural Magic*, on p. 211.

121. Wheatstone, "Reed Organ-Pipes, Speaking Machines," pp. 358–360.

122. Concerning the study of timbre in the nineteenth century, as well as the historiography of acoustics, see Robert J. Silverman, "Instrumentation, Representation, and Perception in Modern Science: Imitating Human Function in the Nineteenth Century" (Ph.D. diss., University of Washington, 1992), chap. 3, "The Torch of Acoustics: Instrumentation, Representation, and the Science of Sound in the Nineteenth Century."

123. Although it was also articulated by Georg Simon Ohm, this acoustical principle should not be confused with the more familiar "Ohm's law" in electromagnetism. See V. Carlton Maley, Jr., *The Theory of Beats and Combination Tones, 1700–1863* (New York: Garland, 1990), originally a Ph.D. diss. (Harvard University, 1967), published by Harvard Dissertations on the History of Science; R. Steven Turner, "The Ohm-Seebeck Dispute, Hermann von Helmholtz, and the Origins of Physiological Acoustics," *British Journal for the History of Science* 10 (1977): 1–24; and Stephan Vogel, "Sensation of Tone, Perception of Sound, and Empiricism: Helmholtz's Physiological Acoustics," in *Hermann von Helmholtz and the Foundations of Nineteenth-Century Science*, ed. David Cahan (Berkeley and Los Angeles: University of California Press, 1993), pp. 259–287.

124. Helmholtz, *Sensations*, p. 61.

125. Helmholtz claimed that the arches of Corti on the cochlea behave like the strings of the undamped piano discussed above (ibid., pp. 128–148). Stephan Vogel characterizes Helmholtz's tuning-fork apparatus as "Ohm's definition of tone cast, as it were, into wood and brass" (Vogel, "Sensation of Tone, Perception of Sound, and Empiricism," p. 274). According to Timothy Lenoir, Helmholtz's theory of auditory perception and his tuning-fork synthesizer provided critical analogues for his theory of visual perception ("Helmholtz and the Materialities of Communication," *Osiris*, 2d ser., 9 [1993]: 185–207).

126. Helmholtz, *Sensations*, p. 109. Rudolph König disagreed with Helmholtz's and Donders's measurements of characteristic pitches. See König, "Sur les notes fixes caractéristiques des diverses voyelles," *Comptes rendus, Académie des sciences, Paris* 70 (1870): 931–933, reprinted in his *Quelques expériences d'acoustique* (Paris, 1882).

127. Helmholtz also used his famous resonators placed in the ear to hear the partial tones and to confirm his estimates of the characteristic pitches (Helmholtz, *Sensations*, p. 110). The determination of the precise strengths of partials for the tuning-fork apparatus was largely empirical (p. 123).

128. Ibid., pp. 127–128.

129. Strutt, *Theory of Sound*, 2:477.

130. D. C. Miller, *The Science of Musical Sounds*, 2d ed. (1916; New York: Macmillan, 1922), p. 246.

131. Strutt, *Theory of Sound*, 2:472.

132. Ibid., p. 474.

133. The earliest ones were conducted by Fleeming Jenkin and J. A. Ewing, in 1878. Fleeming Jenkin and J. A. Ewing, "On the Harmonic Analysis of Certain Vowel Sounds," *Transactions of the Royal Society of Edinburgh* 28 (1876–1878): 745–777. See also Charles R. Cross, "Helmholtz's Vowel Theory and the Phonograph," *Nature* 18 (1878): 93–94; Alexander Graham Bell, "Vowel Theories," *American Journal of Otology* 1 (1879), reprinted in his *The Mechanism of Speech*, 5th ed. (1906; New York: Funk and Wagnalls, 1911), pp. 117–129; Charles R. Cross and George V. Wendell, "On Some Experiments with the Phonograph, Relating to the Vowel Theory of Helmholtz," *Proceedings of the American Academy of Arts and Sciences* 27 (1891–1892): 271–279; and E. W. Scripture, "On the Nature of Vowels," *American Journal of Science* 161 (1901): 302–309.

134. James F. Curtis, "The Rise of Experimental Phonetics," in *History of Speech Education in America*, ed. Karl R. Wallace (New York: Appleton-Century-Crofts, 1954), pp. 348–369, at pp. 355–356.

135. John Q. Stewart, "An Electrical Analogue of the Vocal Organs," *Nature* 110 (1922): 311–312, at 312.

136. William Henry Preece and Augustus Stroh, "Studies in Acoustics. I. On the Synthetic Examination of Vowel Sounds," *Proceedings of the Royal Society of London* 28 (1878–1879): 358–366, at 363.

137. Ibid.

138. Ibid., p. 364.

139. Ellis's comments appear in the translator's additions to Helmholtz's *Sensations*, app. 20, sec. M, p. 543.

140. The disparity between the acoustic theories of Helmholtz and König (principally, their disagreement on the significance of the phase differences among partial tones, in their estimates of timbre), as well as the instruments they devised consequently, is treated in detail in Silverman, "Instrumentation, Representation, and Perception," chap. 3, pt. 3. König's most important publication on the topic was his "Remarques sur le timbre," which is found in his collection of essays, *Quelques expériences d'acoustique*. The essay was originally published in the *Annalen der Physik* in 1881.

141. Alfred Eichorn, "Die Vocalsirene, eine neue Methode der Nachahmung von Vocalklängen," *Annalen der Physik* 39 (1890): 148–154.

142. Rudolph König, "Die Wellensirene," *Annalen der Physik* 57 (1896): 339–388, at 382–386. D. C. Miller criticized the ability of the wave siren to reproduce the tones corresponding to the waveforms cut on its bands (*The Science of Musical Sounds*, pp. 244–245).

143. Alexander John Ellis, *The Alphabet of Nature* (London: S. Bagster; Bath: Isaac Pitman, Phonographic Institution, 1845); Ellis, *Essentials of Phonetics* (London: Pitman, 1848). Ellis's 1848 *Essentials of Phonetics* was written entirely in the typefaces of his universal alphabet.

144. Ellis, *Alphabet*, p. 25.

145. Ibid.

146. Richard Potter, "On the English Sounds of the Vowel-Letters of the Alphabet, on Their Production by Instruments, and on the Natural Musical Sequence of the Vowel-Sounds," read April 28, 1873, published in *Proceedings of the Cambridge Philosophical Society* 2 (1874–1876): 306–308, at 306.

147. Richard John Lloyd, "Speech Sounds: Their Nature and Causation," *Phonetische Studien* 3 (1890): 251–278, at 274.

148. Ibid., pp. 274–275.

149. Ibid., p. 275.

150. Richard Paget, *Human Speech* (New York: Harcourt, Brace; London: Kegan Paul, Trench, Trubner, 1930).

151. Georges René Marie Marage, "Théorie de la formation des voyelles," *Séances de la Société Française de Physique*, 1900, pp. 109–147, at p. 112. Also see his *Petit manuel de physiologie de la voix à l'usage des chanteurs et des orateurs* (1911; Tours: Deslis, n.d.).

152. E. W. Scripture, "Report on the Construction of a Vowel Organ," *Smithsonian Miscellaneous Collections* 47 (1905): 360–364.

153. Miller, *The Science of Musical Sounds*, p. 244.

154. Chapuis and Gélis, *Le monde des automates*, p. 210; Chapuis and Droz, *Automata*, pp. 324–325.

155. Carroll, *The Great Chess Automaton*, p. 12.

156. Richard Altick, *The Shows of London* (Cambridge: Harvard University Press, Belknap Press, 1978), is an outstanding source for information about the amusements displayed in London in the eighteenth and nineteenth centuries, including speaking figures. Also see the "talking head" in Tissandier's *Popular Scientific Recreations*, pp. 135–137. Niemann also briefly mentions a speaking machine built in 1806 by a Berlin sculptor named Posch, who seems to have made it as an amusement ("Sprechenden Figuren," pp. 27–28). Chapuis and Gélis mention Posch also (*Le monde des automates*, 2:209) as well as a speaking head made by Robertson (Étienne Gaspard Robert; see chap. 3) who was to have exhibited it in London along with his mechanical trumpet player, in 1828 (*Le monde des automates*, 2:210).

157. William Wordsworth, *Prelude*, bk. 7, lines 706–721. The "invisible girl" was a popular trick in which the voice of a girl would be heard by spectators, but the speaker could not be seen. She would be in an adjoining chamber and her voice conducted through a speaking tube. See Altick, *Shows of London*, p. 353.

158. On Faber, see Niemann, "Sprechenden Figuren," pp. 28–30; Altick, *Shows of London*, pp. 353–356. A picture of Faber and his machine is given in "The Euphonia," *Illustrated London News* 9 (1846): 96. The appearance of the Euphonia does not seem to have been an imitation of Kempelen's chess player. Altick explains that the automaton magicians customarily wore Turkish attire (*Shows of London*, p. 68). However, Faber may have been influenced indirectly by Kempelen's machine, via the "Talking Turk" in E.T.A. Hoffmann's 1814 story "Die Automate." In his introduction to *The Best Tales of Hoffmann* (New York: Dover, 1967), E. F. Bleiler suggests that this fictional automaton may have been inspired by the chess player, which was being exhibited in Europe by Maelzel at the time Hoffmann was writing (pp. xxi–xxii).

159. "Our Weekly Gossip," *Athenaeum*, 1846, p. 765, described a performance by the Euphonia: "When we entered the room, we found him singing to a select

society; and we believe any portion of the gratification which the latter experienced was not derived from the beauty of the voice."

160. "The Speaking Machine," *Punch* 11 (1846): 83.

161. Ibid.

162. John Hollingshead, *My Lifetime*, 2 vols. (London: Sampson, Low, Marston, 1895), 1:68–69.

163. P. T. Barnum, *Struggles and Triumphs; or, The Life of P. T. Barnum, Written by Himself*, ed. George S. Bryan, 2 vols. (New York: Knopf, 1927), 1:400. The new "Faber" looked to Joseph Henry and the Smithsonian for financial assistance in 1871 and received five hundred dollars in aid from Alexander Graham Bell in 1885. See Robert V. Bruce, *Bell: Alexander Graham Bell and the Conquest of Solitude* (1973; Ithaca: Cornell University Press, 1990), pp. 82, 295. Contrarily, but without a great deal of evidence, A. H. Saxon's biography rated the Euphonia as a splendid money maker. Saxon, *P. T. Barnum: The Legend and the Man* (New York: Columbia University Press, 1989), pp. 150–151.

164. Cited in Chapuis and Gélis, *Le monde des automates*, p. 206.

165. "Faber's Sprechmaschine," *Annalen der Physik* 58 (1843): 175–176; Gariel, "Machine parlante de M. Faber," *Journal de physique* 8 (1879): 274–275. Faber's machine is also noted in Alfred Marshall Mayer's *Sound: A Series of Simple, Entertaining, and Inexpensive Experiments in the Phenomena of Sound for the Use of Students of Every Age* (London: Macmillan, 1891), pp. 147–148. Mayer also made a talking machine that said "mama" from an orange and a toy trumpet. He adorned it with a bonnet, a peanut nose, and black bean eyes (pp. 146–147).

166. [Oliver Wendell Holmes], "The Great Instrument," *Atlantic* 12 (1863): 637–647.

167. Ibid., p. 639.

168. Ibid., p. 646. Holmes cites a Dr. Burney's estimate of the pipes: "As to the *vox humana*, which is so celebrated, it does not at all resemble a human voice, though a very good stop of the kind; but the world is very apt to be imposed upon by names; the instant a common hearer is told that an organist is playing upon a stop which resembles the human voice, he supposes it to be very fine, and never inquires into the propriety of the name, or exactness of the imitation. However, with respect to our own feelings, we must confess, that, of all the stops which we have yet heard, that have been honored with the appellation of *vox humana*, no one in the treble part has ever reminded us of anything human, so much as the cracked voice of an old woman of ninety, or, in the lower parts, of Punch singing through a comb."

George Ashdown Audsley's *The Art of Organ Building*, 2 vols. (1905; New York: Dover, 1965) gives this description: "The stop is voiced with the view of producing tones in imitation of the human voice; hence its name: but even the best results that have hitherto been obtained fall far short of what is to be desired. . . . The tonal and imitative effects of the Vox Humana depend to a large extent on the position it occupies, and the manner in which it is treated in an Organ; and its imitative quality is also greatly affected by the acoustical properties of the building in which the Organ is located. Of all the stops of the Organ, the Vox Humana is the one to which distance lends the greatest charm" (1:574).

169. Paget, *Human Speech*, pp. 233–239.

170. Bruce, *Bell*, pp. 4–5.

171. Ibid., p. 5.

172. Ibid.

173. Alexander Graham Bell, "Prehistoric Telephone Days," *National Geographic* 41 (1922): 223–241, at 223.

174. Ibid., p. 229. Also see Bruce, *Bell*, pp. 110–124. Bruce says that Bell worked on an "ear phonautograph" with an ear obtained for him by Clarence Blake.

175. Bruce, *Bell*, pp. 50–51; Bell, "Prehistoric," pp. 231–233.

176. Bell saw the Euphonia again, in Boston in 1871—this time operated by the new "Faber" (Bruce, *Bell*, pp. 81–82). Neither occurrence is mentioned in his "Prehistoric Telephone Days."

177. Bell, "Prehistoric," pp. 233–235.

178. Ibid., pp. 235–237; Bruce, *Bell*, pp. 36–37.

179. Bruce, *Bell*, pt. 2. Also see Bernard S. Finn, "Alexander Graham Bell's Experiments with the Variable-Resistance Transmitter," *Smithsonian Journal of History* 1 (1966): 1–16.

180. Living things reproduce themselves, but can artificial instruments reproduce themselves? Can a human duplicate himself or herself by art, or only by nature? These are questions worthy of Aristotle. John von Neumann showed mathematically that self-replicating automata are possible. John von Neumann, *Theory of Self-Reproducing Automata*, edited and completed by Arthur Banks (Urbana: University of Illinois Press, 1966); John G. Kemeny, "Man Viewed as a Machine," *Scientific American* 192 (1955): 58; and Freeman Dyson, *Disturbing the Universe* (New York: Harper and Row, 1979), chap. 18, "Thought Experiments."

Chapter Nine
Conclusion

1. H. Helmholtz, "On the Interaction of Natural Forces," in *Popular Lectures on Scientific Subjects*, trans. E. Atkinson (New York: Appleton, 1885), 153–193, at p. 154.

2. Ronald Dworkin has taken this approach with regard to "hard cases" in the law. He argues that the resolution of such cases involves more than redefining the edge of applicability of a legal rule; rather, these impasses lead judges and scholars to reexamine core principles of jurisprudence. Ronald Dworkin, *Law's Empire* (Cambridge: Harvard University Press, Belknap Press, 1986), esp. pp. 41–43.

3. "Experience" (1842), in Ralph Waldo Emerson, *The Complete Works*, Concord Edition, ed. Edward Waldo Emerson, 12 vols. (Boston: Houghton Mifflin, 1903–1904), 3:75–76.

4. "Automata," trans. Major Alexander Ewing, in *The Best Tales of Hoffmann*, ed. E. F. Bleiler (New York: Dover, 1967), pp. 71–103. Numerous literary authors exploited the symbolic value of automata. See Lieselotte Sauer, *Marionetten, Maschinen, Automaten. Der Künstliche Mensch in der deutschen und englischen Romantik* (Bonn: Bouvier Verlag, 1983); Patricia S. Warrick, *The Cybernetic Imagination in Science Fiction* (Cambridge: MIT Press, 1980).

5. Hoffman, *Best Tales*, p. 81.

6. Ibid., p. 96.

7. Ibid., p. 99.

8. Holmes was also intrigued by the mechanisms of cerebration and human locomotion. See Holmes, "Mechanism in Thought and Morals" (1870) in *The Works of Oliver Wendell Holmes*, 15 vols. (Boston: Houghton Mifflin, 1892–1896), 8:260–314. His "The Human Wheel, Its Spokes and Felloes," *Atlantic* 11 (1863): 567–580,

discusses the mechanism of talking and walking automata. It mentions Maelzel's talking doll, as well as artificial limbs—an unpleasant, yet omnipresent subject, since Holmes was writing in the midst of the American Civil War.

9. Johann Wolfgang von Goethe, *Goethe*, 12 vols. (New York: Suhrkamp Publishers, 1983), 12:11 and 14.

10. Lorraine Daston and Peter Galison, in "The Image of Objectivity" (*Representations* 40 [1992]: 81–128, at 82), label this kind of objectivity "mechanical." See also Zeno G. Swijtink, "The Objectification of Observation: Measurement and Statistical Methods in the Nineteenth Century," in *The Probabilistic Revolution*, vol. 1, *Ideas in History*, ed. Lorenz Krüger, Lorraine J. Daston, and Michael Heidelberger (Cambridge: MIT Press, 1989), pp. 261–285, esp. 266–268.

11. Helmholtz, *Popular Lectures on Scientific Subjects*, "The Eye as an Optical Instrument," pp. 197–228, at 197–199, and "The Sensation of Sight," pp. 228–316, at 258–261.

12. Thomas L. Hankins, *Science and the Enlightenment* (Cambridge: Cambridge University Press, 1985), pp. 6–8. Daston and Galison, "The Image of Objectivity," p. 122.

13. "Probabilité," in *Encyclopédie, ou dictionnaire raisonné des sciences, des arts et des métiers*, ed. Jean D'Alembert and Denis Diderot, 35 vols. (Paris: Briasson, David l'aîné, Le Breton, Durand, 1751–1780), 13:393b.

14. Robert Boyle, *Considerations about the Reconcileableness of Reason and Religion* (1675), in *Works of the Honorable Robert Boyle* (London, 1772), 4:182. Quoted from Lorraine Daston, *Classical Probability in the Enlightenment* (Princeton: Princeton University Press, 1988), p. 63. The term "moral certainty" contains a degree of ambiguity. "Moral" comes from the Latin "mos" (plural "mores"), meaning "custom." "Moral certainty" uses the word in this normative sense. It claims to satisfy a customary standard of judgment as in law. But the word "moral" also has a prescriptive sense of good and evil, which is not entirely separable from its normative sense. That is because our customary standards of judgment are based on our recognition of good and evil.

In the elaboration of probability theory in the eighteenth century, the word "moral" came to mean taking into account the circumstances of the person making the judgment. Thus a rich man might enter into a wager that a poor man would refuse. The "mathematical" certainty would be the same for both players, but the poor man would have a lower "moral" certainty (ibid., pp. 70–71).

BIBLIOGRAPHY

Aarsleff, Hans. *From Locke to Saussure: Essays on the Study of Language and Intellectual History*. Minneapolis: University of Minnesota Press, 1982.

Abbot, Austin. "The Eye and the Camera." *Harper's Magazine* 39 (1869): 476–482.

Abrams, Meyer Howard. "Coleridge's 'A Light in Sound': Science, Metascience, and Poetic Imagination." *Proceedings of the American Philosophical Society* 116, no. 6 (1972): 458–476.

———. *The Correspondent Breeze: Essays on English Romanticism*. New York: W. W. Norton and Company, 1984.

———. *The Mirror and the Lamp: Romantic Theory and the Critical Tradition*. Oxford: Oxford University Press, 1953.

———. *Natural Supernaturalism: Tradition and Revolution in Romantic Literature*. New York: Norton, 1971.

Achinstein, Peter, and Owen Hannaway, eds. *Observation, Experiment, and Hypothesis in Modern Physical Science*. Cambridge: MIT Press, 1985.

Acta Academiae Scientarum Imperialis Petropolitanae. Part 2, *Histoire*, 1781, pp. 9–10. [Announcement of prize contest for the reproduction of vowel sounds. A summary of Kratzenstein's essay is contained on pp. 13–15 of this volume.]

Adams, George. *Micrographia illustrata; or, The Knowledge of the Microscope Explain'd: Together with an Account of a New Invented Universal Microscope*. London: The author, 1746.

Adorno, Theodor. "The Curves of the Needle." *October*, no. 55 (1990): 49–55.

———. "The Form of the Phonograph Record." *October*, no. 55 (1990): 56–61.

———. "Opera and the Long-Playing Record." *October*, no. 55 (1990): 62–66.

Alburger, Mary Anne. *Scottish Fiddlers and Their Music*. London: Victor Gollancz, 1983.

Alembert, Jean d'. *Mélanges de littérature, d'histoire et de philosophie*. 5 vols. Amsterdam: Zacharie Chatelain & Fils, 1770.

Alpers, Svetlana. *The Art of Describing: Dutch Art in the Seventeenth Century*. Chicago: University of Chicago Press, 1983.

Altick, Richard D. *The Shows of London*. Cambridge: Harvard University Press, Belknap Press, 1978.

Amman, Jean Coenrad. *The Talking Deaf Man*. Facsimile of the 1694 ed. Menston, England: Scolar, 1972.

Aristotle. *Posterior Analytics*. Translated and edited by Jonathan Barnes. Oxford: Clarendon Press, 1975.

Arnheim, Rudolph. *Visual Thinking*. Berkeley and Los Angeles: University of California Press, 1969.

Ashworth, William B. "Natural History and the Emblematic World View." In *Reappraisals of the Scientific Revolution*, edited by David C. Lindberg and Robert S. Westman, pp. 303–332. Cambridge: Cambridge University Press, 1990.

Audsley, George Ashdown. *The Art of Organ Building*. 1905. 2 vols. New York: Dover, 1965.

"Automata." In *Encyclopedia of World Art*. 1960. 2:182–193.

Bach, Jose Alfredo. "Athanasius Kircher and His Method: A Study in the Relations of the Arts and Sciences in the Seventeenth Century." Ph.D. diss., University of Oklahoma, 1985.

Backer, Augustin de, and Carlos Sommervogel. *Bibliothèque de la Compagnie de Jésus*. 12 vols. Brussels: O. Schepens, 1890–[1932].

Bacon, Francis. *The Advancement of Learning and New Atlantis*. Edited by Arthur Johnston. Oxford: Clarendon Press, 1974.

———. *A Selection of His Works*. Edited by Sidney Warhaft. Toronto: Macmillan, 1965.

———. *Sylva Sylvarum or Natural History*. 7th ed. London: William Lee, 1658.

———. *The Works of Francis Bacon*. Edited by James Spedding, Robert Leslie Ellis, and Douglas Denon Heath. 14 vols. London: Longman & Co., 1857–1874.

Baker, Henry. "An Account of Mr. Leeuwenhoek's Microscopes." *Philosophical Transactions of the Royal Society* 41, pt. 2 (1740): 516.

———. *The Microscope Made Easy*. London: R. Dodsley, 1743.

Baldwin, Martha. "Athanasius Kircher and the Magnetic Philosophy." Ph.D. diss., University of Chicago, 1987.

———. "Magnetism and the Anti-Copernican Polemic." *Journal for the History of Astronomy* 16 (1985): 155–174.

Barnes, John. *Catalogue of the Barnes Museum of Cinematography*. Vol. 2. Cornwall, 1970.

———. "Philip Carpenter 1776–1833." *New Magic Lantern Journal* 3, no. 2 (1984): 8–9.

———. "The Projected Image: A Short History of Magic Lantern Slides." *New Magic Lantern Journal* 3, no. 3 (1985): 2–7.

Barnouw, Erik. "Magicians Created 'Living Pictures' Then Rued the Day." *Smithsonian* 12, no. 4 (1981): 114–116, 118, 120, 122, 124, 126–128.

Barnum, P. T. *Struggles and Triumphs; or, The Life of P. T. Barnum Written by Himself*. Edited by George S. Bryan. 2 vols. New York: Knopf, 1927.

Baudelaire, Charles. *Art in Paris, 1845–1865*. Translated and edited by Jonathan Mayne. London: Phaidon, 1965.

———. *Variétés critiques*. 2 vols. Paris: G. Crès, 1924.

Baxandall, Michael. *Painting and Experience in Fifteenth Century Italy: A Primer in the Social History of Pictorial Style*. 1972. 2d ed. Oxford: Oxford University Press, 1988.

———. *Patterns of Intention: On the Historical Interpretation of Pictures*. New Haven: Yale University Press, 1985.

Bayle, Pierre. *The Dictionary Historical and Critical*. 5 vols. London, 1736. Facsimile ed. New York: Garland, 1984.

Bedini, Silvio A. "The Role of Automata in the History of Technology." *Technology and Culture* 5 (1964): 24–42.

Beer, J. B. "Coleridge and Boehme's 'Aurora.'" *Notes and Queries*, May 1963, pp. 183–187.

Bell, Alexander Graham. "Prehistoric Telephone Days." *National Geographic* 41 (1922): 223–241.

———. "Vowel Theories." *American Journal of Otology* 1 (1879). Reprinted in his *The Mechanism of Speech*, pp. 117–129. 1906. 5th ed. New York: Funk and Wagnalls, 1911.

Bell, Charles. "Of the Organs of the Human Voice." *Philosophical Transactions of the Royal Society*, 1832, pp. 299–320.

Benjamin, Walter. *Gesammelte Schriften*. Unter Mitwirkung von Theodor W. Adorno und Gershom Scholem; herausgegeben von Rolf Tiedemann und Hermann Schweppenhauser. 7 vols. Frankfurt am Main: Suhrkamp Verlag, ca. 1972–.

Bennett, James A. "Cosmology and the Magnetical Philosophy, 1640–1680." *Journal for the History of Astronomy* 12 (1981): 165–177.

———. "A Viol of Water or a Wedge of Glass." In Gooding, Pinch, and Schaffer, *The Uses of Experiment*, pp. 105–114.

Benoit, Maurice. "La musique des couleurs et l'audition colorée." *Mercure de France* 165 (1923): 392–402.

Bertrand, M. "Le Père Castel." *Le Correspondant*, n.s., 29 (1868): 1067–1084.

Bidney, Martin. "The Aeolian Harp Reconsidered: Music of Unfulfilled Longing in Tjutchev, Morike, Thoreau, and Others." *Comparative Literature Studies* 22 (1985): 329–343.

Biggs, Norman L., E. Keith Lloyd, and Robin J. Wilson. *Graph Theory, 1736–1936*. Oxford: Clarendon Press, 1976.

Bilaniuk, Olexa Myron. "Lichtenberg, Georg Christoph." In *Dictionary of Scientific Biography*, edited by Charles Gillispie, 8:320–323.

"Binocular Vision." *Edinburgh Review* 108 (1858): 437–473.

Biographie universelle et portative . . . Paris: F. G. Levrault, 1834.

Biot, Jean-Baptiste. *Traité de physique expérimentale et mathématique*. 4 vols. Paris: Deterville, 1816.

Birch, Thomas. *The History of the Royal Society of London*. 4 vols. London: A. Millar, 1756–1757.

Bishop, John. "Experimental Researches into the Physiology of the Human Voice." *Philosophical Magazine*, 3d ser., 9 (1836): 201–209, 269–277, and 342–348.

Blake, Clarence J. "The Use of the Membrana Tympani as Phonautograph and Logograph." *Archives of Ophthalmology and Otology* 5 (1876): 108–113.

Blanchet, Jean. *L'art du chant*. Paris: Dessaint et Saillant, 1755.

Bloomfield, Robert. "Nature's Music." 1801. In *Works*. 8 vols. in 3. London: Longman, Hurst, Rees, Orme and Brown, 1819–1826.

Bonner, Stephen. "Aeolian Harp." In *The New Grove Dictionary of Music and Musicians*, edited by Stanley Sadie, 1:115–117. London: Macmillan, 1980.

Bonner, Stephen, and M. G. Davies, eds. *Aeolian Harp*. 4 vols. Duxford, Cambridge: Bois de Boulogne, 1970–1974.

Borell, Merriley. "Extending the Senses: The Graphic Method." *Medical Heritage* 2 (1986): 114–121.

Boring, Edwin G. *Sensation and Perception in the History of Experimental Psychology*. New York: Appleton-Century-Crofts, 1942.

Born, Wolfgang. "Early Peep-shows and the Renaissance Stage." *Connoisseur* 107 (1941): 67–71, 161–164, 180.

Boswell, Jeanetta. "Three Poets and the Aeolian Harp: Emerson, Thoreau, and Melville." *University Forum, Fort Hays State University*, no. 31 (September 1983): 5.

Boutan, A., and J. d'Alméïda. *Cours élémentaire de physique*. Paris: Dunod, 1862.

Bowers, Brian. *Sir Charles Wheatstone. F.R.S., 1802–1875*. London: Her Majesty's Stationery Office, 1975.

Brachner, Alto, et al. *G. F. Brander 1713–1783: Wissenschaftliche Instrumente aus seiner Werkstatt*. Munich: Deutsches Museum, 1983.

Brauen, Fred. "Athanasius Kircher (1602–1680)." *Journal of the History of Ideas* 43 (1982): 129–134.

Braun, Marta. *Picturing Time: The Work of Étienne-Jules Marey (1830–1904)*. Chicago: University of Chicago Press, 1992.

Brent, Joseph. *Charles Sanders Peirce: A Life*. Bloomington: Indiana University Press, 1993.

Brewster, David. "Account of a Binocular Camera, and of a Method of Obtaining Drawings of Full Length and Colossal Statues, and of Living Bodies, Which Can Be Exhibited as Solids by the Stereoscope." *Transactions of the Royal Scottish Society of Arts* 3 (1851) 259–264.

———. "Binocular Vision and the Stereoscope." *North British Review* 17 (1852): 165–204. Published anonymously.

———. "Description of a Binocular Camera." *Report of the British Association. Transactions of the Sections*, 1849, p. 5.

———. *The Kaleidoscope, Its History, Theory and Construction with Its Application to the Fine and Useful Arts*. London: J. Murray, 1858.

———. *Letters on Natural Magic Addressed to Sir Walter Scott, Bart*. London: J. Murray, 1832.

———. "On the Form of Images Produced by Lenses and Mirrors of Different Sizes." *Report of the British Association. Transactions of the Sections*, 1852, pp. 3–7.

———. "Photography." *North British Review* 7 (1847): 465–504. Published anonymously.

———. *The Stereoscope: Its History, Theory, and Construction with Its Application to the Fine and Useful Arts and to Education*. London: J. Murray, 1856.

———. *A Treatise on Optics*. The Cabinet Encyclopedia of Natural Philosophy. London: Longman et al., 1831.

———. *A Treatise on the Kaleidoscope*. Edinburgh: Archibald Constable, 1819.

Brown, Harold I. "Galileo on the Telescope and the Eye." *Journal of the History of Ideas* 46 (1985): 487–501.

Bruce, Robert V. *Bell: Alexander Graham Bell and the Conquest of Solitude*. 1973. Ithaca: Cornell University Press, 1990.

Buffon, George-Louis Leclerc, comte de. "Initial Discourse to the Histoire Naturelle." In *From Natural History to the History of Nature: Readings from Buffon and His Critics*, translated and edited by John Lyon and Phillip R. Sloan. Notre Dame: University of Notre Dame Press, 1981.

Bugge, Thomas. *Science in France in the Revolutionary Era*. Edited by Maurice P. Crosland. Society for the History of Technology, 7. Cambridge: MIT Press, 1969.

Burtt, E. A. *The Metaphysical Foundations of Modern Physical Science*. Rev. ed. Atlantic Highlands, N.J.: Humanities, 1952.

Burwick, Frederick. *Approaches to Organic Form: Permutations in Science and Culture*. Dordrecht: D. Reidel, 1987.

Busch, Wilhelm. "Die deutsche Fachsprache der Mathematik: Ihre Entwicklung und ihre wichtigsten Erscheinungen mit besonderer Rucksicht auf Johann Heinrich Lambert." *Giessener Beitrage zur deutschen Philologie*, no. 30. Giessen: Wilhelm Schmitz Verlag, 1933. Reprint, Amsterdam: Swets & Zeitlinger N.V., 1968.

Butler, Marilyn. "Romanticism in England." In *Romanticism in National Context*, edited by Roy Porter and Mikulas Teich, pp. 37–67. Cambridge: Cambridge University Press, 1988.

Cagniard de la Tour, Charles. "Voix humaine." *L'Institut* 5, no. 212 (1837): 179–180.

Cahan, David. "From Dust Figures to the Kinetic Theory of Gases: August Kundt and the Changing Nature of Experimental Physics in the 1860s and 1870s." *Annals of Science* 47 (1990): 151–172.

———. "Pride and Prejudice in the History of Physics: The German Speaking World, 1740–1945." *Historical Studies in the Physical and Biological Sciences* 19 (1988): 173–191.

———, ed. *Hermann von Helmholtz and the Foundations of Nineteenth-Century Science*. Berkeley and Los Angeles: University of California Press, 1993.

Cajori, Florian. *A History of Physics*. New York: Macmillan, 1899.

Canguilhem, Georges. "The Role of Analogies and Models in Biological Discovery." In *Scientific Change*, edited by Alastair C. Crombie, pp. 507–520. New York: Basic Books, 1963.

Cantor, Geoffrey N. *Optics after Newton: Theories of Light in Britain and Ireland, 1704–1840*. Manchester, England: Manchester University Press, 1983.

———. "Revelation and the Cyclical Cosmos of John Hutchinson." In *Images of the Earth: Essays in the History of the Environmental Sciences*, edited by L. J. Jordanova and Roy S. Porter, pp. 3–22. BSHS Monographs. Chalfont St. Giles: The British Society for the History of Science, 1979.

Carroll, Charles Michael. *The Great Chess Automaton*. New York: Dover, 1975.

Cartwright, Lisa. "'Experiments of Destruction': Cinematic Inscriptions of Physiology." *Representations* 40 (1992): 129–152.

Cassirer, Ernst. *Das Erkenntnisproblem in der Philosophie und Wissenschaft der neueren Zeit*. Berlin: Verlag Bruno Cassirer, 1922.

Castel, Louis-Bertrand. "Clavecin pour les yeux, avec l'art de peindre les sons, et toutes sortes de pièces de musique, Lettre écrite de Paris le 20 février 1725 par le R. P. Castel, Jesuite, à M. Decourt, à Amiens." *Mercure de France*, November 1725, pp. 2552–2577.

———. "Démonstration géométrique du clavecin pour les yeux et pour tous les sens, avec l'éclaircissement de quelques difficultez, et deux nouvelles observations." *Mercure de France*, February 1726, pp. 277–292.

———. "Difficultez sur le clavecin oculaire, avec leurs réponses." *Mercure de France*, March 1726, pp. 455–463.

———. "Lettre à M. C***, sur l'existence d'un milieu entre le naturel et le surnaturel, qu'il appelle artificiel." *Journal de Trévoux* (also known as *Mémoires pour l'histoire des sciences et des beaux arts*), December 1722, pp. 2072–2097.

———. "Nouvelles expériences d'optique et d'acoustique: addressées à M. le Président de Montesquieu." *Journal de Trévoux*, August 1735, pp. 1444–1482, 1619–1666; September, pp. 1807–1839; October, pp. 2018–2053; November, pp. 2335–2372; December, pp. 2642–2768.

———. *L'optique des couleurs, fondée sur les simples observations, et tournée surtout à la pratique de la peinture, de la teinture et des autres arts coloristes*. Paris: Briasson, 1740.

———. "Réflexions sur la nature et la source du sublime dans le discours: sur le vrai philosophie de discours poétique, et sur l'analogie qui est la clef des découvertes." *Mercure de France*, June 1733, pp. 1309–1322, and *Journal de Trévoux*, October 1733, pp. 1747–1762.

———. *Le vrai système de physique générale de M. Isaac Newton . . .* Paris: C.-F. Simon, 1743.

Caus, Salomon de. *Les raisons des forces mouvantes avec diverses machines tant utiles que plaisantes*. 1615. Amsterdam: Frits Knuf, 1973.

Chadarevian, Soraya de. "Graphical Method and Discipline: Self-Recording Instruments in Nineteenth-Century Physiology." *Studies in History and Philosophy of Science* 24 (1993): 267–291.

Chambers's Encyclopaedia. American Revised Edition. Philadelphia: J. B. Lippincott, 1883.

Champeix, Robert. *Savants méconnus, inventions oubliées* . . . Paris: Dunod, 1966.

Chapuis, Alfred. *Les automates dans les oeuvres d'imagination*. Neuchâtel: Éditions du Griffon, 1947.

Chapuis, Alfred, and Edmond Droz. *Automata: A Historical and Technical Study*. Translated by Alec Reid. Neuchâtel: Éditions du Griffon, 1958.

———. *Les automates, figures artificielles d'hômmes et d'animaux*. Neuchâtel: Éditions du Griffon, 1949.

Chapuis, Alfred, and Édouard Gélis. *Le monde des automates*. 2 vols. Paris, 1928.

Chauvin, Stephanus. *Lexicon philosophicum*. 1713. Reprint, Stern-Verlag Janssen & Co., 1967.

Chevremont, F. *Jean-Paul Marat: borné de son portrait; esprit politique accompagné de sa vie scientifique, politique et privée*. 2 vols. Paris: Chez l'auteur, 1880.

Chladni, Ernst Florens Friedrich. *Traité d'acoustique*. Paris: Courcier, 1809. Translation of *Die Akustik* (1802).

Chouillet, Jacques. *La formation des idées esthétiques de Diderot, 1745–1763*. Paris: Librairie Armand Colin, 1973.

Chouillet-Roche, Anne-Marie. "Le clavecin oculaire du Père Castel." *Dix-huitième siècle* 8 (1976): 141–166.

Christensen, Thomas. "Music Theory as Scientific Propaganda: The Case of d'Alembert's *Élémens de musique*." *Journal of the History of Ideas* 50 (1989): 409–427.

Clark, Latimer. "On an Arrangement for Taking Stereoscopic Pictures with a Single Camera." *Photographic Journal* 1 (1853–1854): 57–59.

Clark, Stuart. "The Scientific Status of Demonology." In Vickers, *Occult and Scientific Mentalities in the Renaissance*, pp. 351–374.

Clarke, Desmond M. *Occult Powers and Hypotheses: Cartesian Natural Philosophy under Louis XIV*. Oxford: Clarendon Press, 1989.

Clay, R. S. "The Stereoscope." *Transactions of the Optical Society* 29 (1927–1928): 149–166.

Clifford, William Kingdon. *Mathematical Papers*. Edited by Robert Tucker. London: Macmillan and Co., 1882.

Clulee, Nicholas H. "At the Crossroads of Magic and Science: John Dee's Archemastrie." In Vickers, *Occult and Scientific Mentalities in the Renaissance*, pp. 57–71.

Cohen, Albert. *Music in the French Royal Academy of Sciences: A Study in the Evolution of Musical Thought*. Princeton: Princeton University Press, 1991.

Cohen, H. F. *Quantifying Music: The Science of Music at the First Stage of the Scientific Revolution, 1580–1650*. The University of Western Ontario Series in Philosophy of Science, 23. Dordrecht: D. Reidel, 1984.

Cohen, I. Bernard. *Revolution in Science*. Cambridge: Harvard University Press, Belknap Press, 1985.

Cohen, John. *Human Robots in Myth and Science*. London: George Allen and Unwin, 1966.

Coleridge, Samuel Taylor. *Aids to Reflection, and the Confessions of an Inquiring Spirit*. London: George Bell and Sons, 1901.

———. *The Best of Coleridge*. Edited by Earl Leslie Griggs. New York: The Ronald Press, 1934.

———. *Collected Letters of Samuel Taylor Coleridge*. Edited by Earl Leslie Griggs. 6 vols. Oxford: Oxford University Press, 1956–1971.

———. *The Complete Works*. 7 vols. New York: Harper & Brothers, 1853.

———. *The Friend*. Edited by Barbara E. Rooke. In *The Collected Works of Samuel Taylor Coleridge*, vol. 4, pt. 1. London: Routledge & Kegan Paul, 1969.

———. *Marginalia*. Edited by George Whalley. Vol. 12 of *Collected Works*. Princeton: Princeton University Press, 1980.

———. *The Notebooks*. Edited by Kathleen Coburn. 4 vols. New York: Pantheon Books, 1957–.

Colie, Rosalie L. *"Some Thankfulnesse to Constantine": A Study of English Influence upon the Early Works of Constantijn Huygens*. The Hague: Nijhoff, 1956.

Collegium Anglorum Societatus Jesu Leodii. *Florus Anglo-Bavaricus*. Liège: Ex officina Typographica Guilielmi Henrici Streel, 1686.

Collingwood, R. G. *The Idea of History*. Oxford: Clarendon Press, 1946.

Condorcet, Marie-Jean-Antoine-Nicolas de. *Oeuvres de Condorcet*. Edited by A. Condorcet-O'Connor and F. Arago. 12 vols. Paris, 1847.

Cooke, Conrad William. *Automata Old and New*. London: Chiswick Press, 1893.

Correspondance littéraire, philosophique et critique, par Grimm, Diderot, Raynal, Meister, etc. Edited by Maurice Tourneux. 16 vols. Paris: Garnier, 1877–1882.

Crary, Jonathan. *Techniques of the Observer: On Vision and Modernity in the Nineteenth Century*. Cambridge: MIT Press, 1990.

Crawford, William. *The Keepers of Light: A History and Working Guide to Early Photographic Processes*. Dobbs Ferry, N.Y.: Morgan and Morgan, 1979.

Crombie, A. C. "The Mechanistic Hypothesis and the Scientific Study of Vision: Some Optical Ideas as a Background to the Invention of the Microscope." In *Historical Aspects of Microscopy*, edited by S. Bradbury and G. L'E. Turner, pp. 3–112. Cambridge: Royal Microscopical Society, 1967.

Cross, Charles R. "Helmholtz's Vowel Theory and the Phonograph." *Nature* 18 (1878): 93–94.

Cross, Charles R., and George V. Wendell. "On Some Experiments with the Phonograph, Relating to the Vowel Theory of Helmholtz." *Proceedings of the American Academy of Arts and Sciences* 27 (1891–1892): 271–279.

Crowther, J. G. *Founders of British Science*. London: Crescent, 1960.

Cudworth, Ralph. *True Intellectual System of the Universe, Part I*. For Richard Royston, 1678.

Cunningham, Andrew, and Nicholas Jardine, eds. *Romanticism and the Sciences*. Cambridge: Cambridge University Press, 1990.

Currier, Thomas F., and Eleanor M. Tilton. *A Bibliography of Oliver Wendell Holmes*. New York: New York University, 1953.

Curtis, James F. "The Rise of Experimental Phonetics." In *History of Speech Education in America*, edited by Karl R. Wallace, pp. 348–369. New York: Appleton-Century-Crofts, 1954.

Dagognet, François. *Étienne-Jules Marey: A Passion for the Trace*. Translated by Robert Galeta and Jeanine Herman. New York: Zone Books, 1992.

Daguin, P. A. *Traité élémentaire de physique théorique et expérimentale*. 4th ed. 4 vols. Paris: Delagrave; Toulouse: Paul Privat, 1878.

Darius, John. *Beyond Vision*. New York: Oxford University Press, 1984.

Darnton, Robert. *The Great Cat Massacre and Other Episodes in French Cultural History*. New York: Basic Books, 1984.

———. *The Literary Underground of the Old Regime*. Cambridge: Harvard University Press, 1982.

Darrah, William C. *The World of Stereographs*. Gettysburg: W. C. Darrah, 1977.

Darwin, Erasmus. *The Letters of Erasmus Darwin*. Edited by Desmond King-Hele. Cambridge: Cambridge University Press, 1981.

———. *The Temple of Nature; or, The Origin of Society*. London: J. Johnson, 1803.

Daston, Lorraine. *Classical Probability in the Enlightenment*. Princeton: Princeton University Press, 1988.

Daston, Lorraine, and Peter Galison. "The Image of Objectivity." *Representations* 40 (1992): 81–128.

Dauben, Joseph W. "Marat: His Science and the French Revolution." *Archives Internationales d'Histoire des Sciences* 22 (1969): 235–261.

Daumas, Maurice. *Les instruments scientifiques aux XVIIe et XVIIIe siècles*. Paris: Presses Universitaires de France, 1953.

———. *Scientific Instruments of the Seventeenth and Eighteenth Centuries*. Translated by Mary Holbrook. New York: Praeger, 1972.

Davis, Audrey B. *Medicine and Its Technology*. Westport, Conn.: Greenwood, 1981.

de Clercq, P. R., ed. *Nineteenth-Century Scientific Instruments and Their Makers*, Amsterdam: Rodopi, 1985.

De la Rue, Warren. "Report on the Present State of Celestial Photography in England." *Report of the British Association*, 1859, pp. 130–153.

Dear, Peter. "Jesuit Mathematical Science and the Reconstitution of Experience in the Early Seventeenth Century." *Studies in History and Philosophy of Science* 18 (1987): 133–175.

———. *Mersenne and the Learning of the Schools*. Ithaca: Cornell University Press, 1988.

———. "Narratives, Anecdotes, and Experiments: Turning Experience into Science in the Seventeenth Century." In *The Literary Structure of Scientific Argument: Historical Studies*, edited by Peter Dear, pp. 135–163. Philadelphia: University of Pennsylvania Press, 1991.

———. "Totius in Verba: Rhetoric and Authority in the Early Royal Society." *Isis* 76 (1985): 145–161.

Debus, Allen G. "Robert Fludd and the Use of Gilbert's *De magnete* in the Weaponsalve Controversy." *Journal of the History of Medicine* 19 (1964): 389–417.

Dechales, Claude François Milliet. *Cursus seu mundus mathematicus*. Lyons, 1674.

Delaporte, François. *Nature's Second Kingdom: Explorations of Vegetality in the Eighteenth Century*. Translated by Arthur Goldhammer. Cambridge: MIT Press, 1982.

Della Porta, Giambattista. *Natural Magick*. 1st ed., Naples, 1558. Reprint of the 1st English ed., London: Thomas Young and Samuel Speek, 1658. Edited by Derek J. Price. New York: Basic Books, 1957.

Descartes, René. *Oeuvres*. Edited by Charles Adam and Paul Tannery. 13 vols. Paris: L. Cerf, 1897–1913.

———. *Treatise of Man*. French text with translation by Thomas Steele Hall. Cambridge: Harvard University Press, 1972.

Deschanel, A. Privat. *Elementary Treatise on Natural Philosophy*. Translated and edited by J. D. Everett. New York: D. Appleton, 1873.

Devaux, Pierre. *Automates et automatisme*. Paris: Presses Universitaires de France, 1944.

Devlin, Christopher. *Poor Kit Smart*. London: Rupert Hart-Davis, 1961.

Dibner, Bern, and Stillman Drake. *A Letter from Galileo*. Burndy Library Publications, 24. Norwalk, Conn.: Burndy Library, 1967. Contains Dibner, "Galileo the Innovator," and Drake, "A Long-lost Letter from Galileo to Peiresc on a Magnetic Clock."

Dickinson, H. W., and Rhys Jenkins. *James Watt and the Steam Engine*. Oxford: Clarendon Press, 1927.

Dictionary of Scientific Biography. Charles Gillispie, editor-in-chief. 14 vols. plus Supplement. New York: Scribner, 1970–1980.

Diderot, Denis. "Clavecin oculaire." In *Encyclopédie, ou Dictionnaire raisonné des sciences*, 3:511–512.

———*Oeuvres complètes*. Édition critique et annotée, publieé sous la direction de Herbert Dieckmann, Jean Fabre et Jacques Proust; avec les soins de Jean Varloot. 25 vols. Paris: Hermann, 1975–.

———. *Oeuvres philosophiques*. Edited by Paul Vernière. Paris: Garnier Frères, 1961.

Dix, Robin C. "The Harps of Memnon and Aeolus: A Study in the Propagation of an Error." *Modern Philology* 85 (1988): 288–293.

Dodart, Denis. "Mémoire sur les causes de la voix de l'homme, et de ses différens tons." *Mémoires de l'Académie royale des sciences*, 1700, pp. 238–287.

Dolby, R.G.A. "Reflections on Deviant Science." In *On the Margins of Science: The Social Construction of Rejected Knowledge*, edited by Roy Wallis, pp. 9–47. Sociological Review Monograph, 27. Keele: University of Keele, 1979.

Donders, F. C. "Zur Klangfarbe der Vocale." *Annalen der Physik* 199 (1864): 527–528.

Dortous de Mairan, Jean-Jacques. "Sur la propagation du son dans les différents tons qui le modifient." *Académie royale des sciences, Paris, Mémoires*, 1737, pp. 34–45.

Dostrovsky, Sigalia. "Early Vibration Theory: Physics and Music in the Seventeenth Century." *Archive for History of Exact Sciences* 14 (1975): 169–218.

Doyon, André, and Lucien Liaigre. *Jacques Vaucanson, mécanicien de génie*. Paris: Presses Universitaires de France, 1966.

———. "Méthodologie comparée du biomécanisme et de la mécanique comparée." *Dialectica* 40 (1956): 292–323.

Drake, Stillman. *Galileo Studies: Personality, Tradition, and Revolution*. Ann Arbor: University of Michigan Press, 1970.

———. "Renaissance Music and Experimental Science." *Journal of the History of Ideas* 31 (1970): 483–500.

Draper, John William. "On the Process of Daguerreotype and Its Application to Taking Portraits from Life." *Philosophical Magazine*, 3d ser., 17 (1840): 217–225.

Duck, Michael J. "Newton and Goethe on Colour: Physical and Physiological Considerations." *Annals of Science* 45 (1988): 507–519.

Dudley, Homer, and T. H. Tarnoczy. "The Speaking Machine of Wolfgang von Kempelen." *Journal of the Acoustical Society of America* 22 (1950): 151–166.

Dworkin, Ronald. *Law's Empire*. Cambridge: Harvard University Press, Belknap Press, 1986.

Dyson, Freeman. *Disturbing the Universe*. New York: Harper and Row, 1979.

Eadweard Muybridge: The Stanford Years, 1872–1882. Stanford: Stanford Museum of Art, 1972. Rev. ed., 1973.

Eamon, William. "Arcana Disclosed: The Book of Secrets Tradition and the Development of Experimental Science." *History of Science* 22 (1984): 111–150.

———. *Science and the Secrets of Nature: Books of Secrets in Medieval and Early Modern Culture*. Princeton: Princeton University Press, 1994.

———. "Technology and Magic in the Late Middle Ages and in the Renaissance." *Janus* 70 (1983): 171–212.

Earle, Edward W., ed. *Points of View: The Stereograph in America—A Cultural History*. Rochester: Visual Studies Workshop, 1979.

Eastlake, Elizabeth. "Photography." *Quarterly Review* 101 (1857): 442–468. Published anonymously.

Eichorn, Alfred. "Die Vocalsirene, eine neue Methode der Nachahmung von Vocalklängen." *Annalen der Physik* 39 (1890): 148–154.

Ellis, Alexander John. *The Alphabet of Nature*. London: S. Bagster; Bath: Isaac Pitman, Phonographic Institution, 1845.

———. *Essentials of Phonetics*. London: Pitman, 1848.

Elsky, Martin. "Bacon's Hieroglyphs and the Separation of Words and Things." *Philological Quarterly* 63 (1984): 449–460.

Emerson, Ralph Waldo. *The Complete Works*. Concord Edition. Edited by Edward Waldo Emerson. 12 vols. Boston: Houghton Mifflin, 1903–1904.

———. *Five Essays on Man and Nature*. Edited by Robert E. Spiller. New York: Appleton-Century-Crofts, 1954.

———. *Journals*. Edited by Edward Waldo Emerson and Waldo Emerson Forbes. 10 vols. Boston: Houghton Mifflin, 1909–1914.

Encyclopédie, ou dictionnaire raisonné des sciences, des arts et des métiers. Edited by Jean d'Alembert and Denis Diderot. 35 vols. Paris: Briasson, David l'aîné, Le Breton, Durand, 1751–1780.

Erhardt-Siebold, Erika von. "Harmony of the Senses in English, German, and French Romanticism." *PMLA* 47 (1932): 577–592.

———. "The Heliotrope Tradition." *Osiris* 3 (1933): 22–46.

———. "Some Inventions of the Pre-Romantic Period and Their Influence upon Literature." *Englische Studien* 66 (1931–1932): 347–363.

Eschinardi, Francesco. *Centuriae opticae*. Rome, 1664.

Euler, Leonhard. "Emendatio laternae magicae ac microscopii solaris." *Nova Commentarii Academiae Scientiarum Imperialis Petropolitanae* 3 (1750/51): 363–380. In *Opera omnia*, ser. 3, 2:22–37. Turici: Orell Fussli, 1962.

———. *Letters of Euler on Different Subjects in Natural Philosophy Addressed to a German Princess*. Translated by Henry Hunter. Notes by David Brewster. Reprint of 1833 ed. 2 vols. in 1. 1768, 1772. New York: Arno, 1975.

———. *Lettres à une Princesse d'Allemagne*. In *Opera omnia*, ser. 3, vol. 12. Turici: Orell Fussli, 1960.

"The Euphonia." *Illustrated London News* 9 (1846): 96.

Evans, Henry Ridgely. *Edgar Allan Poe and Baron von Kempelen's Chess-Playing Automaton*. Kenton, Ohio: International Brotherhood of Magicians, 1939.

Evans, Hughes. "Losing Touch: The Controversy over the Introduction of Blood Pressure Instruments into Medicine." *Technology and Culture* 34 (1993): 784–807.

Evelyn, John. *The Diary of John Evelyn*. Edited by E. S. de Beer. 6 vols. Oxford: Clarendon Press, 1955.

Explanation of the Ocular Harpsichord upon Shew to the Public. London: Printed for S. Hooper and A. Morley, 1757.

"Faber's Sprechmaschine." *Annalen der Physik* 58 (1843): 175–176.

Fayet, Joseph. *La révolution française et la science 1789–1795*. Paris: Librairie M. Riviere, 1960.

Feingold, Mordechai. "The Occult Tradition in the English Universities of the Renaissance: A Reassessment." In Vickers, *Occult and Scientific Mentalities in the Renaissance*, pp. 73–94.

Ferrein, Antoine. "De la formation de la voix de l'homme." *Mémoires de l'Académie royale des sciences*, 1741, pp. 409–432.

Field, J. V. "What Is Scientific about a Scientific Instrument?" *Nuncius* 3, no. 2 (1988): 3–26.

Finn, Bernard S. "Alexander Graham Bell's Experiments with the Variable-Resistance Transmitter." *Smithsonian Journal of History* 1 (1966): 1–16.

———. "Laplace and the Speed of Sound." *Isis* 55 (1964): 7–19.

Fletcher, John E. "Astronomy in the Life and Correspondence of Athanasius Kircher." *Isis* 61 (1970): 52–67.

———. "Athanasius Kircher: A Man under Pressure." In *Athanasius Kircher und seine Beziehungen zum gelehrten Europa seiner Zeit*, edited by John Fletcher, pp. 1–15. Weisbaden: Kommission bei Otto Harrassowitz, 1988

———. "Athanasius Kircher and the Distribution of His Books." *The Library* 23 (1968): 108–117.

———. "A Brief Survey of the Unpublished Manuscripts of Athanasius Kircher." *Manuscripta* 13 (1969): 150–160.

———. "Claude Fabri de Peiresc and the Other French Correspondents of Athanasius Kircher (1602–1680)." *Australian Journal of French Studies* 9 (1972): 250–273.

———. "Johann Marcus Marci Writes to Athanasius Kircher." *Janus* 59 (1972): 95–118.

Fludd, Robert. *Tomus Secundus De Supernaturali, Naturali, Praeternaturali et contranaturali Microcosmi historia in tractatus tres distributa*. Oppenheim: Johann Theodor de Bry, 1619.

Fontenelle, Bernard le Bovier de. "Rapport sur la lumière et les couleurs d'après le 'Mémoire' de Malebranche." In Nicolas Malebranche, *Oeuvres complètes*, edited by André Robinet, 2:510. 22 vols. Paris: Librairie J. Vrin, 1958–.

Forman, Paul. "Independence, Not Transcendence, for the Historian of Science." *Isis* 82 (1991): 71–86.

Fourier, Joseph B. J. "Éloge de J.A.C. Charles." *Académie des sciences, Paris, Mémoires* 8 (1829): lxxiii–lxxxviii.

Fournié, Édouard. *Physiologie de la voix et de la parole*. Paris: Adrien Delhaye, 1866.

France, Anatole. "L'Elvire de Lamartine," from *Pages d'histoire et de littérature*. In *Oeuvres complètes*, 24:267–333. 26 vols. Paris: Calmann-Lévy, 1934.

Frängsmyr, Tore, J. L. Heilbron, and Robin E. Rider. *The Quantifying Spirit in the Eighteenth Century*. Berkeley and Los Angeles: University of California Press, 1990.

Frank, Robert G., Jr. "The Telltale Heart: Physiological Instruments, Graphic Methods, and Clinical Hopes, 1854–1914." In *The Investigative Enterprise: Experimental Physiology in Nineteenth Century Medicine*, edited by William Coleman and Frederic L. Holmes, pp. 211–290. Berkeley and Los Angeles: University of California Press, 1988.

Franklin, Allan. *The Neglect of Experiment*. Cambridge: Cambridge University Press, 1986.

Franklin, Benjamin. *The Works of Benjamin Franklin*. Edited by Jared Sparks. Rev. ed. 10 vols. Philadelphia: Childs & Peterson, 1840.

Franssen, Maarten. "The Ocular Harpsichord of Louis-Bertrand Castel: The Science and Aesthetics of an Eighteenth-Century *Cause célèbre*." *Tractrix* 3 (1991): 15–77.

Friedrich, Joseph, Freiherr zu Racknitz. *Über den Schachspieler des Herrn von Kempelen und dessen Nachbildung*. Leipzig and Dresden, 1789.

Frizot, Michel. *Avant la cinématographie, la chronophotographie: temps, photographie et mouvement autour de E.-J. Marey*. Beaune: Association des amis de Marey, 1984.

Fryer, David M., and John C. Marshall. "The Motives of Jacques de Vaucanson." *Technology and Culture* 20 (1979): 257–269.

Funkhouser, H. Gray. "Historical Development of the Graphical Representation of Statistical Data." *Osiris*, 1st ser., 3 (1937): 269–404.

Galilei, Galileo. *Dialogue Concerning the Two Chief World Systems*. Translated by Stillman Drake. Berkeley and Los Angeles: University of California Press, 1953.

———. *Discourses Concerning Two New Sciences*. Translated by Henry Crew and Alfonso de Salvio. New York: Dover, 1954.

———. *Discoveries and Opinions of Galileo*. Edited and translated by Stillman Drake. New York: Anchor/Doubleday, 1957.

———. *The Galileo Affair: A Documentary History*. Edited by Maurice A. Finocchiaro. Berkeley and Los Angeles: University of California Press, 1989.

———. *Le opere di Galileo Galilei*. Edizione nazionale. 20 vols. in 21. Florence: Barbera, 1890–1909.

———. *Siderius Nuncius, or, The Sidereal Messenger*. Translated by Albert van Helden. Chicago: University of Chicago Press, 1989.

———. *Two New Sciences*. Translated by Stillman Drake. Madison: University of Wisconsin Press, 1974.

Galison, Peter L. "Bubble Chambers and the Experimental Workplace." In Achinstein and Hannaway, *Observation, Experiment, and Hypothesis in Modern Physical Science*, pp. 309–373.

———. "Descartes' Comparisons." *Isis* 75 (1984): 311–326.

———. "History, Philosophy, and the Central Metaphor." *Science in Context* 2 (1988): 192–212.

———. *How Experiments End*. Chicago: University of Chicago Press, 1987.

Galison, Peter, and Alexi Assmus. "Artificial Clouds, Real Particles." In Gooding, Pinch, and Schaffer, *The Uses of Experiment*, pp. 225–274.

Ganot, Adolphe. *Elementary Treatise on Physics*. Translated and edited by E. Atkinson. 13th ed. New York: William Wood, 1890.

Gariel. "Machine parlante de M. Faber." *Journal de physique* 8 (1879): 274–275.

Gassendi, Pierre. *The Mirrour of True Nobility and Gentility being the Life of the Renowned Nicolaus Claudius Fabricius Lord of Peiresk, Senator of the Parliament at Aix*. Translated by W. Rand. London: Printed by J. Streater for Humphrey Moseley, 1657.

Gehler, Samuel Traugott. *Physikalisches Worterbuch*. Leipzig: E. B. Schwickert, 1825.

Gernsheim, Helmut and Alison. *The History of Photography from the Earliest Use of the Camera Obscura in the Eleventh Century up to 1914*. London: Oxford University Press, 1955.

Gilbert, William. *On the Magnet*. Translated by S. P. Thomson. Edited by Derek J. Price. New York: Basic Books, 1958.

Gill, A. T. "Early Stereoscopes." *Photographic Journal* 109 (1969): 546–559, 606–614, 641–651.

Gillispie, Charles Coulston. See *Dictionary of Scientific Biography*.

———. *The Edge of Objectivity*. Princeton: Princeton University Press, 1960.

———. *The Montgolfier Brothers and the Invention of Aviation, 1783–1784: With a Word on the Importance of Ballooning for the Science of Heat and the Art of Building Railroads*. Princeton: Princeton University Press, 1983.

Glick, Thomas F. "On the Influence of Kircher in Spain." *Isis* 62 (1971): 379–381.

Goclenius, Rudolphus. *Lexicon philosophicum quo tanquam clave philosophiae fores aperiuntur*. 1613. Reprint, Hildesheim: Georg Olms Verlag, 1980.

Godwin, Joscelyn. "Athanasius Kircher and the Occult." In *Athanasius Kircher und seine Beziehungen zum gelehrten Europa seiner Zeit*, edited by John Fletcher, pp. 17–36. Wiesbaden: Kommission bei Otto Harrassowitz, 1988.

———. *Athanasius Kircher: A Renaissance Man and the Quest for Lost Knowledge*. London: Thames and Hudson, 1979.

———. *Robert Fludd: Hermetic Philosopher and Surveyor of Two Worlds*. London: Thames and Hudson, 1979

Goethe, Johann Wolfgang von. *Faust*. Translated by Walter Kaufman. New York: Doubleday/Anchor, 1961.

———. *Goethe*. 12 vols. New York: Suhrkamp Publishers, 1983.

——— *Die Schriften zur Naturwissenschaft*. Hrsg. im Auftrage der Deutschen Akademie der Naturforscher zu Halle. 11 vols. in 18. Weimar: H. Böhlaus, 1947–.

———. *Scientific Studies*. Translated and edited by Douglas Miller. New York: Suhrkamp Publishers, 1988.

Gold, Liza Hannah. "Automata and the Origins of the Mechanical Philosophy." Senior thesis, Harvard University, 1981.

Golinski, Jan. "Precision Instruments and the Demonstrative Order." *Osiris* 9 (1994): 30–47.

———. "The Secret Life of an Alchemist." In *Let Newton Be! A New Perspective on His Life and Works* edited by John Fauvel, Raymond Flood, Michael Shortland, and Robin Wilson, pp. 146–176. Oxford: Oxford University Press, 1988.

Gombrich, E. H. *Art and Illusion*. 1960. Princeton: Princeton University Press, 1984.

———. "Image and Code: Scope and Limits of Conventionalism in Pictorial Representation." In *Image and Code*, edited by Wendy Steiner. Ann Arbor: Michigan Studies in the Humanities, 1981.

Gombrich, E. H. "Mirror and Map: Theories of Pictorial Representation." *Philosophical Transactions of the Royal Society* (B) 270 (1975): 119–149.

———. *The Sense of Order: A Study in the Psychology of Decorative Art*. Oxford: Phaidon Press, 1979.

———. "Standards of Truth: The Arrested Image and the Moving Eye." *Critical Inquiry* 7 (1980): 237–273.

———. "The 'What' and the 'How': Perspective Representation and the Phenomenal World." In *Logic and Art: Essays in Honor of Nelson Goodman*, edited by Richard Rudner and Israel Scheffler, pp. 129–149. Indianapolis: Bobbs-Merrill, 1972.

Gooding, David. *Experiment and the Making of Meaning: Human Agency in Scientific Observation and Experiment*. Dordrecht: Kluwer, 1990.

———. " 'Magnetic Curves' and the Magnetic Field: Experimentation and Representation in the History of a Theory." In Gooding, Pinch, and Schaffer, *The Uses of Experiment*, pp. 183–223.

Gooding, David, Trevor Pinch, and Simon Schaffer, eds. *The Uses of Experiment: Studies in the Natural Sciences*. Cambridge: Cambridge University Press, 1989.

Goodman, Nelson. *Languages of Art*. 1968. 2d ed. Indianapolis: Hackett, 1976.

Gosser, H. Mark. "Kircher and the Laterna Magica—a Reexamination." *Journal of the SMPTE* 90 (1981): 972–978.

Gouk, Penelope. "The Harmonic Roots of Newtonian Science." In *Let Newton Be! A New Perspective on His Life and Works*, edited by John Fauvel, Raymond Flood, Michael Shortland, and Robin Wilson, pp. 101–125. Oxford: Oxford University Press, 1988.

———. "The Role of Acoustics and Music Theory in the Scientific Work of Robert Hooke." *Annals of Science* 37 (1980): 573–605.

Graves, Robert P. *Life of Sir William Rowan Hamilton*. 3 vols. Dublin: Hodges, Figgis and Co., 1882–1891.

Gravesande, Willem Jacob van 's. *Mathematical Elements of Natural Philosophy Confirmed by Experiments, or An Introduction to Sir Isaac Newton's Philosophy*. Translated by J. T. Desaguliers. 2d ed. 2 vols. London: J. Senex and W. Taylor, 1721–1726.

———. *Oeuvres philosophiques et mathématiques*. Edited by Jean Nic. Seb. Allamand. 2 pts. in 1 vol. Amsterdam: M. M. Rey, 1774.

———. *Physices elementa mathematica, experimentis confirmata; sive introductio ad philosophiam Newtonianam*. Geneva: Apud Henricum-Albertum Gosse & Soc., 1721.

Gray, J. J., and Laura Tilling. "Johann Heinrich Lambert, Mathematician and Scientist, 1728–1777." *Historia Mathematica* 5 (1978): 13–41.

Green, Harvey. " 'Pasteboard Masks': The Stereograph in American Culture 1865–1910." In Earle, *Points of View*, pp. 109–115.

Greenacre, Derek. "Great Heavens: Some Unusual Slides." *New Magic Lantern Journal* 5 (1988): 6.

Greene, Mott T. "History of Geology." *Osiris*, 2d ser., 1 (1985): 97–116.

Grigson, Geoffrey. *The Harp of Aeolus and Other Essays on Art, Literature and Nature*. London: George Routledge & Sons, 1947.

Grmek, M. D. "Dodart, Denis." In *Dictionary of Scientific Biography*, 4:135–136.

———. "Ferrein, Antoine." In *Dictionary of Scientific Biography*, 4:589–590.

Guerlac, Henry. "Can There Be Colors in the Dark? Physical Color Theory before Newton." *Journal of the History of Ideas* 47 (1986): 3–20.

———. "The Word 'Spectrum': A Lexicographic Note with a Query." *Isis* 56 (1965): 206–207.

Guest, Harriet. *A Form of Sound Words: The Religious Poetry of Christopher Smart*. Oxford: Clarendon Press, 1989.

Gunther, R. T. *Early Science in Oxford*. 15 vols. Oxford: Printed for the Author, 1931.

Haakfort, Casper. "Newton's Optics: The Changing Spectrum of Science." In *Let Newton Be! A New Perspective on His Life and Works*, edited by John Fauvel, Raymond Flood, Michael Shortland, and Robin Wilson, pp. 81–100. Oxford: Oxford University Press, 1988.

Hacking, Ian. "Do We See through a Microscope?" *Pacific Philosophical Quarterly* 62 (1981): 305–322.

———. *Representing and Intervening*. Cambridge: Cambridge University Press, 1983.

Hackmann, W. D. "The Relationship between Concept and Instrument Design in Eighteenth-Century Experimental Science." *Annals of Science* 26 (1979): 205–224.

———. "Scientific Instruments: Models of Brass and Aids to Discovery." In Gooding, Pinch, and Schaffer, *The Uses of Experiment*, pp. 31–65.

———. "Underwater Acoustics and the Royal Navy, 1893–1930." *Annals of Science* 26 (1979): 255–278.

Hahn, Roger. *The Anatomy of a Scientific Institution: The Paris Academy of Sciences, 1666–1803*. Berkeley and Los Angeles: University of California Press, 1971.

Hall, Marie Boas. "Salomon's House Emergent: The Early Royal Society and Cooperative Research." In *The Analytic Spirit: Essays in the History of Science in Honor of Henry Guerlac*, edited by Harry Woolf, pp. 177–194. Ithaca: Cornell University Press, 1981.

Hammond, John H. *The Camera Obscura: A Chronicle*. Bristol: Adam Hilger, 1981.

Hankins, Thomas L. *Jean d'Alembert: Science and the Enlightenment*. Oxford: Clarendon Press, 1970.

———. "Newton's 'Mathematical Way' a Century after the Principia." In *Some Truer Method: Reflections on the Heritage of Newton*, edited by Frank Durham and Robert D. Purrington, pp. 89–112. New York: Columbia University Press, 1990.

———. *Science and the Enlightenment*. Cambridge: Cambridge University Press, 1985.

Hardie, Walter. "Description of a New Pseudoscope." *Philosophical Magazine*, 4th ser., 5 (1853): 442–446.

———. "The Telestereoscope." *Philosophical Magazine*, 4th ser., 15 (1858): 156–157.

Harding, Anthony John. *Coleridge and the Inspired Word*. Kingston and Montreal: McGill-Queens University Press, 1985.

Harman, P. M. *Energy, Force, and Matter: The Conceptual Development of Nineteenth-Century Physics*. Cambridge: Cambridge University Press, 1982.

Harris, John. *Lexicon Technicum or, an Universal English Dictionary of Arts and Sciences Explaining Not Only the Terms of Art but the Arts Themselves*. London: Dan. Brown, Tim. Goodwin . . . , 1704.

Harris, Lawrence Ernest. *The Two Netherlanders: Humphrey Bradley and Cornelis Drebbel*. Leiden: E. J. Brill, 1961.

Harvey, William. *Exercitatio anatomica de motu cordis et sanguinis in animalibus.* With English translation and annotations by Chauncey D. Leake. Baltimore: Charles C. Thomas, 1928.

Hawkes, Terence. *Structuralism and Semiotics.* Berkeley and Los Angeles: University of California Press, 1977.

Hawthorne, Nathaniel. "The Artist of the Beautiful." In his *Mosses from an Old Manse,* pp. 504–536. Boston: Houghton Mifflin, 1846.

———. "Dr. Heidegger's Experiment." In his *Twice-Told Tales,* pp. 258–271. 1837. Boston: Houghton Mifflin, 1882.

Hecht, Hermann. "The History of Projecting Phantoms, Ghosts and Apparitions." *New Magic Lantern Journal* 3, no. 1 (1984): 2–6, and 3, no. 2 (1984): 2–6.

Heilbron, John L. *Electricity in the Seventeenth and Eighteenth Centuries: A Study of Early Modern Physics.* Berkeley and Los Angeles: University of California Press, 1979.

———. "Introductory Essay." In Frängsmyr, Heilbron, and Rider, *The Quantifying Spirit,* pp. 1–23.

Heimann, P. M., and J. E. McGuire. "Newtonian Forces and Lockean Powers: Concepts of Matter in Eighteenth-Century Thought." *Historical Studies in the Physical Sciences* 3 (1971): 233–306.

Helmholtz, Hermann von. "On the Interaction of Natural Forces." In his *Popular Lectures on Scientific Subjects,* translated by E. Atkinson, pp. 153–193. New York: Appleton, 1885.

———. *On the Sensations of Tone as a Physiological Basis for the Theory of Music.* 1863 (1st German ed.). Translated by A. J. Ellis. 1885 (1st English ed.). 2d English ed., based on the 4th German ed. of 1887. New York: Dover, 1954.

———. "On the Telestereoscope." *Philosophical Magazine,* 4th ser., 15 (1858): 19–24. From *Annalen der Physik* no. 9 (1857).

———. *Physiological Optics.* Vol. 1, 1856; vol. 2, 1860; vol. 3, 1866. Edited and translated by James P. C. Southall. 3 vols. New York: Dover, 1901.

———. *Popular Lectures on Scientific Subjects.* Translated by E. Atkinson. New York: Appleton, 1885.

———. "Über die Klangfarbe der Vocale." *Annalen der Physik* 184 (1859): 280–290. Translated in *Philosophical Magazine,* 4th ser., 19 (1860): 81–88.

Helsinger, Elizabeth K. *Ruskin and the Art of the Beholder.* Cambridge: Harvard University Press, 1982.

Hendricks, Gordon. *Eadweard Muybridge: The Father of the Motion Picture.* New York: Grossman, 1975.

Hennig, John. "Goethe and the Jesuits." *Thought* 24 (1949): 449–665.

Henry, John. "Newton, Matter and Magic." In *Let Newton Be! A New Perspective on His Life and Works,* edited by John Fauvel, Raymond Flood, Michael Shortland, and Robin Wilson, pp. 127–146. Oxford: Oxford University Press, 1988.

———. "Occult Qualities and the Experimental Philosophy: Active Principles in Pre-Newtonian Matter Theory." *History of Science* 24 (1986): 335–381.

———. "Robert Hooke, the Incongruous Mechanist." In *Robert Hooke: New Studies,* edited by Michael Hunter and Simon Schaffer, pp. 149–180. Woodbridge: The Boydell Press, 1989.

Hepworth, T. C. "The Evolution of the Magic Lantern." *Chambers Journal,* 6th ser., 1 (1897): 213–215.

Herschel, John F. W. *Outlines of Astronomy*. London: Longmans, Green, and Co., 1875.

———. "Sound." *Encyclopaedia Metropolitana* 4 (1830): 747–824.

———. "The Stereoscopic Angle." *Photographic News* 1 (1858): 110.

Hesse, Mary B. "Hooke's Philosophical Algebra." *Isis* 57 (1966): 67–83.

Hewison, Robert. *John Ruskin: The Argument of the Eye*. Princeton: Princeton University Press, 1976.

Himes, Charles F. "Contributions to the Subject of Binocular Vision." *Journal of the Franklin Institute* 62 (1871): 263–270, 340–348, 413–419; 63 (1872): 141–144, 350–356.

Hindle, Brooke. *Emulation and Innovation*. New York: New York University Press, 1981.

Hine, William L. "Athanasius Kircher and Magnetism." In *Athanasius Kircher und seine Beziehungen zum gelehrten Europa seiner Zeit*, edited John Fletcher, pp. 72–97. Wiesbaden: Otto Harrassowitz, 1988.

———. "Marin Mersenne: Renaissance Naturalism and Renaissance Magic." In Vickers, *Occult and Scientific Mentalities in the Renaissance*, pp. 165–176.

———. "The Mersenne-Kircher Correspondence on Magnetism." *Proceedings of the Western Society for French History* 10 (1982): 106–117.

Hobbes, Thomas. *Leviathan or The Matter, Forme, and Power of a Commonwealth, Ecclesiasticall and Civill*. London, 1651.

Hoefer, Jean Chrétien, ed. *Nouvelle biographie générale*. 46 vols. Paris: Firmin Didot Frères, 1852–1866.

Hoff, Hebbel E., and L. A. Geddes. "The Beginnings of Graphic Recording." *Isis* 53 (1962): 287–310.

———. "Graphic Recording before Carl Ludwig: An Historical Summary." *Archives internationales d'histoire des sciences* 12 (1959): 1–25.

———. "Graphic Registration before Ludwig: The Antecedents of the Kymograph." *Isis* 50 (1959): 5–21.

———. "A Historical Perspective on Physiological Monitoring: Chaveau's Projecting Kymograph and the Projecting Physiograph." *Cardiovascular Research Center Bulletin* 14 (1975): 3–35.

———. "A Historical Perspective on Physiological Monitoring: Sherrington's Mammalian Laboratory and Its Antecedents." *Cardiovascular Research Center Bulletin* 13 (1974): 19–39.

———. "The Technological Background of Physiological Discovery: Ballistics and the Graphic Method." *Journal of the History of Medicine* 15 (1960): 345–363.

Hoffmann, E.T.A. *The Best Tales of Hoffmann*. Edited by E. F. Bleiler. New York: Dover, 1967.

———. *Sämtliche poetische Werke*. Edited by Hannsludwig Geiger. 4 vols. Weisbaden: Emil Vollmer Verlag, 1972.

Hofmann, Johann Jacob. *Lexicon universale historiam sacram et profanum*. 4 vols. Leyden: Jacob Hackium, 1698.

Hollingshead, John. *My Lifetime*. 2 vols. London: Sampson, Low, Marston, 1895.

Holmes, Oliver Wendell. "Doings of the Sunbeam." *Atlantic* 12 (1863): 1–15. Published anonymously.

———. "The Great Instrument." *Atlantic* 12 (1863): 637–647. Published anonymously.

Holmes, Oliver Wendell. "History of the 'American Stereoscope.'" *Philadelphia Photographer* 6 (1869): 1–3.

———. "The Human Wheel, Its Spokes and Felloes." *Atlantic* 11 (1863): 567–580. Published anonymously.

———. "Mechanism in Thought and Morals." In *The Works of Oliver Wendell Holmes*, 8:260–314. 15 vols. Boston: Houghton Mifflin, 1892–1896.

———. "The Stereoscope and the Stereograph." *Atlantic* 3 (1859): 738–748. Published anonymously.

———. "Sun-Painting and Sun-Sculpture." *Atlantic* 8 (1861): 13–29. Published anonymously.

"The 'Holmes' Stereoscope." *Philadelphia Photographer* 6 (1869): 23–25.

Hooke, Robert. "Contrivance to Make the Picture of any Thing Appear on a Wall." *Philosophical Transactions of the Royal Society of London* 3 (August 17, 1668): 741–743.

———. *The Diary of Robert Hooke, 1672–1680.* Edited by Henry W. Robinson and Walter Adams. London: Taylor and Francis, 1935.

———. *Micrographia: or Some Physiological Descriptions of Minute Bodies Made by Magnifying Glasses with Observations and Inquiries Thereupon.* London, 1665. Reprint, New York: Dover, 1961.

———. *Posthumous Works.* 2d ed. Introduced by T. M. Brown. [London]: F. Cass, [1971].

Hughes, Thomas P. "Model Builders and Instrument Makers." *Science in Context* 2 (1988): 59–75.

Humbert, Pierre. *Un amateur: Peiresc, 1580–1637.* Paris: Desclée de Brouwer, 1933.

Hunt, F. V. *Origins in Acoustics: The Science of Sound from Antiquity to the Age of Newton.* New Haven: Yale University Press, 1978.

Hutchison, Keith. "Idiosyncrasy, Achromatic Lenses, and Early Romanticism." *Centaurus* 34 (1991): 125–171.

———. "What Happened to Occult Qualities in the Scientific Revolution?" *Isis* 73 (1982): 233–253.

Huygens, Christiaan. *Oeuvres complètes.* 22 vols. The Hague: Société Hollandaise des Sciences/Nijhoff, 1888–1950.

———. *Traité de la lumière.* Facsimile of the 1690 ed. London: Dawsons of Pall Mall, 1966.

———. *Treatise on Light.* Translated by S. P. Thompson. Reprint, New York: Dover, 1952.

Hyde, Ralph. *Panoramania!* London: Trefoil Publications in association with Barbican Art Gallery, 1988.

Iliffe, Rob. "'In the Warehouse': Privacy, Property and Priority in the Early Royal Society." *History of Science* 30 (1992): 29–68.

"Inanimate Reason." *Monthly Review*, 1st ser., 70 (1784): 307–308. [On Windisch's account of Kempelen's chess player.]

Isherwood, Robert M. *Farce and Fantasy: Popular Entertainment in Eighteenth-Century Paris.* Oxford: Oxford University Press, 1986.

Jardine, Lisa. *Francis Bacon: Discovery and the Art of Discourse.* Cambridge: Cambridge University Press, 1974.

Jasper, David. *Coleridge as Poet and Religious Thinker: Inspiration and Revelation.* London: Macmillan, 1985.

Jenkin, Fleeming, and J. A. Ewing. "On the Harmonic Analysis of Certain Vowel

Sounds." *Transactions of the Royal Society of Edinburgh* 28 (1876–1878): 745–777.

Jones, William. *Physiological Disquisitions; or, Discourses on the Natural Philosophy of the Elements . . .* London: J. Rivington and Sons, 1781.

Joseph, Steven F. "Wheatstone and Fenton: A Vision Shared." *History of Photography* 9 (1985): 305–309.

———. "Wheatstone's Double Vision." *History of Photography* 8 (1984): 329–332.

Jungmarker, Gunnar. "Kattklaver och voterings-instrument Verklighet och fantasi." *Artes*, no. 5 (1982): 116–125.

Jungnickel, Christa, and Russell McCormmach. *Intellectual Mastery of Nature: Theoretical Physics from Ohm to to Einstein.* 2 vols. Vol. 1, *The Torch of Mathematics, 1800–1870.* Vol. 2, *The Now Mighty Theoretical Physics, 1870–1925.* Chicago: University of Chicago Press, 1986.

Kahn, Douglas. "Track Organology." *October*, no. 55 (1990): 68–78.

Káláman, Bolla, ed. "In Memoriam Farkas Kempelen: Articles Written in Commemoration of the 250th Anniversary of Farkas Kempelen's Birth." *Magyar Fonetikai Füzetek—Hungarian Papers in Phonetics* 13 (1984). This volume is in Hungarian with English summaries.

Kassler, Jamie Croy. "Man—a Musical Instrument: Models of the Brain and Mental Functioning before the Computer." *History of Science* 22 (1984): 59–92.

———. "The Systematic Writings on Music of William Jones (1726–1800)." *Journal of the American Musicological Society* 26 (1973): 92–107.

Kastner, Georges. *La harpe d'Éole et la musique cosmique: études sur les rapports des phénomènes sonores de la nature avec la science et l'art.* Paris: G. Brandus, Dufour et Cie, 1856.

Kaufman, Thomas DaCosta. *The Mastery of Nature: Aspects of Art, Science, and Humanism in the Renaissance.* Princeton: Princeton University Press, 1993.

Kemeny, John G. "Man Viewed as a Machine." *Scientific American* 192 (1955): 58.

Kempelen, Wolfgang von. *Le mécanisme de la parole, suivi de la déscription d'une machine parlante.* Vienna: J. V. Degen, 1791.

———. *Mechanismus der menschlichen Sprache nebst der Beschreibung seiner sprechenden Maschine.* Vienna: J. V. Degen, 1791. Reprinted with biographical and bibliographical material, Stuttgart-Bad Cannstatt: Friedrich Frommann, 1970.

Kepler, Johannes. *Gesammelte Werke.* Edited by Walther van Dyck and Max Caspar. 17 vols. Munich: C. H. Bech, 1937–1975.

Kestler, Johann Stephan. *Physiologia kircheriana experimentalis.* Amsterdam: Janssonio-Waesbergiana, 1680.

King-Hele, Desmond. *Doctor of Revolution: The Life and Genius of Erasmus Darwin.* London: Faber & Faber, 1977.

———. *Erasmus Darwin.* New York: Scribner, 1963.

———. *Erasmus Darwin and the Romantic Poets.* London: Macmillan, 1986.

Kircher, Athanasius. *Ars magna lucis et umbrae.* Rome: Sumptibus Hermanni Scheuss, 1646. 2d ed. Amsterdam, 1671.

———. *Athanasii Kircherie Soc. Jesu China monumentis, qua sacris qua profanis, nec non variis naturae & artis spectaculis, aliarumque rerum memorabilium argumentis illustrata, auspiciis Leopoldi Primi roman, imper.* Antwerp: Apud Jacobum a Meurs, 1667.

———. *Magnes, sive de arte magnetica.* Rome: Sumptibus H. Scheus, 1641.

Kircher, Athanasius. *Musurgia universalis*. 2 vols. Rome: Francisci Corbelletti, 1650. Facsimile ed. Hildesheim: Olms, 1970.

———. *Phonurgia nova sive conjugium mechanico-physicum naturae pananymphia phonosophia concinnatum*. Kempten: R. Dreherr, 1673.

———. *Polygraphia nova et universalis ex combinatoria arte detecta*. Rome: Typographia Varesij, 1663.

———. *Turris Babel*. Amsterdam: Ex officina Jansonio-Waesbergiana, 1679.

Klein, Adrian Bernard. *Colour-Music: The Art of Light*. 2d ed. London: Crosby Lockwood and Son, 1930.

Knapp, Lewis Mansfield. *Tobias Smollett, Doctor of Men and Manners*. Princeton: Princeton University Press, 1949.

Knobloch, Eberhard. "Musurgia Universalis: Unknown Combinatorial Studies in the Age of Baroque Absolutism." *History of Science* 17 (1979): 258–275.

Knowlson, James. *Universal Language Schemes in England and France, 1600–1800*. Toronto: University of Toronto Press, 1975.

Koelbing, Huldrych M. "Newton's and Goethe's Colour Theories—Contradictory or Complementary Approaches?" In *Theory and Experiment: Recent Insights and New Perspectives on Their Relation*, edited by Diderik Bateus and Jean Paul van Bendegem, pp. 189–205. Studies in Epistemology, Logic, Methodology, and Philosophy of Science, 195. Dordrecht: D. Reidel, 1988.

König, Karl Rudolph. *Catalogue des appareils d'acoustique*. Paris: 1889.

———. *Quelques expériences d'acoustique*. Paris, 1882.

———. "Die Wellensirene." *Annalen der Physik* 57 (1896): 339–388.

Krafft, Georg. *Sermones in solemni academiae scientiarum imperialis* . . . St. Petersburg, 1742.

Kratzenstein, Christian Gottlieb. "Sur la naissance & la formation des voyelles." *Journal de Physique*, 1782, pp. 358–380.

Kruger, Johannes Gottlob. "De novo musices quo oculi delectantur genere." *Miscellanea Berolinensia* 7 (1743): 354.

Kuhn, Albert J. "Glory or Gravity: Hutchinson vs. Newton." *Journal of the History of Ideas* 22 (1961): 303–322.

Kutschmann, Werner. "Scientific Instruments and the Senses: Towards an Anthropological Historiography of the Natural Sciences." *International Studies in the Philosophy of Science* 1 (1986): 106–123.

La Mettrie, Julien Offray de. *Oeuvres philosophiques*. New ed. 2 vols. Berlin, 1774.

Lambert, Johann Heinrich. *Beyträge zum Gebrauch der Mathematik und derren Anwendungen*. 3 vols. in 4. Berlin: Im Verlage des Buchladens der Realschule, 1765–1772.

———. "Essai de taxéometrie, ou sur la mesure de l'ordre." *Akademie der Wissenschaften, Berlin, Nouveau mémoires*, 1770, pp. 327–342; 1773, pp. 347–368.

———. *Johann Heinrich Lamberts deutscher gelehrter Briefwechsel*. Edited by Johann Bernoulli. 5 vols. in 7. Berlin: Bey dem Herausgeber, 1781–1787.

———. *Philosophische Schriften*. 7 vols. Facsimile ed. Hildesheim: Georg Olms Verlagsbuchhandlung, 1965–.

———. *Pyrometrie oder vom Maasse des Feuers und der Wärme*. Berlin: Bey Hande und Spener, 1779.

Latour, Bruno. "Drawing Things Together." In Lynch and Woolgar, *Representation*, pp. 19–68.

———. *Science in Action*. Cambridge: Harvard University Press, 1987.

Latour, Bruno, and Steve Woolgar. *Laboratory Life: The Social Construction of Scientific Facts*. Introduced by Jonas Salk. Sage Library of Social Research, 80. Beverly Hills: Sage Publications, 1979.

Lavoisier, Antoine Laurent. *Traité élémentaire de chimie*. Paris: Gauthiers-Villars, 1937.

Law, William. *A Serious Call to a Devout and Holy Life*. Introduced by Norman Sykes. London, 1967.

Lawrence, Christopher. "Physiological Apparatus in the Wellcome Museum. 1. The Marey Sphygmograph." *Medical History* 22 (1978): 196–200.

Layer, Harold A. "Stereoscopy: An Analysis of Its History and Its Import to Education and the Communication Process." Ed.D. diss., Indiana University, 1970.

Le Grand, Homer E., ed. *Experimental Inquiries: Historical, Philosophical and Social Studies of Experimentation in Science*. Dordrecht: Kluwer, 1990.

———. "Is a Picture Worth a Thousand Experiments?" In Le Grand, *Experimental Inquiries*, pp. 241–270.

LeConte, John. "On the Influence of Musical Sounds on the Flame of a Jet of Coal-Gas." *American Journal of Science* 75 (1858): 62–67.

LeConte, Joseph. *Sight: An Exposition of the Principles of Monocular and Binocular Vision*. 1881. Rev. ed. New York: Appleton, 1897.

Leighton, John. "Binocular Photographs." *Photographic Journal* 1 (1853–1854): 211–212.

Lemay, J. A. Leo. *Ebenezer Kinnersley: Franklin's Friend*. Philadelphia: University of Pennsylvania Press, 1964.

Lenoir, Timothy. "Practice, Reason, Context: The Dialogue between Theory and Experiment." *Science in Context* 2 (1988): 3–22.

Lesch, John E. "Systematics and the Geometrical Spirit." In Frängsmyr, Heilbron, and Rider, *The Quantifying Spirit*, pp. 73–111.

Levere, Trevor Harvey. *Poetry Realized in Nature: Samuel Taylor Coleridge and Early Nineteenth-Century Science*. Cambridge: Cambridge University Press, 1981.

Levin, Thomas Y. "For the Record: Adorno on Music in the Age of Its Technological Reproducibility." *October*, no. 55 (1990): 23–47.

Lichtenberg, H. *Göttingen Taschen-kalendar*. Göttingen: Johann Christian Dietrich, 1792.

Liesegang, Franz Paul. *Dates and Sources: A Contribution to the History of the Art of Projection and to Cinematography*. Translated by Hermann Hecht. London: The Magic Lantern Society of Great Britain, 1986.

Lindberg, David C. *Theories of Vision from Al-Kindi to Kepler*. Chicago: University of Chicago Press, 1976.

———. "The Theory of Pinhole Images from Antiquity to the Thirteenth Century." *Archive for History of Exact Sciences* 5 (1968): 154–176.

Lindsay, R. Bruce, ed. *Acoustics: Historical and Philosophical Development*. Stroudsburg, Pa.: Dowden, Hutchinson, & Ross, 1972.

Lissajous, Jules Antoine. "Mémoire sur l'étude optique des mouvements vibratoire." *Annales de chimie*, 3rd ser., 51 (1857): 147–231.

Lloyd, Richard John. "Speech Sounds: Their Nature and Causation." *Phonetische Studien* 3 (1890): 251–278.

Locke, John. *An Essay Concerning Human Understanding*. London: Printed by Elizabeth Holt for Thomas Basset, 1690. Edited by Alexander Campbell Fraser. 2 vols. Oxford: Clarendon Press, 1894.

Lonsdale, Roger H. *Dr. Charles Burney, A Literary Biography*. Oxford: Clarendon Press, 1965.

Loudon, James. "A Century of Progress in Acoustics." *Science*, n.s., 14 (1901): 987–995.

Ludwig, Carl. "Beiträge zur Kenntniss des Einflusses der Respirationsbewegungen auf den Blutlauf im Aortensysteme." *Archiv für Anatomie, Physiologie, und wissenschaftliche Medizin*, 1847, pp. 242–302.

Lynch, Michael, and Steve Woolgar, eds. *Representation in Scientific Practice*. Cambridge: MIT Press, 1990.

Macksey, Richard, ed. *The Structuralist Controversy: The Languages of Criticism and the Sciences of Man*. Baltimore: The Johns Hopkins University Press, 1971.

MacLean, Ian. "The Interpretation of Natural Signs: Cardano's *De subtilitate* versus Scaliger's *Exercitationes*." In Vickers, *Occult and Scientific Mentalities in the Renaissance*, pp. 231–252.

Magni, Valeriano. *Demonstratio ocularis. Loci sine locato: Corporis successive moti in vacuo: Luminis nulli corpori inhaerentis*. Warsaw: 1647.

Malebranche, Nicolas. *Nicolas Malebranche: The Search after Truth*. Translated by Thomas M. Lennon and Paul J. Olscamp. *Elucidations*. Translated with philosophical commentary by Thomas M. Lennon. Columbus: Ohio State University Press, 1980.

Maley, V. Carlton, Jr. *The Theory of Beats and Combination Tones, 1700–1863*. New York: Garland, 1990. Originally a Ph.D. diss. (Harvard University, 1967), published by Harvard Dissertations in the History of Science.

Mannoni, Laurent. "The Magic Lantern Makers of France." *New Magic Lantern Journal* 5, no. 2 (1987): 3–7.

Marage, Georges René Marie. "Études des cornets acoustiques par la photographie des flammes de Koenig." *Séances de la Société Française de Physique*, 1897, pp. 74–85.

———. "La méthode graphique dans l'étude des voyelles." *Comptes rendus, Académie des sciences, Paris* 128 (1899): 425–427.

———. *Petit manuel de physiologie de la voix à l'usage des chanteurs et des orateurs*. 1911. Tours: Deslis, n.d.

———. "Photographie des flammes de Koenig." *Comptes rendus, Académie des sciences, Paris* 124 (1897): 811–813.

———. "Théorie de la formation des voyelles." *Séances de la Société Française de Physique*, 1900, pp. 109–147.

Marat, Jean-Paul. *Recherches physiques sur le feu*. Paris: C. A. Jombert, 1780.

Marder, William and Estelle. *Anthony: The Man, the Company, the Cameras; An American Photographic Pioneer; 140 Year History of a Company from Anthony to Ansco to GAF*. Plantation, Fla.: Pine Ridge Publishing Co., 1982.

Marey, Étienne-Jules. *Animal Mechanism*. New York: Appleton, 1873.

———. *La circulation du sang à l'état physiologique et dans les maladies*. Paris: G. Masson, 1881.

———. "Mesures à prendre pour l'uniformisation des méthodes et le contrôle des instruments employés en physiologie." *Comptes rendus, Académie des sciences, Paris* 127 (1898): 375–381.

———. *La méthode graphique dans les sciences expérimentales*. 1878. 2d ed. Paris: G. Masson, 1885.

Marion, F. *The Wonders of Optics*. Translated by Charles W. Quin. London: Sampson Low, Son, and Marston, 1868.

Marks, Lawrence E. "On Colored-Hearing Synesthesia: Cross-modal Translations of Sensory Dimensions." *Psychological Bulletin* 82 (1975): 303–331.

Martin, Benjamin. *A New and Compendious System of Optics*. London: James Hodges, 1740.

———. *The Young Gentleman and Lady's Philosophy*. 3d ed. London, 1781.

Mascher, J. F. "On Taking Daguerreotypes without a Camera." *Journal of the Franklin Institute* 59 (1855): 344–347.

———. "On the Cause of Distortions in Photographic Pictures—A Disclaimer." *Journal of the Franklin Institute* 60 (1855): 65–66.

"Mascher's Stereoscopic Books." *Scientific American* 11 (1856): 228.

Masteller, Richard N. "Western Views in Eastern Parlors: The Contribution of the Stereograph Photographer to the Conquest of the West." *Prospects* 6 (1981): 55–71.

Matteson, Robert S. "Emerson and the Aeolian Harp." *South Central Bulletin* (South Central Modern Language Association) 24, no. 1 (February 1964): 4–9.

Matthiessen, F. O. *American Renaissance: Art and Expression in the Age of Emerson and Whitman*. New York: Oxford University Press, 1941.

Mayer, Alfred Marshall. "On a New Form of Lantern-Galvanometer." *American Journal of Science and the Arts* 103 (1872): 414–418.

———. *Sound: A Series of Simple, Entertaining, and Inexpensive Experiments in the Phenomena of Sound for the Use of Students of Every Age*. London: Macmillan, 1891.

Mayr, Otto. *Authority, Liberty, and Automatic Machinery in Early Modern Europe*. Baltimore: The Johns Hopkins University Press, 1986.

McCloy, Shelby Thomas. *French Inventions of the Eighteenth Century*. Lexington: University of Kentucky Press, 1952.

McCormmach, Russell. *Night Thoughts of a Classical Physicist*. 1982. New York: Avon, 1983.

McDermott, Jeanne A. "The Kaleidoscope, Magic in a Tube, Is Enjoying Revival." *Smithsonian* 13 (November 1982): 98–108.

McMullin, Ernan. "The Conception of Science in Galileo's Work." In *New Perspectives on Galileo*, edited by Robert E. Butts and Joseph C. Pitt, pp. 209–257. The University of Western Ontario Series in Philosophy of Science, 14. Dordrecht: D. Reidel, 1978.

———. "Conceptions of Science in the Scientific Revolution." In *Reappraisals of the Scientific Revolution*, edited by David C. Lindberg and Robert S. Westman, pp. 27–92. Cambridge: Cambridge University Press, 1990.

Megill, Allan Dickson. "The Enlightenment Debate on the Origin of Language and Its Historical Background." Ph.D. diss., Columbia University, 1975.

Melville, Herman. *The Works of Herman Melville. Standard Edition*. 16 vols. London: Constable and Co., 1922–1924.

Merritt, T. L. "Stereoscopic Angles." *Notes and Queries*, 1st ser., 8 (1853): 109–110.

Mersenne, Marin. *Correspondance du Père Marin Mersenne, religieux minime*. Edited by Paul Tannery and C. de Waard. 13 vols. Paris: G. Beauchesne et fils, 1932–1977.

"Mical." In *Biographie universelle, ancienne et moderne*, 28:517–519. Paris: Michaud, 1811–1828.

Micraelius, Johannes. *Lexicon philosophicum terminorum philosophis usitatorum.* 1662. Reprint, Stern-Verlag Janssen & Co., 1966.

Middleton, W. E. Knowles. "Athanasius Kircher, Buffon, and the Burning-Mirrors." *Isis* 62 (1971): 533–543.

———. *A History of the Thermometer and Its Use in Meteorology*. Baltimore: The Johns Hopkins University Press, 1966.

Millard, André. *Edison and the Business of Invention*. Baltimore: The Johns Hopkins University Press, 1990.

Millburn, John R. *Benjamin Martin: Author, Instrument Maker, and "Country Showman."* Leyden: Noordhoff, 1976.

Miller, Dayton Clarence. *Anecdotal History of the Science of Sound: To the Beginning of the Twentieth Century*. New York: Macmillan, 1935.

———. *The Science of Musical Sounds*. 1916. 2d ed. New York: Macmillan, 1922.

Miller, Ron. "The Manifestation of Occult Qualities in the Scientific Revolution." In *Religion, Science and Worldview: Essays in Honor of Richard S. Westfall*, edited by Margaret J. Osler and Paul Lawrence Farber, pp. 185–216. Cambridge: Cambridge University Press, 1985.

Moigno, F. "Phonautographe et fixation graphique de la voix." *Cosmos*, 1859, pp. 314–320.

Molyneux, William. *Dioptrica nova*. London: Benj. Tooke, 1692.

Monchamps, Georges. *Galilée et la Belgique: essai historique sur les vicissitudes du système de Copernic en Belgique*. Saint-Trond: G. Moreaus-Schouberects, 1892.

Montesquieu, Charles de Secondat, baron de. *Oeuvres complètes, publiées sous la direction de M. André Masson*. 3 vols. Paris: Éditions Nagêl, [1950]–1955.

Morrison-Low, A. D., and J.R.R. Christie, eds. *'Martyr of Science': Sir David Brewster, 1781–1868*. Edinburgh: Royal Scottish Museum, 1984.

Morse, Edgar W. "Natural Philosophy, Hypotheses and Impiety: Sir David Brewster Confronts the Undulatory Theory of Light." Ph.D. diss., University of California, Berkeley, 1972.

Müller, Johannes. *Elements of Physiology*. Translated by William Baly. 2d ed. 2 vols. London: Taylor and Walton, 1839.

Mulligan, Lotte. "'Reason,' 'Right Reason,' and 'Revelation' in Mid-seventeenth-century England." In Vickers, *Occult and Scientific Mentalities in the Renaissance*, pp. 375–401.

Munto, William. "Scott, Sir Walter." In *Encyclopedia Britannica*, 24:469–475. 11th ed. Cambridge: Cambridge University Press, 1911.

Musschenbroek, Pieter van. *Essai de physique*. Leyden: S. Luchtmans, 1739.

Myers, Greg. "Every Picture Tells a Story: Illustrations in E. O. Wilson's *Sociobiology*." In Lynch and Woolgar, *Representation*, pp. 231–265.

N. E. Thing Enterprises. *Magic Eye III—Visions: A New Dimension in Art*. Kansas City: Andrews and McMeel, 1994.

Nathan, Isaac. *Musurgia vocalis*. 2d ed. London: Fentum, 1836.

Newton, Isaac. *The Correspondence of Isaac Newton*. Edited by H. W. Turnbull et al. 7 vols. Cambridge: Cambridge University Press, 1959–1977.

———. *Mathematical Principles of Natural Philosophy and His System of the World*. Edited by Florian Cajori. Berkeley and Los Angeles: University of California Press, 1960.

———. *The Optical Papers of Isaac Newton.* Edited by Alan E. Shapiro. Vol. 1, *The Optical Lectures 1670–1672.* Cambridge: Cambridge University Press, 1984–.

Nicolson, Marjorie Hope. *Newton Demands the Muse: Newton's Opticks and the Eighteenth-Century Poets.* Westport, Conn.: Greenwood Press, 1946.

Niemann, W. "Sprechenden Figuren. Ein Beitrag zur Vorgeschichte der Phonographen." *Geschichtsblätter für Technik und Industrie* 7 (1920): 2–30.

Nietzsche, Friedrich. "On Truth and Falsity in Their Ultramoral Sense." 1873. In *The Complete Works of Friedrich Nietzsche,* translated by Maximilian A. Mügge, edited by Oscar Levy. 18 vols. New York: Macmillan, 1924.

Nollet, Jean Antoine. *Leçons de physique expérimentale.* 6 vols. Amsterdam and Leipzig: Arkste'e & Merkus, 1754–1765.

Noulet, Emilie. "Le Père Castel et le 'clavecin oculaire.'" *La Nouvelle NRF* 1 (1953): 553–559.

Novalis, Friedrich von Hardenberg. *Werke in einem Band.* Edited by Uwe Lassen. 3d ed. Hamburg, 1966.

Oersted, H. C. *Correspondance avec divers savants.* Edited by H. C. Harding. 2 vols. in 1. Copenhagen: H. Aschehoug & Co., 1920.

———. *The Soul of Nature.* Translated by Leonora and Joanna B. Horner. London: Dawsons, 1966.

O'Malley, C. D. *Andreas Vesalius of Brussels 1514–1564.* Berkeley and Los Angeles: University of California Press, 1964.

O'Malley, Glenn. "Shelley's 'Air-Prism': The Synesthetic Scheme of *Alastor.*" *Modern Philology* 55 (1958): 178–187.

Oldenburg, Henry. *The Correspondence of Henry Oldenburg.* Edited by A. Rupert Hall and Marie Boas Hall. 13 vols. Madison: University of Wisconsin Press, 1965–1986.

Ord-Hume, Arthur W.J.G. *Clockwork Music: An Illustrated History of Mechanical Musical Instruments from the Musical Box to the Pianola, from Automaton Lady Virginal to Orchestrion.* London: Allen and Unwin, 1973.

"Our Weekly Gossip."*Athenaeum*, 1846, p. 765. [Discussion of Faber's speaking machine.]

Ozanam, Jacques. *Recréations mathématiques et physiques.* Paris, 1694.

Paganelli, Arturo, Giancarlo Cassina, and Fiorella Rocca. "The Botanical Garden of the University of Padua." *The Herbalist* (Concord, Mass.), no. 49 (1983): 72–78.

Paget, Richard. *Human Speech.* New York: Harcourt, Brace; London: Kegan Paul, Trench, Trubner, 1930.

Paley, William. *Natural Theology: or, Evidences of the existence and attributes of the Deity, collected from the appearances of nature.* London: R. Faulder, 1802.

Palisca, Claude V. "Scientific Empiricism in Musical Thought." In *Seventeenth Century Science and the Arts,* edited by H. H. Rhys, pp. 91–137. Princeton: Princeton University Press, 1961.

Panofsky, Erwin. *Galileo as Critic of the Arts.* The Hague: Nijhoff, 1954.

Park, Katherine, and Lorraine J. Daston. "Unnatural Conceptions: The Study of Monsters in Sixteenth- and Seventeenth-Century France and England." *Past and Present* 92 (August 1981): 20–54.

Park, Katharine, Lorraine J. Daston, and Peter L. Galison. "Bacon, Galileo, and Descartes on Imagination and Analogy." *Isis* 75 (1984): 287–326. Contains "Bacon's 'Enchanted Glass,'" by Park (290–302); "Galilean Analogies: Imagination at the Bounds of Sense," by Daston (302–310); and "Descartes' Comparisons: From the Invisible to the Visible," by Galison (311–326).

Paroy, Jean-Philippe-Guy Le Gentil, marquis de. *Mémoires du Comte de Paroy: Souvenirs d'un défenseur de la famille royale pendant la Révolution (1789–1797)*. Paris: Plon, 1895.

Peacock, George. *Life of Thomas Young*. London: J. Murray, 1855.

Peirce, Charles Sanders. *Collected Papers of Charles Sanders Peirce*. Edited by Charles Hartshorne and Paul Weiss. 7 vols. Cambridge: Harvard University Press, 1933.

———. *Values in a Universe of Chance: Selected Writings of C. S. Peirce*. Edited by Philip Wiener. Garden City, N.Y.: Doubleday, 1958.

Peiresc, Nicolas Claude Fabri de. "Une horloge qui monstre les heures à l'ombre dans une chambre fermée." Bibliothèque inguimbertine, Carpentras. MS 1864, fol. 215 et seq.

———. *Lettres de Peiresc*. Edited by Philippe Tamizey de Larroque. 7 vols. Paris: Imprimerie Nationale, 1888–1898.

Pellisov, Charles-Émile. "Andeutungen zur Begrundung einer Theorie der Aeolsharfe." *Annalen der Physik und Chemie* 19 (1830): 237ff.

Pepys, Samuel. *The Diary of Samuel Pepys*. Edited by Henry B. Wheatley. 8 vols. London: G. Bell and Sons, [1892].

Perregaux, Charles, and F. Louis Perrot. *Les Jaquet-Droz et Leschot*. Neuchâtel: Attinger Frères, 1916.

Perriault, Jacques. *Mémoires de l'ombre et du son: une archéologie de l'audio-visuel*. Paris: Flammarion, 1981.

Pfaundler, L., ed. *Müller-Pouillet's Lehrbuch der Physik und Meteorologie*. 9th ed. 3 vols. in 4. Brunswick: F. Vieweg, 1886–1898.

"The Photographic Portrait." *The Crayon* 4 (1857): 154–155.

Pietrasancta, Silvestro. *De symbolis heroicis libri IX*. Antwerp: Ex officina Plantiniana B. Moreti, 1634.

Pirenne, M. H. *Optics, Painting, and Photography*. Cambridge: Cambridge University Press, 1970.

Pisko, Franz Josef. *Die neuren Apparate der Akustik*. Vienna: Carl Gerold's Sohn, 1865.

Playfair, William. *The Commercial and Political Atlas, Representing by Copperplate Charts, the Progress of the Commerce, Revenues, Expenditure, and Debts of England, During the Whole of the Eighteenth Century*. 3d ed. London: J. Wallis, 1801.

———. *An Inquiry into the Permanent Causes of the Decline and Fall of Powerful and Wealthy Nations, Illustrated by Four Engraved Charts*. 2d ed. London, 1807.

"Playfair, William." In *Dictionary of National Biography*, edited by Leslie Stephens, 15:1300. New York: Macmillan, 1885–1901.

Plomp, Reiner. "Beats of Mistuned Consonances." *Journal of the Acoustical Society of America* 42 (1969): 462–474.

———. "Timbre as a Multidimensional Attribute of Complex Tones." In *Frequency Analysis and Periodicity Detection in Hearing*, edited by R. Plomp and G. F. Smoorenburg, pp. 397–411. Leiden: A. W. Sijthoff, 1970.

Plomp, Reiner, and H.J.M. Steeneken. "Effect of Phase on the Timbre of Complex Tones." *Journal of the Acoustical Society of America* 46 (1969): 409–421.

Plummer, H. C. "Color Music—A New Art Created with the Aid of Science." *Scientific American* 112 (1915): 343, 350–351.

Poe, Edgar Allan. *The Complete Illustrated Stories and Poems of Edgar Allan Poe*. London: Chancellor, 1988.

Potter, Richard. "On the English Sounds of the Vowel-Letters of the Alphabet, on Their Production by Instruments, and on the Natural Musical Sequence of the Vowel-Sounds." Read April 28, 1873. *Proceedings of the Cambridge Philosophical Society* 2 (1874–1876): 306–308.

Preece, William Henry, and Augustus Stroh. "Studies in Acoustics. I. On the Synthetic Examination of Vowel Sounds." *Proceedings of the Royal Society of London* 28 (1878–1879): 358–366.

Price, Derek J. de Solla. "Automata and the Origins of Mechanism and Mechanistic Philosophy." *Technology and Culture* 5 (1964): 9–24.

Prickett, Stephen. *Romanticism and Religion: The Tradition of Coleridge and Wordsworth in the Victorian Church*. Cambridge: Cambridge University Press, 1976.

Pumfrey, Stephen. "Ideas above His Station: A Social Study of Hooke's Curatorship of Experiments." *History of Science* 29 (1991): 1–44.

Quandt, Christian Friedrich. "Versuche und Vermutungen über die Aeolsharfe." *Lausizische Monatschrift*, 1795, pp. 277–278.

Quigley, Martin, Jr. *Magic Shadows: The Story of the Origin of Motion Pictures*. Washington, D.C.: Georgetown University Press, 1948.

Ravetz, J. "The Representation of Physical Quantities in Eighteenth-Century Mathematical Physics." *Isis* 52 (1961): 7–20.

Rees, Abraham. *The Cyclopaedia; Or Universal Dictionary of Arts, Sciences, and Literature*. 39 vols. London: Longman, Hurst, Rees, Orme, & Brown, 1802–1820.

Rees, Graham. "Francis Bacon's Semi-Paracelsian Cosmology." *Ambix* 22 (1975): 81–101.

Reid, Anthony. *Southeast Asia in the Age of Commerce 1450–1680*. Vol. 1, *The Lands below the Winds*. New Haven: Yale University Press, 1988.

Reilly, Conor. *Athanasius Kircher, S.J., Master of a Hundred Arts, 1602–1680*. Studia Kircheriana, 1. Wiesbaden: Edizioni del Mondo, 1974.

Reynolds, Osborne. "Review of J. J. Thomson's *The Motion of Vortex Rings* (London: Macmillan, 1883)." *Nature* 29 (April 1883): 193–195.

Ribe, Neil M. "Goethe's Critique of Newton: A Reconsideration." *Studies in History and Philosophy of Science* 16 (1985): 315–335.

Richerand, A. *Elements of Physiology*. Translated by G.L.M. De Lys. 3d ed. London: Thomas and George Underwood, 1819.

Rider, Robin E. "Measure of Ideas, Rule of Language: Mathematics and Language in the Eighteenth Century." In Frängsmyr, Heilbron, and Rider, *The Quantifying Spirit*, pp. 113–140.

———. *The Show of Science*. The Friends of the Bancroft Library, Keepsakes, 31. Berkeley: The Friends of the Bancroft Library, University of California, 1983.

Rienstra, Miller Howard. "Giovanni Battista Della Porta and Renaissance Science." Ph.D. diss., University of Michigan, 1963.

Rivarol, Antoine. *Oeuvres complètes*. 1818. 5 vols. Geneva: Slatkine Reprints, 1868.

Roberts, Lissa. "A Word and the World: The Significance of Naming the Calorimeter." *Isis* 82 (1991): 198–222.

Robertson, Étienne-Gaspard. *Mémoires recréatifs, scientifiques et anecdotiques d'un physicien-aeronaute*. Reprint of vol. 1. Langres: Café Clima Editeur, 1985.

Robinson, David. "Robinson on Robertson." *New Magic Lantern Journal* 4 (1986): 4–13. [Also appears as the *Ten Year Book of the Magic Lantern Society of Great Britain*.]

Roget, Peter Mark. *Animal and Vegetable Physiology Considered with Reference to Natural Theology*. 2 vols. London: William Pickering, 1834.

Rohr, Moritz von. *Die Binocularen Instrumente*. Berlin: Springer, 1920.

Rosenberg, John D. *The Darkening Glass: A Portrait of Ruskin's Genius*. New York: Columbia University Press, 1961.

Rosenberger, Ferdinand. *Die Geschichte der Physik*. Vol. 1, 1882; vol. 2, 1884; vol. 3, 1887–1890. 3 vols. in 2. Hildesheim: Georg Olms, 1965.

Rosner, Lisa. "Eighteenth-Century Medical Education and the Didactic Model of Experiment." In *The Literary Structure of Scientific Argument: Historical Studies*, edited by Peter Dear, pp. 182–194. Philadelphia: University of Pennsylvania Press, 1991.

Rossi, Paolo. *Philosophy, Technology, and the Arts in the Early Modern Era*. Translated by Salvator Attanasio. Edited by Benjamin Nelson. New York: Harper & Row, 1970.

Roth, Nancy. "'First Stammerings of the Heart': Ludwig's Kymograph." *Medical Instrumentation* 12 (1978): 348.

Rousselot, P. J. *Principes de phonétique expérimentale*. New ed. 2 vols. Paris: Didier, 1924.

Royston, Erica. "A Note on the History of the Graphical Presentation of Data." *Biometrika* 43 (1956): 241–247.

Rubens, Peter Paul. *The Letters of Peter Paul Rubens*. Edited and translated by Ruth Saunders Magurn. Cambridge: Harvard University Press, 1955.

———. *P. P. Rubens als Boekillustrator, Stadt Antwerpen Museum Plantin-Moretus, 7 May–4 Jul. 1977.*

———. *Rubens and the Book: Title Pages by Peter Paul Rubens*. Edited by Julius S. Held. Catalog to an exhibition held at Chapin Library, Williams College, Williamstown, Mass., May 2–31, 1977.

Rudwick, Martin J. S. "The Emergence of a Visual Language for Geological Science, 1760–1840." *History of Science* 14 (1976): 149–195.

Ruskin, John. *The Complete Works of John Ruskin*. Edited by E. T. Cook and Alexander Wedderburn. 39 vols. London: G. Allen, 1903–1912.

Ryan, W. F. "Limelight on Eastern Europe: The Great Dissolving Views at The Royal Polytechnic." *New Magic Lantern Journal* 4 (1986): 48–55.

Sachs, Julius von. *History of Botany (1530–1860)*. Translated by Henry E. F. Garnsey. Revised by Isaac Bayley Balfour. Oxford: Clarendon Press, 1890.

Saettler, Paul. *A History of Instructional Technology*. New York: McGraw-Hill, 1968.

Sargent, Rose-Mary. "Scientific Experiment and Legal Expertise: The Way of Experience in Seventeenth-Century England." *Studies in History and Philosophy of Science* 20 (1989): 19–45.

Sauer, Lieselotte. *Marionetten, Maschinen, Automaten. Der Künstliche Mensch in der deutschen und englischen Romantik*. Bonn: Bouvier Verlag, 1983.

Savart, Félix. "Mémoire sur la voix humaine." *Annales de chimie*, 2d ser., 30 (1825): 64–87.

———. "Notes sur la sensibilité de l'organe de l'ouïe." *Annales de chimie* 44 (1830): 337–352.

———. "Notes sur la limite de la perception des sons graves." *Annales de chimie* 47 (1831): 64–74.

Saxon, A. H. *P. T. Barnum: The Legend and the Man*. New York: Columbia University Press, 1989.

Schaffer, Simon. "Glass Works: Newton's Prisms and the Uses of Experiment." In Gooding, Pinch, and Schaffer, *The Uses of Experiment*, pp. 67–104.

———. "Natural Philosophy." In *The Ferment of Knowledge: Studies in the Historiography of Eighteenth-Century Science*, edited by George Rousseau and Roy Porter, pp. 55–91. Cambridge: Cambridge University Press, 1980.

———. "Natural Philosophy and Public Spectacle in the Eighteenth Century." *History of Science* 21 (1983): 1–43.

Schier, Donald S. *Louis Bertrand Castel, Anti-Newtonian Scientist*. Cedar Rapids, Iowa: The Torch Press, 1941.

Schiller, Friedrich. *Poems of Schiller*. Boston: S. E. Cassino and Co., 1884.

Schneebeli, Heinrich. "Expériences avec le phonautographe." *Archives des sciences physiques et naturelles*, n.s., 4 (1878): 78–83.

Schneider, Duane B. "Coleridge's Light-Sound Theory." *Notes and Queries*, May 1963, pp. 182–183.

Schofield, Robert E. *The Lunar Society of Birmingham: A Social History of Provincial Science in Eighteenth-Century England*. Oxford: Clarendon Press, 1963.

———. *Mechanism and Materialism: British Natural Philosophy in an Age of Reason*. Princeton: Princeton University Press, 1970.

Schott, Gaspar. *Magia universalis naturae et artis, sive recondita naturalium et artificialium rerum scientia*. 4 vols. Würzburg, 1657–1659.

———. *Magica hydraulico-pneumatica*. Frankfurt: J. G. Schonwetteri, 1657.

Scott de Martinville, Édouard-Léon. *Histoire de la sténographie depuis les temps anciens jusqu'à nos jours; ou précis historique et critique des divers moyens qui ont été proposés ou employés pour rendre l'écriture aussi rapide que la parole*. Paris: Charles Tondeur, 1849.

———. "Inscription automatique des sons de l'air au moyen d'une oreille artificielle." *Comptes rendus, Académie des sciences, Paris,* 53 (1861): 108–111.

———. *Les noms de baptême et les prénoms*. 1857. 2d ed. Paris: Alexandre Houssiaux, 1858.

———. *Le problème de la parole s'écrivant elle-même*. Paris: Scott, 1878.

Scriba, Christoph J. "Lambert, Johann Heinrich." In *Dictionary of Scientific Biography*, 7:595–600.

Scripture, Edward Wheeler. *The Elements of Experimental Phonetics*. New York: Charles Scribner's Sons; London: Edward Arnold, 1902.

———. "On the Nature of Vowels." *American Journal of Science* 161 (1901): 302–309.

———. "Report on the Construction of a Vowel Organ." *Smithsonian Miscellaneous Collections* 47 (1905): 360–364.

Sebeok, Thomas A. "'Semiotics' and Its Congeners." In *Frontiers in Semiotics*, edited by John Deely, Brooke Williams, and Felicia E. Kruse, pp. 255–263. Bloomington: Indiana University Press, 1986.

Sepper, Dennis L. *Goethe contra Newton: Polemics and the Project for a New Science of Color*. Cambridge: Cambridge University Press, 1988.

Serene, Eileen. "Demonstrative Science." In *The Cambridge History of Later Medieval Philosophy*, edited by Norman Kretzmann, Anthony Kenny, and Jon Pinborg, pp. 496–517. Cambridge: Cambridge University Press, 1982.

Shapin, Steven. "The House of Experiment in Seventeenth-Century England." *Isis* 79 (1988): 373–404.

———. "Who Was Robert Hooke?" In *Robert Hooke: New Studies*, edited by Michael Hunter and Simon Schaffer, pp. 253–285. Woodbridge: The Boydell Press, 1989.

Shapin, Steven, and Simon Schaffer. *Leviathan and the Air-Pump: Hobbes, Boyle, and the Experimental Life*. Princeton: Princeton University Press, 1985.

Shapiro, Barbara J. *John Wilkins, 1614–1672: An Intellectual Biography*. Berkeley and Los Angeles: University of California Press, 1969.

———. *Probability and Certainty in Seventeenth-Century England: A Study of the Relationships between Natural Science, Religion, History, Law, and Literature*. Princeton: Princeton University Press, 1983.

Shelley, Percy Bysshe. *The Complete Poetical Works*. Edited by Neville Rogers. 4 vols. Oxford: Clarendon Press, 1975.

———. *Shelley's Literary and Philosophical Criticism*. Edited by John Shawcross. Oxford: Oxford University Press, 1909.

Sherman, Paul. "The Wise Silence: Sound as the Agency of Correspondence in Thoreau." *New England Quarterly* 22 (1949): 511–527.

Sherwood, Merriam. "Magic and Mechanics in Medieval Fiction." *Studies in Philology* 44 (1947): 567–592.

Sheynin, O. B. "J. H. Lambert's Work on Probability." *Archive for History of Exact Science* 7 (1971): 244–256.

———. "Origin of the Theory of Errors." *Nature*, no. 5052 (August 27, 1966): 1003–1004.

Shields, Margaret C. "The Early History of Graphs in Physical Literature." *American Physics Teacher* 5 (1937): 68–71.

———. "James Watt and Graphs." *American Physics Teacher* 6 (1938): 162.

Silverman, Robert J. "Instrumentation, Representation, and Perception in Modern Science: Imitating Human Function in the Nineteenth Century." Ph.D. diss., University of Washington, 1992.

Slaughter, Mary. *Universal Languages and Scientific Taxonomy in the Seventeenth Century*. Cambridge: Cambridge University Press, 1982.

Smart, Christopher. *Poems*. Edited by Robert Brittain. Princeton: Princeton University Press, 1950.

———. *The Poetical Works*. Vol. 1, *Jubilate Agno*. Edited by Karina Williamson. Oxford: Clarendon Press, 1980.

Smith, Pamela H. *The Business of Alchemy: Science and Culture in the Holy Roman Empire*. Princeton: Princeton University Press, 1994.

Smollett, Tobias. *The Adventures of Ferdinand Count Fathom*. Edited by Jerry C. Beasley and O. M. Brack, Jr. Athens: University of Georgia Press, 1988.

Snellen, H. A., ed. *E. J. Marey and Cardiology*. Rotterdam: Kooyker, 1980.

Snorrason, E. *C. G. Kratzenstein: professor physices experimentalis Petropol. et Havn. and His Studies on Electricity during the Eighteenth Century*. *Acta Historica Scientiarum Naturalium et Medicinalium* 29 (1974).

Snyder, Joel. "Picturing Vision." *Critical Inquiry* 6 (1980): 499–526.

Snyder, Joel, and Neil Walsh Allen. "Photography, Vision, and Representation." *Critical Inquiry* 2 (1975): 143–169.

Sopka, Katherine Russell. "John Tyndall: International Popularizer of Science." In *John Tyndall, Essays on a Natural Philosopher*, edited by W. H. Brock, N. D. McMillan, and R. C. Mollan, pp. 193–203. Dublin: Royal Dublin Society, 1981.

"The Speaking Machine." *Punch* 11 (1846): 83.

Sprat, Thomas. *History of the Royal Society*. Edited by Jackson I. Cope and Harold Whitmore Jones. Reprint of 1667 ed. St. Louis: Washington University Press, 1958.

Stallo, J. B. *General Principles of the Philosophy of Nature*. Boston: W. M. Crosby and H. P. Nichols, 1848.

"The Stereoscope." *Illustrated London News* 20 (1852): 229–230.

"The Stereoscope." *National Magazine* 12 (1858): 49–54.

"Stereoscope." In *Chambers's Encyclopaedia*. American Revised Edition. Philadelphia: J. B. Lippincott, 1883.

"The Stereoscope, Pseudoscope, and Solid Daguerreotypes." *Illustrated London News* 20 (1852): 77–78.

"Stereoscopic Journeys." *Eclectic Magazine* 40 (1857): 560–561.

Stevens, Walter LeConte. "Rudolph Koenig." *Science*, n.s., 14 (1901): 724–727.

———. "Sketch of Rudolph Koenig." *Popular Science Monthly* 37 (1890): 545–550.

———. "The Stereoscope: Its History." *Popular Science Monthly* 21 (1882): 37–53.

Stewart, John Q. "An Electrical Analogue of the Vocal Organs." *Nature* 110 (1922): 311–312.

Straker, Stephen. "The Eye Made 'Other': Dürer, Kepler and the Mechanization of Light and Vision." In *Science, Technology and Culture in Historical Perspective*, edited by L. A. Knafla, M. Staum, and T.H.E. Travers, pp. 7–25. University of Calgary Studies in History, 1. Calgary: University of Calgary Press, 1976.

———. "What Is the History of Theories of Perception the History Of?" In *Religion, Science, and Worldview: Essays in Honor of Richard S. Westfall*, edited by M. J. Osler and P. L. Farber, pp. 245–273. Cambridge: Cambridge University Press, 1985.

Strickland, Stuart Walter. "Circumscribing Science: Johann Wilhelm Ritter and the Physics of Sidereal Man." Ph.D. diss., Harvard University, 1992.

Strouhal, V. "Über eine besondere Art der Tonerregung." *Annalen der Physik und Chemie* 5 (1878): 216.

Strutt, John William, 3d Baron Rayleigh. *The Theory of Sound*. Vol. 1, 1877; vol. 2, 1878. 2d ed. Vol. 1, 1894; vol. 2, 1896. 2d rev. ed. 2 vols. New York: Dover, 1945.

Sturm, Johann Christoph. *Collegium experimentale, sive curiosum*. 2 vols. in 1. Nuremberg: Wolfgangi Mauritii Endteri & Johannis Andrae Endteri, 1676–1685.

Sutton, M. A. "Sir John Herschel and the Development of Spectroscopy in Britain." *British Journal for the History of Science* 7 (1974): 42–60.

Swijtink, Zeno G. "The Objectification of Observation: Measurement and Statistical Methods in the Nineteenth Century." In *The Probabilistic Revolution*, edited by Lorenz Krüger, Lorraine J. Daston, and Michael Heidelberger, vol. 1, *Ideas in History*, pp. 261–285. Cambridge: MIT Press, 1989.

Sylvester, J. J. "Chemistry and Algebra." In his *Collected Mathematical Papers*, vol. 3, no. 14. 4 vols. Cambridge: Cambridge University Press, 1904–1912.

Sylvester, J. J. "On an Application of the New Atomic Theory to the Graphical Representation of the Invariants and Covariants of Binary Quantics—with Three Appendices." *American Journal of Mathematics* 1 (1878): 64–128.

Taft, Robert. *Photography and the American Scene: A Social History, 1839–1889.* 1938. New York: Dover, 1964.

Taton, René, ed. *Enseignement et diffusion des sciences en France au xviiie siècle.* Histoire de la pensée, 11. Paris: Hermann, 1964.

Taylor, E.G.R. *The Mathematical Practitioners in Tudor and Stuart England.* Cambridge: Cambridge University Press for the Institute of Navigation, 1954.

———. *The Mathematical Practitioners of Hanoverian England.* Cambridge: Cambridge University Press, 1966.

Telemann, Georg Philipp. *Beschreibung der Augen-orgel oder des Augen-clavicimbels.* Hamburg, 1739. Reprinted in Lorenz Christoph Mizler von Kolor, *Musikalische Bibliothek oder gründliche Nachricht von alten und neuen musikalischen Schrifften und Büchern . . .*, 4 vols. Leipzig, 1739–1754, 2 (1742): 262–266.

Ten Doesschate, G. *Perspective: Fundamentals, Controversials, History.* Nieuwkoop: B. De Graaf, 1964.

Thackray, Arnold. *Atoms and Powers: An Essay on Newtonian Matter-theory and the Development of Chemistry.* Cambridge: Harvard University Press, 1970.

Thompson, Benjamin, Count Rumford. "New Experiments upon Gun-powder, with Occasional Observations and Practical Inferences . . ." *Royal Society of London, Philosophical Transactions* 71, pt. 2 (1781): 229–328.

Thompson, Silvanus P. "Koenig's Experiments in Acoustics." *Nature* 43 (1891): 203–206, 275–278.

———. "The Researches of Dr. R. Koenig on the Physical Basis of Musical Sounds." *Nature* 43 (1891): 199–203, 224–227, 249–253.

———. "Rudolph Koenig." *Nature* 64 (1901): 630–632.

Thomson, James. *The Castle of Indolence.* In *The Castle of Indolence and Other Poems*, edited by Alan Dugald McKillop. Lawrence: University of Kansas Press, 1961.

———. *The Castle of Indolence.* In *Liberty, The Castle of Indolence and Other Poems*, edited by James Sambrook. Oxford: Clarendon Press, 1986.

Thoreau, Henry David. *The Journal of Henry David Thoreau.* Edited by Bradford Torrey and Francis H. Allen. 14 vols. in 2. New York: Dover, 1962.

———. *A Week on the Concord and Merrimack Rivers.* New York: New American Library, 1961.

Tierie, Gerrit. *Cornelis Drebbel (1572–1633).* Amsterdam: H. J. Paris, 1932.

Tietjens, Oskar Karl Gustav, and Ludwig Prandtl. *Applied Hydro- and Aerodynamics.* Translated by Jacob Pieter Den Hartog. New York: McGraw-Hill, 1934.

Tilling, Laura. "Early Experimental Graphs." *British Journal for the History of Science* 8 (1975): 193–213.

Tilton, Eleanor M. *Amiable Autocrat: A Biography of Dr. Oliver Wendell Holmes.* New York: Henry Schuman, 1947.

Tissandier, Gaston. *A History and Handbook of Photography.* Edited by J. Thomson. 2d rev. ed. Reprint of 1878 ed. New York: Arno, 1973.

———. *Popular Scientific Recreations.* London: Ward, Lock, 1883.

Todhunter, Isaac. *A History of the Theory of Elasticity and the Strength of Materials from Galilei to Lord Kelvin.* Vol. 1, 1886; vol. 2. 1893. 2 vols. in 3. New York: Dover, 1960.

Toledo, Amalia. *Contribution à l'histoire de l'enseignement de projections lumineuses, l'abbé Moigno (1872–1880)*. Diplôme de l'Ecole des hautes études en sciences sociales, Paris, 1976.

Torlais, Jean. "Un prestidigitateur célèbre chef de service d'électrothérapie au XVIIIe siècle, Ledru dit Comus (1731–1807)." *Histoire de la médecine* 5 (1953): 13–25.

Tourneux, Maurice, ed. *Correspondance littéraire, philosophique et critique, par Grimm, Diderot, Raynal, Meister, etc.* 16 vols. Paris: Garnier, 1877.

Trachtenberg, Alan. *Reading American Photographs: Images as History, Mathew Brady to Walker Evans*. New York: Hill and Wang, 1989.

Truesdell, Clifford A. "The Rational Mechanics of Flexible or Elastic Bodies, 1638–1788." In Leonhard Euler, *Opera omnia*, ser. 2, vol. 11, pt. 2. Zurich: Orell Fussli, 1960.

———. "The Theory of Aerial Sound, 1687–1788." In Leonhard Euler, *Opera omnia*, ser. 2, vol. 13, pt. 2. Zurich: Orell Fussli, 1960.

Tucker, Jennifer. "Bees, Corks, Monsters, and Mites: The Debut of the Compound Microscope and Retooling of Discovery in Seventeenth-Century England." Senior thesis, Stanford University, 1988.

Turner, Gerard L'Etrange. "Animadversions on the Origin of the Microscope." In *The Light of Nature: Essays in the History and Philosophy of Science Presented for A. C. Crombie*, edited by J. D. North and J. J. Roche, pp. 193–207. Dordrecht: Nijhoff Publishers, 1985.

———. "The Microscope as a Technical Frontier in Science." In *Historical Aspects of Microscopy*, edited by S. Bradbury and G. L'E. Turner, pp. 175–199. Cambridge: W. Heffer and Sons, 1967.

———. *Nineteenth-Century Scientific Instruments*. London and Berkeley: Sotheby/University of California Press, 1983.

Turner, Peter, and Richard Wood. *Peter Emerson: Photographer of Norfolk*. Boston: David R. Goodine, 1974.

Turner, R. Steven. "Consensus and Controversy: Helmholtz on the Visual Perception of Space." In Cahan, *Hermann von Helmholtz and the Foundations of Nineteenth-Century Science*, pp. 154–204.

———. "The Ohm-Seebeck Dispute, Hermann von Helmholtz, and the Origins of Physiological Acoustics." *British Journal for the History of Science* 10 (1977): 1–24.

Tyndall, John. *Lectures on Light Delivered in the United States in 1872–73*. New York: D. Appleton and Company, 1873.

———. *Sound*. Rev. 3d ed. New York: Appleton, 1908.

Ullmann, Dieter. "Chladni und die Entwicklung der experimentellen Akustik um 1800." *Archive for History of Exact Sciences* 31 (1984): 35–52.

Valentini, Michael Bernhard. *Museum museorum; oder, Vollständige schaü-bühne aller materialien und specereyen, nebst deren natürlichen beschreibung, election, nutzen und gebrauch* . . . 2d ed. 3 vols. in 2. Frankfurt am Main: J. D. Zummer und J. A. Jungen, 1714.

Valesius, Georgius de Sepibus. *Romanii collegiis societatus jesu musaeum*. Amsterdam: Janssonio-Waesbergiana, 1678.

Van Helden, Albert. "The Birth of the Modern Scientific Instrument, 1550–1700." In *The Uses of Science in the Age of Newton*, edited by John G. Burke, pp. 49–84. William Andrews Clark Memorial Library, Los Angeles: University of California Press, 1983.

Van Helden, Albert. "The Invention of the Telescope." *Transactions of the American Philosophical Society* 67 (1977): pt. 4.

———. "The Telescope in the Seventeenth Century." *Isis* 65 (1974): 38–58.

Van Nooten, S. I. "Contributions of Dutchmen in the Beginnings of Film Technology." *Journal of the SMPTE* 81 (1972): 116–123.

Vaucanson, Jacques de. *Le mécanisme du fluteur automate; An Account of the Mechanism of an Automaton or Image Playing on the German-Flute*. Translated by J. T. Desaguliers. Buren: Frits Knuf, 1979.

Vellay, Charles. *La correspondance de Marat*. Paris: Charpentier et Fasquelle, 1908.

Vetter, Théodore. "Le Cat, Claude-Nicolas." In *Dictionary of Scientific Biography*, 8:114–116.

Vickers, Brian. "Analogy versus Identity: The Rejection of Occult Symbolism, 1580–1680." In his *Occult and Scientific Mentalities in the Renaissance*, pp. 95–163.

———, ed. *Occult and Scientific Mentalities in the Renaissance*. Cambridge: Cambridge University Press, 1984.

Vogel, Stephan. "Sensation of Tone, Perception of Sound, and Empiricism: Helmholtz's Physiological Acoustics." In Cahan, *Hermann von Helmholtz and the Foundations of Nineteenth-Century Science*, pp. 259–287.

Voltaire. *Correspondence*. Edited by Theodore Besterman. 135 vols. Geneva: Institut et Musée Voltaire, 1953–1977.

———. *Oeuvres complètes*. Edited by Louis Moland. 52 vols. Paris: Garnier frères, 1877–1885.

von Neumann, John. *Theory of Self-Reproducing Automata*. Edited and completed by Arthur Banks. Urbana: University of Illinois Press, 1966.

Voskuil, J. "The Speaking Machine through the Ages." *Transactions of the Newcomen Society* 26 (1947–1949): 259–267.

Wade, Nicholas J., ed. *Brewster and Wheatstone on Vision*. London: Academic Press, 1983.

Wagenaar, W. A. "The True Inventor of the Magic Lantern: Kircher, Walgenstein, or Huygens?" *Janus* 66 (1979): 193–207.

Walker, D. P. *Spiritual and Demonic Magic from Ficino to Campanella*. Notre Dame: University of Notre Dame Press, 1958.

———. *Studies in Musical Science in the Late Renaissance*. London: Warburg Institute; Leiden: E. J. Brill, 1978.

Waller, Mary D. *Chladni Figures: A Study in Symmetry*. London: D. Bell, 1961.

Waller, Richard. "The Life of Dr. Robert Hooke." In *The Posthumous Works of Robert Hooke*. London: S. Smith and B. Walford, 1705.

Wallis, Roy, ed. *On the Margins of Science: The Social Construction of Rejected Knowledge*, Sociological Review Monograph, 27. Keele, England: University of Keele, 1979.

Walters, Alice Nell. "Tools of Enlightenment: The Material Culture of Science in Eighteenth-Century England." Ph.D. diss., University of California at Berkeley, 1992.

Ward, John, and Sara Stevenson. *Printed Light: The Scientific Art of William Henry Fox Talbot and David Octavius Hill with Robert Adamson*. Edinburgh: H.M.S.O., 1986.

Warner, Deborah Jean. "The Landscape Mirror and Glass." *Antiques* 105 (1974): 158–159.

———. "Mathematical and Philosophical Instruments in Early Modern England." Unpublished MS.

———. "What Is a Scientific Instrument, When Did It Become One, and Why?" *British Journal for the History of Science* 23 (1990): 83–93.

———, ed. *Pike's Illustrated Catalogue of Scientific and Medical Instruments.* 1848. 2d ed. New York: Benjamin Pike, 1856. Facsimile of the 2d ed. 2 vols. Dracut, Mass.: The Antiquarian Scientist; San Francisco: Jeremy Norman, 1984.

Warrick, Patricia S. *The Cybernetic Imagination in Science Fiction.* Cambridge: MIT Press, 1980.

Wasserman, Earl R. "Nature Moralized: The Divine Analogy in the Eighteenth Century." *ELH* 20 (1953): 39–76.

Weinhold, Adolf Ferdinand. *Physikalische Demonstrationen: Anleitung zum Experimentieren im Unterricht an Gymnasium, Realgymnasium, Realschulen und Gewerbschulen.* Leipzig: Quandt & Handel, 1905.

Wellek, Albert. "Das Doppelempfinden im abendländischen Altertum und Mittelalter." *Archiv für die Gesamte Psychologie* 80 (1931): 120–166.

———. "Farbenharmonie und Farbenklavier: ihre Entstehungsgeschichte im 18. Jahrhundert." *Archiv für die Gesamte Psychologie* 94 (August–December 1935): 347–375.

———. "Renaissance- und Barock-Synästhesie." *Deutsche Vierteljahrsschrift für Literaturwissenschaft und Geistesgeschichte* 9 (1931): 534–584.

———. "Zur Geschichte und Kritik der Synästhesie-Forschung." *Archiv für die Gesamte Psychologie* 79 (1931): 325–384.

Westfall, Richard S. "Science and Patronage: Galileo and the Telescope." *Isis* 76 (1985): 11–30.

Westman, Robert S. "Nature, Art, and Psyche: Jung, Pauli, and the Kepler-Fludd Polemic." In Vickers, *Occult and Scientific Mentalities in the Renaissance*, pp. 177–229.

Westrum, Ron. "Knowledge about Sea-Serpents." In Wallis, *On the Margins of Science*, pp. 293–314.

Wetzels, Walter D. "Art and Science: Organicism and Goethe's Classical Aesthetics." In *Approaches to Organic Form: Permutations in Science and Culture*, edited by Frederick Burwick, pp. 71–85. Boston Studies in the Philosophy of Science, 105. Dordrecht: D. Reidel, 1987.

———. *Johann Wilhelm Ritter: Physik im Wirkungsfeld der deutschen Romantik.* Berlin: Walter de Gruyter, 1973.

Wheatstone, Charles. "On the Various Attempts Which Have Been Made to Imitate Human Speech by Mechanical Means." *Report of the British Association. Transactions of the Sections*, 1835, p. 14.

———. *The Scientific Papers of Sir Charles Wheatstone.* London: Taylor and Francis, 1879.

Whewell, William. *History of the Inductive Sciences.* 3d ed. 2 vols. New York: Appleton, 1901.

———. *The Philosophy of the Inductive Sciences, Founded upon Their History.* Introduced by John Herivel. 2d ed. 2 vols. Facsimile of the 1847 ed. New York: Johnson, 1967.

Whitehead, Alfred North. *Science and the Modern World.* New York: Mentor Books, 1958.

Wilde, C. B. "Hutchinsonianism, Natural Philosophy and Religious Controversy in Eighteenth-Century Britain." *History of Science* 18 (1980): 1–24.

Wilkins, John. *An Essay Towards a Real Character, and a Philosophical Language.* 1668. Menston, England: Scolar, 1968.

———. *Mathematicall Magick. Or, The Wonders that may be Performed by Mechanicall Geometry.* London: Printed by M. F. for S. Gellibrand, 1648.

Williams, L. Pearce. "De la Rue, Warren." In *Dictionary of Scientific Biography,* 4:18–19.

———. *Michael Faraday: A Biography.* New York: Basic Books, 1965.

Willis, Robert. "On the Mechanism of the Larynx." *Transactions of the Cambridge Philosophical Society* 4 (1833): 323–352.

———. "On the Vowel Sounds, and on Reed Organ-Pipes." *Transactions of the Cambridge Philosophical Society* 3 (1830): 231–268.

Wilson, Catherine. "Visual Surface and Visual Symbol: The Microscope and the Occult in Early Modern Science." *Journal of the History of Ideas* 49 (1988): 85–108.

Wimsatt, W. K., Jr. "Poe and the Chess Automaton." *American Literature* 11 (1939): 138–151.

Windisch, Charles Gottlieb de. *Lettres sur le Joueur d'Echecs de M. de Kempelen.* Translated by Chrétien de Mechel. Basel: Mechel, 1783.

Winkler, Mary G., and Albert Van Helden. "Representing the Heavens: Galileo and Visual Astronomy." *Isis* 83 (1992): 195–217.

Wolters, Gereon. *Basis und Deduktion. Studien zur Entstehung und Bedeutung der Theorie der axiomatischen Methode bei J. H. Lambert (1728–1777).* Berlin: Walter de Gruyter, 1980.

———. "Some Pragmatic Aspects of the Methodology of Johann Heinrich Lambert." In *Change and Progress in Modern Science,* edited by Joseph C. Pitt, pp. 133–170. University of Western Ontario Series in Philosophy of Science, 27. Dordrecht: D. Reidel, 1985.

Wood, Alexander B. *A Textbook of Sound: Being an Account of the Physics of Vibrations with Special Reference to Recent Theoretical and Technical Developments.* 1930. 2d ed. New York: The Macmillan Company, 1941.

———. *Thomas Young, Natural Philosopher.* Cambridge: Cambridge University Press, 1954.

Wordsworth, William. *The Complete Poetical Works.* Edited by Andrew J. George. Boston: Houghton Mifflin, 1932.

———. *The Prelude or Growth of a Poet's Mind.* Edited by Ernest De Selincourt. 2d ed. Oxford: Clarendon Press, 1959.

Wright, Lewis. *Optical Projection: A Treatise on the Use of the Lantern in Exhibition and Scientific Demonstration.* London: Longmans, Green, and Co., 1901.

Yates, Frances A. *The Rosicrucian Enlightenment.* London: Routledge and Kegan Paul, 1972.

Yeo, Eileen, and E. P. Thompson. *The Unknown Mayhew.* New York: Pantheon Books, [1971].

Yolton, John W. *Perceptual Acquaintance from Descartes to Reid.* Minneapolis: University of Minnesota Press, 1984.

Young, Matthew. *An Enquiry into the Principal Phaenomena of Sounds and Musical Strings.* Dublin: G. Robinson, 1784.

Young, Thomas. *A Course of Lectures on Natural Philosophy and the Mechanical Arts*. 2 vols. London: J. Johnson, 1807.

———. *Miscellaneous Works of the Late Thomas Young*. Edited by George Peacock. 3 vols. London: J. Murray, 1855. Reprint, New York: Johnson, 1972.

Zahn, Johann. *Oculus artificialis teledioptricus sive telescopium*. 3 vols. Würzburg: Q. Heyl, 1685–1686.

Zetterberg, Jack Peter. "Echoes of Nature in Salomon's House." *Journal of the History of Ideas* 43 (1982): 179–193.

———. "Mathematicall Magick in England 1550–1650." Ph.D. diss., University of Wisconsin, Madison, 1976.

Zilczer, Judith. "'Color Music': Synaesthesia and Nineteenth-Century Sources for Abstract Art." *Artibus et historiae: An Art Anthology* 16 (1987): 1101–1126.

INDEX

About the Authors

Thomas L. Hankins is Professor of History at the University of Washington, and Robert J. Silverman, who holds a Ph.D. from the University of Washington, is completing a law degree at New York University.